THE PUTNAM
AERONAUTICAL REVIEW

Volume One

Edited by John Motum

© 1989 Conway Maritime Press Ltd

This edition first published in Great Britain in 1990
by Putnam Aeronautical Books, an imprint of
Conway Maritime Press Ltd
24 Bride Lane, Fleet Street, London EC4Y 8DR

ISBN 0-85177-837-2 (Conway Maritime Press)

Published and distributed in the United States of America
and Canada by the Naval Institute Press, Annapolis, Maryland 21402.

Library of Congress Catalog Card No. 89-61978

ISBN 0-87021-610-4 (Naval Institute Press)

Manufactured in Great Britain

THE PUTNAM AERONAUTICAL REVIEW

Index of Aircraft: Volume 1

THE PUTNAM AERONAUTICAL REVIEW

ISSUE NUMBER ONE **MAY 1989**

Editor: John Motum

Consultant Editor: John Stroud

Assistant Editor: Linda Jones

Design: Swanston Graphics Ltd

Managing Editor: Julian Mannering

Publisher: William R Blackmore

© Conway Maritime Press Ltd

ISBN 0 85177 525X
ISSN 0955-7822

Published quarterly by:
Conway Maritime Press Ltd
24 Bride Lane, Fleet Street
London EC4Y 8DR

Telephone: 01 583 2412

**Annual subscription for
four issues:**
£16.00 UK, £18.00 overseas,
$33.00 USA and $39.00 Canada.
All post paid.

Typeset by Swanston Graphics Ltd, Derby.
Printed and bound in Great Britain
by Page Bros (Norwich) Ltd.

CONTRIBUTORS

Maurice Allward retired recently from BAe Hatfield, Hertfordshire. From the time he began as a design draughtsman at Hawker, Kingston, Surrey, under Sir Sidney Camm in the Second World War, he has worked on forty-seven different aircraft types and flown in as many again. He has written over fifty aviation books.

Dr Norman Barfield, following completion of an engineering apprenticeship in 1950, has spent a working lifetime on the Brooklands Weybridge, Surrey, site, principally concerned with the engineering development and promotion of commercial aeroplanes.

Sir George Edwards was pioneer and chief designer of the propeller turbine Viscount and the turbojet Valiant, first of the V bombers. He subsequently became design teams leader for the VC10, BAC One-Eleven and the British portion of the Concorde programme.

J R Finnimore was BOAC's aircraft development manager from the VC10's inception in 1958 until 1970, before continuing as the airline's general manager supplies in the Boeing 747 era.

Dr Roberto Gentilli, author and aviation historian, concentrates his research on Italian aviation between the two world wars. His five published books include one on the Savoia Marchetti S.79.

David Smith is an air traffic controller and author, his principal subject being UK aerodromes. His latest book, *Britain's Military Airfields 1939-1945*, is due for publication in the late spring.

John Stroud, in his work as an aviation writer, has been making operational analyses of sectors flown by transport aeroplanes of increasing complexity, up to and including Concorde, during a period of more than 40 years.

Barry Wheeler trained as an aviation journalist on *Flight International* and is currently editor of the British Ministry of Defence's *Joint Services Recognition Journal*.

R A R Wilson is an air historian, following a lifetime of practical experience in aviation including aerial survey work in over twenty countries and airline experience with both BEA and British Airways. He was in the RNVR throughout the Second World War and beyond as a senior observer and air signals officer, with detachments both to the RAF and USAAF on operations.

Front cover: The VC10 prototype which flew for the first time on 29 June 1962. (*Vickers*)
Back cover: The Savoia Marchetti S.79 MN.365, of the Italian team, which flew in the Istres–Damascus–Paris Air Race of August 1937. Clearly visible is its race number, I-13, and the emblem of the three green mice. (*Dr Roberto Gentilli*)

WELCOME

HOW GOOD to welcome, as I do with much pleasure and high expectation, the arrival of a new quarterly journal to be devoted to the broad sweep of aeronautical affairs – past, present and future.

There is, in these days, a massive flow of news and detailed information about the still fast expanding aviation business. But with it we have all too little informed comment and opinion; of authoritative interpretation and guidance on what it is all about – the answers to those three where questions – Where have we been? Where are we now? Where do we go from here?

Every devotee of aviation will look forward to all that the new *Review* will bring to us in these fascinating arenas. I wish the new venture success, and to its readers much edification and enjoyment.

Sir Peter Masefield
Reigate, Surrey

Why We Are Here

As the last decade of the twentieth century approaches, it would be too simple to suggest that the clues to the future of aerospace activity lie rooted in the past. In fact, to the people currently engaged in designing, building and operating aircraft, military or commercial, and to those few who are seeking to bridge the gap between conventional flight and operation in earth orbit, the very suggestion might seem absurd. If Henry Ford really did say 'history is bunk', we may be pretty certain that, were he alive today, he would not be numbered among the readers of this *Review*. What, then, are we trying to do?

Today, it is almost taken for granted that a major new aircraft or an aero engine programme will often involve international technical co-operation and financial risk-sharing across the world rather than just over adjoining national boundaries. Without such co-operation, most new programmes could no longer be started, let alone completed.

The main subject of this issue is the VC10, and 30 years ago, when planning was in progress, the complete programme was undertaken in the United Kingdom. Ultimately, the costs had to be spread over a production run of just fifty-four aircraft involving severe losses to the manufacturing parent, Vickers, and further losses to the taxpayer. At that time, the concept of financial risk-sharing among equipment suppliers had yet to be developed, though wide sub-contracting was in vogue to maintain an even load in the workshops of the main producer, and in some cases to share experience with a future partner. Such objectives were met by the production of Super VC10 fin structures by Sud-Aviation, later part of Aérospatiale — a fragment of the learning curve towards Concorde.

There are many fruitful areas for our *Review* to explore. We aim to set the past record into the context of today. With that precious commodity, hindsight, we shall recognise, trace and analyse the evolutionary paths that have led to the present. Our contributors will be our guides. Not always will they steer us past the thickets of uncertainty first time, for our readers will provide fresh knowledge and other points of view. How different the same aeroplane looks to the creator, the operator, the accountant and even the passenger.

By recognising our heritage, the longer-term implications of today's tactics may be better understood. While failure to adapt to change is to court disaster, the need for vision, which comes through understanding, has never been greater. One current example will be to find solutions to the ordered development of Hotol — the horizontal take-off and landing concept for bridging the gap between conventional air-breathing engines and space flight.

Throughout, we shall recognise that aviation has invariably been a truly international activity since the early days of team endeavour which followed the individual efforts of the pioneers. We acknowledge that there are few monopolies in bright ideas; two examples are introduced in this issue, both concerning engine placement.

The French company Sud-Aviation pioneered the rear-engined formula for commercial airliners with the Caravelle. The first and only long-range aircraft in the West to follow this lead was the British VC10, inspiring the French company's delightful advertisement showing two young boys looking upwards at a small rear-engined speck in the sky while one makes the point, 'Ils ont copié Caravelle'. It will be for our readers to ponder why the VC10 should be the only long-range airliner in the West to depend on rear-mounted engines.

Likewise, in the story of the Istres–Damascus–Paris Air Race of 1937, which at first sight may appear to be a piece of sheer nostalgia, we thought it did no harm to remind ourselves that the early three-engined formula so well expressed in the Fokker and Ford Tri-Motors of the late 1920s and the Junkers-Ju 52/3m of the early 1930s, currently enjoying a wave of popularity, did not in fact end there. In the capable hands of Savoia's designer, Dr Alessandro Marchetti, the apparently outdated formula was still breaking records in the late 1930s. All this was a quarter of a century before engineers at Hatfield were putting three jet engines at the back and calling the result the Trident. That set off the Boeing 727, one of the highest selling commercial aeroplanes in history.

For many decades now, aviation by its very nature has been a specialised endeavour where the tale of any particular programme may be told from the viewpoint of design, manufacture, testing on the ground or in the air, and finally from in-service experience. There are many gaps to fill and you, our readers, will be our allies, feeding back new facts and fresh perspectives to provide a forum of evidence which will be published.

Finally, let us remind ourselves that for many decades aviation writers have made major contributions to the recording of aviation's history and development and even contributed to aircraft safety. Britain has been fortunate in having a fair share of them and we shall be inviting writers from home and abroad to contribute to the *Review*. But we are also determined that those who have been responsible for aircraft development should find space on our pages to describe the background to their work and tell us what influenced their decisions.

We have an ambitious programme. Thank you to our contributors who have brought us to issue one, and thank you to Sir Peter Masefield, the complete aviation man, for your message of welcome to all of us. **John Motum** □

VC10: Personal Overview & Perspective

Sir George Edwards OM, CBE, FRS

If there is one man uniquely qualified to introduce the VC10 it is the author of this perspective. As chief designer at Vickers-Armstrongs (Aircraft) from 1945, he turned to turbine power early for both commercial and military applications. Told here are the salient facts behind the qualities of the VC10 and the basic reasons for its continuance in service with the Royal Air Force today.

They also serve. An RAF VC10 K.Mk.3 trails its centre hose. Flight Refuelling and Vickers-Armstrongs (Aircraft) developed probe and drogue refuelling initially for Valiant V bombers, all of which were capable of transferring or receiving fuel in flight, more than 30 years ago. (*MoD*)

Whatever the traumas that beset us with the VC10 during its early years – and there were plenty – seen today 27 years on from the first flight of the prototype at Weybridge (on 29 June 1962), it continues as a proud and productive symbol of British aviation capability – serving exclusively with what I have always believed to be the most important of all of our customers, the Royal Air Force.

Scarcely a week goes by without we see the VC10 C.Mk.1s of No. 10 Squadron on our television screens literally 'flying the flag' with HM The Queen and other members of the Royal Family, the Prime Minister and senior members of HM Government throughout the world. At the same time, there are the hundred and one other unseen and unsung tasks that these 'multi-mission' transports perform for the benefit of the nation, plus the 'force multiplier' activity of the VC10 K.Mk.2 and 3, strategic tankers of No. 101 Squadron. And both versions will, they say, be flying well into the twenty-first century.

This means that, from our earliest notions in the mid-1950s, the total VC10 saga will eventually span an incredible 50 years and, I am willing to bet, thanks to traditional Weybridge craftsmanship and quality, it will continue to do a first-class and reliable job just as it always has done since first entering service with BOAC (British Overseas Airways Corporation) in April 1964.

How did it start? The reason lies in two aeroplanes before the VC10. These were the V1000 which was an intercontinental jet military transport based on the Valiant (V bomber) design philosophy. We conceived the V1000 in very close co-operation with BOAC. It was the first airliner with what was to become the universal standard in the first generation of big jets, six-abreast seating with a centre aisle, and thus had a large diameter fuselage for those days. It had a very difficult airfield performance to meet – and nearly all the aeroplanes that we have designed in this country have been bedevilled by having to meet

VC10: Early Days

These pictures and the basic make-up of the captions to accompany them were taken from the contemporary Weybridge record, stuck down in a hard covered book in the manner of a family snapshot album.

14 January 1958. The managing directors of Vickers-Armstrongs (Aircraft) Ltd and the British Overseas Airways Corporation sign a contract for thirty-five VC10s. Sir George Edwards (left) with Basil Smallpeice (later Sir Basil) the customer. (*Vickers*)

15 April 1962. At 0830hrs on a bitterly cold and cloudy Sunday morning, the Vickers-owned prototype G-ARTA is rolled out for the first time and poses for photographs taken from a Westland Widgeon helicopter. (*Popperfoto*)

29 June 1962. A fast taxi run to be followed by a first flight, watched and recorded by the world's press. (*BAe*)

15 May 1962. After completing a ground resonance programme that lasted about 4 weeks, the prototype, built with full production jigging and tooling, is ready for engine running. A Cullum detuner is positioned to reduce the noise to neighbouring residents. The following day, one detuner unit is torn from its mounting and blown a distance of about 150ft. Engine runs continue without it. (*BAe*)

27 June 1962. The main gear which VC10 pilots and passengers will learn to love has still to see air beneath the tyres, except when the aircraft is on jacks. (*BAe*)

25 August 1962. Today Ian Macdonald, in an Empire Test Pilots School (ETPS) Meteor, takes the first air-to-air photographs. Many views, as here, with tape over the retracted slats, cannot be used for publicity purposes without retouching. (*BAe*)

a short airfield performance requirement – which is something the Americans never seem to be unduly bothered about.

With a prototype and six production aircraft ordered for RAF Transport Command, we were within about 6 months of the first example being finished in 1955 when the programme was cancelled for short-term political reasons, ostensibly because the RAF needed more large transport aircraft capacity much quicker than we could produce it, and there were Britannias available that were going to be made in politically attractive Northern Ireland.

This cancellation was, in my view, the most serious setback that the British aircraft industry has suffered since the end of the Second World War because it came at precisely the moment when long-range jet operation was about to become big business.

Whereas the military concept of having a strategic airlift force able to shift large numbers of troops and their equipment over long distances at high speed subsequently came into being through the VC10 which the Royal Air Force eventually acquired, this would have been available a lot sooner with the V1000.

As far as the civil market was concerned, the V1000 stood in just the same position for us that the Viscount had a few years earlier – with a performance that was as good as, and indeed better, I know, than the initial Boeing 707 which could not do the transatlantic crossing non-stop, whereas the V1000, with its 15 per cent bigger wing, could. Thus we would have been well set with the Viscount operating on short ranges and the V1000 on long ranges, with modern technology Rolls-Royce engines in both.

The statements that were made in the House of Commons at the time of the cancellation were pretty pathetic, because one said that BOAC was quite content that it was going to manage the North Atlantic route with propeller-turbines throughout the 1960s and there was no question of needing a jet aeroplane. Reality was predictably and quickly quite the opposite.

I think BOAC had really wanted to buy aeroplanes that other major world airlines were operating because it is much safer to be in competition with common equipment than where you have got some exclusive gleaming monster which might be all right, but nobody else has it. This meant that they were leaning towards American aeroplanes pretty well all the time.

Coming to the VC10 itself, it was designed to an exceptionally severe requirement written by, and exclusive to,

BOAC. This notably necessitated operating from the existing airfield at Singapore to Karachi non-stop against a 35 knot headwind, and getting in and out of Kano and Nairobi, hot and high. These were the kind of conditions that the Boeing 707 certainly could not meet, because everywhere that it had been put into operation a substantial extension of the runway had been needed and, of course, as the years went by, all the major airport runways were duly lengthened. So we had to do better than the 707 – not different, but better – and the main job was to improve the lifting capabilities in order to meet these exacting conditions.

That we had achieved this objective was clearly evident on the original first flight of which my one lasting memory was seeing such a tiny Brooklands motor racing circuit under such a big aeroplane and being quietly satisfied that we had made it out of 'our own backyard' (using less than half of the 4,500ft runway and watched by the eight thousand people who had built it) just the way we had intended.

Happy and rewarding, too, was the moment 19 minutes later, after a gentle sweep over Odiham and back over Farnborough in Hampshire, when it landed beautifully in the sure and expert hands of Jock Bryce and Brian Trubshaw and their crew at our flight test airfield at Wisley just four miles away from Brooklands. And then only a few days later being able to get 'hands on' myself, already knowing that it incorporated the characteristically excellent flying qualities of the Vickers lineage from as far back as I could remember.

All of this came flooding back 18 months ago on 7 July 1987 – just a week after the Silver Anniversary of that historic flight – with the same kind of exemplary arrival back at its original home at Brooklands of the Sultan of Oman's VC10 which he so graciously donated to the Brooklands Museum Trust as a lasting monument to the largest aircraft ever put into production in the United Kingdom.

Meeting such a severe airfield performance requirement was the basic reason why we put the four Rolls-Royce Conway engines (the pioneer of the dual-flow or by-pass principle that has since become universal) at the back. In order to get in and out of these small airfields we needed all the lift that we could generate. However, when you have the engines dangling under the wings like a Christmas tree – like the American types always did – you split up the landing flaps into small sections with a loss of lift at the end of each break, whereas if you have a 'clean wing'

you can have completely continuous lift-generating devices on both the leading and trailing edges. The net result was that the VC10 landed about 20 knots slower (and much gentler thanks to Vickers' own landing gear) than the 707 and got itself both into and out of these places with a full load.

The fact that the rear-mounting of the engines also gave the VC10 a quiet interior was a sheer accident, but another big plus. In turn, it provided the fundamental advantage of superior passenger attraction, which no amount of comparative cost formulae could take any account of, but which sustained it throughout its operation.

This, then, was the basis of our response to the BOAC need, which was that all the routes that they wanted to serve with jet aeroplanes, but could not at that time accommodate the 707, would be served by the VC10; hence a large fleet of VC10s was going to be the backbone of the airline's global operation. So we signed a contract for thirty-five with an option to purchase another twenty.

There were certainly a number of people in high places in the airline who at that time saw an elimination of the 707 and their fleet becoming totally VC10, because we had already devised a larger version – the Super VC10 – in which the exceptional airfield performance had enabled us to put on a longer body and to carry a much larger payload into the bigger airfields that were then beginning to appear.

Though the VC10 was obviously in a poor position compared with where the V1000 would have been, because it was now several years later, it still looked a good bet. But as time went by, the pressures inside BOAC to unify its fleet went through a complete reversal and the eventual compromise was that they had a mixed fleet of VC10s and 707s.

It is well-known history that the passengers much preferred the VC10; the passenger load factors, and hence the revenue-earning capacity, were very much higher, and the pilots liked the aeroplane, especially the low landing speed and the smooth Vickers undercarriage. It therefore proved to be very successful, except for those of us who made it, because we lost a lot of money on it, the total quantity that we built being fifty-four, including fourteen for the Royal Air Force. But it was an aeroplane built to a tough requirement which in the fullness of time evaporated.

The Achilles heel of the British aircraft industry over the years has always been the home market; not only was it not very big, but it tended to be specialised because the operators – military as well as civil – had their own particular needs. From their point of view, this was probably quite justified, but tended to get the manufacturer into devising specialised equipment that was not universally acceptable.

A handful of other operators who had basically the same difficult job to do, especially in Africa, bought the VC10: Freddie Laker and his original British United which was later taken over by British Caledonian, Ghana, Nigeria, Middle East and East African; and the Royal Air Force when the same kind of short, high altitude and tropically located airfields, which had so characterised the whole VC10 conception, were again specified when its own strategic jet transport requirement reappraisal was made at the end of the 1950s.

Like all of our output at Weybridge, we built the VC10 to last. In this, more than half the weight of the structure was machined from the solid and we devised the best corrosion protection scheme that anybody has ever had anywhere, even until now.

The net result is that the VC10 has proved to be ideal for the RAF, and long-lasting, and they celebrated 21 years of VC10 operation on 7 July 1987, just 8 days after the 25th anniversary of the prototype first flight. And that made me as proud as they obviously are.

Perhaps more sobering, the VC10 saga provided for me one of the clearest possible vindications of an axiom that I have so often expressed about the British aviation industry: the truth always comes out but so often when it is too late to rectify the situation.

But at least with the VC10 the Royal Air Force has benefited in large measure and they are still delighted with it. Through them, the needs of international diplomacy, the peacetime transport of British servicemen and their families and the vital logistic support of front-line combat and surveillance air-

craft through in-flight refuelling, the VC10 will continue to symbolise British engineering capability and integrity throughout the world in the best way possible, both for the aviation business and the nation as a whole. As I have said so often, without the aircraft industry and the Royal Air Force we certainly would be a second-rate nation. Despite the radical changes in both spheres in recent times, happily that is not the case – and, all these years on, I am gratified that the VC10 is, and long will be, a small but key reason why not. □

29 June 1962. Sharing the load 1. Coming down the steps at Wisley after the first flight, Vickers' chief test pilot G R 'Jock' Bryce, followed by his deputy Brian Trubshaw and Bill Cairns, acting on this occasion as flight engineer. (*BAe*)

Below right: 29 June 1962. Sharing the load 2. At Wisley for the arrival, E E Marshall, chief project engineer, Ken Lawson, chief aerodynamicist, and managing director, Sir George Edwards. (*BAe*)

Above: Big aircraft on a short runway. A routine but always impressive moment at Brooklands, 1962–70. (*BAe*)

VC10: In RAF Service Today

Barry C Wheeler

The Royal Air Force (RAF) operates the VC10 today in the role of a strategic transport for which it was expressly purchased, and in the air-to-air tanker refuelling role to which it has been adapted by the conversion of ex-airline aircraft.

One of the greatest accolades an aircraft designer can receive is to have his idea copied or adopted by his peers. This occurred with the VC10 when the Ilyushin design bureau in the Soviet Union suddenly revealed, in September 1962, a new long-range airliner bearing an almost identical layout to the British aircraft and designated the Il-62. Unlike its inspirational Western counterpart, however, it was never to achieve the military success of the VC10.

Strategic Transport VC10

The origin of the military VC10 goes back to the original Ministry of Aviation Specification C239 issued in 1960 which called for a strategic jet transport to meet an Air Staff Requirement (ASR). Its aim was to take passengers and freight over long distances and replace some of the older aircraft still in regular service with RAF Transport Command, such as the Hastings, Argosy and, later, the Britannia.

The initial order, placed in September 1961 with Vickers-Armstrongs (Aircraft) at Weybridge, Surrey, then part of the British Aircraft Corporation (BAC), was for five aircraft, subsequently increased to eleven in August 1962. These were to be Standard aircraft with military modifications. In July 1964, three Super VC10 production line positions reserved for the British Overseas Airways Corporation (BOAC) were made over to the RAF, but to the same standard as the earlier orders. Known as the Vickers Type 1106, the RAF aircraft were designated VC10 C.Mk.1s.

Due to the types of operation envisaged for these aircraft (such as freight-carrying, troop transport and medical evacuation) and the areas of the world where they would be flown, the design was modified from that of the two civil versions, the Standard and the Super VC10. The resulting aircraft had the fuselage of the former combined with the wings, fin, undercarriage and engines of the latter. The fin was wet, allowing an extra 1,355 gallons of fuel to be carried, and provision was made for a nose-mounted in-flight refuelling probe to extend further the aircraft's range. For its freight-carrying role, the aircraft incorporated a large, upward-hingeing cargo door in the port side of the forward fuselage. The door measured 84 by 140in and was originally developed for use by British United Airways. Freight lashing points were located in the strengthened floor, and cargo pallets

The graceful lines of a VC10 C.Mk.1 photographed from the rear ramp of an RAF Hercules by Cpl Maurice Lockey. (*RAF*)

and/or small vehicles up to a total weight of 59,000lb could be carried – a useful capability, as later events proved.

In addition to the four crew (pilot, co-pilot, navigator and flight engineer), the VC10 passenger aircraft could (and still does) accommodate up to one hundred and fifty fully equipped troops or one hundred and twenty in airline-type seating or up to seventy-eight stretcher cases with six medical and two cabin attendants.

The first VC10 for the RAF (serial XR806) took off from Weybridge on 26 November 1965 and flew the short distance to the BAC test airfield, Wisley, Surrey, for acceptance trials. These proved to be relatively trouble-free, after which the aircraft was transferred to the Aeroplane and Armament Experimental

In the early days of VC10 operations, the RAF made regular flights to Hong Kong using aircraft marked Transport Command (above) before changing to Air Support Command (below).

Establishment, Boscombe Down, for Ministry of Aviation (MoA) assessment.

No. 10 Squadron, a recently disbanded Victor bomber squadron, was chosen to operate the new aircraft and reformed under the command of Wg Cdr M G Beavis. The squadron received the first VC10 at its Oxfordshire base at RAF Brize Norton in July 1966. As the aircraft was already proven in airline service, it did not require a lengthy work-up programme, but route-proving, in August of that year, took the aircraft on its first long overseas trip to Hong Kong. The squadron began full scheduled VC10 operations in April 1967

and 4 months later the unit flew the inaugural service of the newly formed Air Support Command.

Today, No. 10 Squadron still flies the VC10 C.Mk.1, the thirteen aircraft available to the unit being named appropriately after RAF holders of the Victoria Cross (see the table on p12). The fourteenth VC10 purchased by the MoA (serial XR809) was subsequently leased to Rolls-Royce for development of the RB.211 turbofan engine. Re-registered as G-AXLR, it completed a series of exhaustive trials with the port-side Conway engines replaced by a single RB.211. Returned to the RAF in 1976, it was eventually reduced to spares at RAF Kemble, Gloucestershire, as its reconversion to military standard would have been very costly.

With a maximum range, unrefuelled, of some 6,200nm, the VC10s operate with near airline regularity a series of scheduled routes linking the United Kingdom with the remaining British overseas bases such as Belize, Cyprus and Gütersloh, West Germany. They also serve destinations in the USA and Canada as well as making non-scheduled flights to the Middle East, Africa, the Far East and other distant locations. The squadron is also frequently called upon to fly VVIPs to destinations overseas for official visits. In May 1972, a VC10 became the first British military aircraft to visit China since 1946 and, more recently, Moscow has witnessed the arrival of this majestic aircraft. The unit also contributed aircraft for the British withdrawal from Aden in 1968, and during the Falklands conflict in 1982 it provided a vital casualty evacuation service from Ascension Island to the United Kingdom. No. 10 Squadron's operations go largely unsung, but they continue throughout the year: supporting exercises, responding to disaster relief calls and, of course, passenger-, troop- and freight-carrying around the clock. A replacement for these versatile aircraft has yet to be identified, but the RAF's new fleet of TriStar tanker/transport aircraft will assist in the years ahead with some of the missions that have until now been the prerogative of the VC10s.

Tanker VC10

The other role of the VC10 in RAF service is that of an airborne tanker. Since the 1950s, air-to-air refuelling of military aircraft has steadily assumed a greater importance. At an early stage, the RAF standardised the flexible trailing hose and probe method developed by Sir Alan Cobham's Flight Refuelling company. The Vickers Valiant was the RAF's first operational tanker aircraft,

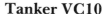

followed by the Victor, both types originally intended to support the operations of the V Force and short-range jet aircraft. With the advent of increased air defence forces (with Phantoms and Lightnings) and the greater need to support tactical elements (with Harriers and Jaguars) in the 1970s, additional tankers were needed, resulting in ASR 406. This was formulated early in 1978 and studies put in hand to determine the feasibility of converting a number of redundant civil VC10s for the air-to-air refuelling role. The idea was sound and a contract to convert nine aircraft was placed with the Weybridge-Bristol Division of the British Aerospace Group in May 1979. The design proceeded under Specification K294DP and the conversions themselves were completed at Filton, Bristol, in the Brabazon aircraft assembly hall.

The nine aircraft were all civil machines: five Standard VC10s originally built for BOAC and subsequently operated by Gulf Air of the Middle East, and four Super VC10s delivered to East African Airways between 1966 and 1970. The Service designations for the two different versions were VC10 K.Mk.2 for the Standards (which took the new Type Number 1112 in the British Aerospace Weybridge Design Engineering Organisation sequence) and VC10 K.Mk.3 for the four Supers (Type 1164).

For operational and engineering reasons, the RAF wanted a high level of common characteristics among its VC10 fleet, which included the transports of No.10 Squadron. While there was no question of changing the fuselage lengths, the VC10 C.Mk.1 engine – the Super's Conway Mk 550B (21,800lb static thrust) – was fitted to both tanker versions. Thrust reversers were fitted only on the outboard engines.

The conversion itself involved the fitting of a Flight Refuelling Mk 32/2800 pod under each wing, capable of trailing 50ft of refuelling hose. A strengthened wing structure was provided to carry the pods, which had to be positioned so as to clear the aircraft's large trailing edge flaps and to give adequate separation from the fuselage/engines for receiver aircraft. Inside the fuselage, a long additional five cell tank was installed in the old passenger cabin to carry extra fuel. Of double-skinned metal construction with internal flexible bags, the inner skin of these fuselage cells is designed to support the bladder cells and the outer skin is designed to withstand the maximum cabin pressure differential of 9lb per sq in. To ensure that fuel vapour does not enter the cabin, the space between the two skins is vented to atmosphere. Both supply and vent pipes can be rapidly switched between cells or to the main fuel system. To refuel the fuselage tank there is a riser pipe in the main wheel bay which feeds by gravity from front to rear – valves at the base of each cell preventing reverse flow – and from the rear cell to the aircraft centre section tank which retains its function as the primary transfer tank.

The VC10's third refuelling point is the rear fuselage installation containing an FR Mk.17B hose drum unit (HDU) capable of dispensing 500 Imperial gallons per minute via an 81ft trailing hose. This equipment, like the pods, is remotely controlled from the flight deck.

The aircraft's operational flexibility is further enhanced by having its own in-flight fuel receiving equipment. This consists of a 9ft long nose-mounted probe which feeds the main fuel system via a supply line running along the left-hand lower fuselage. Thus the VC10 tanker can conduct strategic operations at great ranges with tanker-to-tanker aircraft missions, the oil capacity in the engines being the only limiting factor.

Following conversion, the first of the modified aircraft made its initial flight from Filton on 22 June 1982. It was finished in a green and grey camouflage pattern, although later aircraft

UK support for fellow NATO member Norway means regular exercises requiring air support from Strike Command transports. Here, American and British trucks load equipment from a VC10 while a Hercules taxies past in the background. (*MoD*)

Cluster bomb units (CBUs) being unloaded through the side cargo door of *James Nicholson VC*, alias XV107.

received a hemp colour scheme to reduce visual identification and vulnerability when parked on the concrete hardstandings where most of the VC10-sized aircraft spend their days when not in the air or undergoing maintenance. The VC10 K.Mk.2 ZA141 was subsequently damaged during a high-speed yawing manoeuvre while undertaking BAe pre-acceptance trials. After repairs, which included a replacement tail unit, the aircraft was transferred to Boscombe Down for extensive trials to ensure its compatibility with the various types of RAF aircraft it would refuel. To operate the nine aircraft, the RAF reformed No. 101 Squadron, a bomber unit of long standing which was established in 1917 and had been a Vulcan B.2 squadron until disbandment in 1982. In a move to reduce costs, the new unit was based at Brize Norton where the 'truckies' of No. 10 Squadron were located and where existing VC10 spares and servicing equipment would make for greater operational efficiency. The Tanker Training Flight (TTF) received the first aircraft in August 1983 to begin the 10 week training courses for the new crews which mostly came from other 'heavies' in the Service such as the Victor and VC10 transports.

Under the command of Wg Cdr David Hurrell, No. 101 Squadron was officially formed on 1 May 1984 and began operations with the first K.Mk.2s. Such was the urgency for more tanker support for RAF front-line forces following the Falklands conflict that by August, 3 months later, 1,000 hours had already been achieved by the TTF and the squadron on the available aircraft. During this time the VC10s had supported Jaguar and Tornado overseas deployments and had assisted in Operation Tiger Tail, the collection of No. 74 Squadron's refurbished F-4J Phantoms

from the USA. The first flight of the K.Mk.3 (serial ZA148) was made on 3 July 1984 and by March 1986, the unit had received its ninth and final aircraft.

The tanker has the same flight crew of four as the VC10 C.Mk.1 except that the flight engineer sits on the right-hand side of the flight deck, has the air-to-air refuelling operating facilities and a closed-circuit television (CCTV) system (the camera for which is mounted under the rear fuselage and can be trained on any of the three refuelling points).

Refuelling from a VC10 depends very much on the type of receiver aircraft. Larger aircraft such as the Nimrod or another VC10 are limited to

The asymmetric engine installation of the fourteenth VC10 for the RAF which was transferred to Rolls-Royce. The single Rolls-Royce RB.211 on the port side was air tested prior to its use on the Lockheed TriStar. (*Rolls-Royce*)

using the central line drogue as this position produces less buffet than the underwing units and delivers a higher rate of fuel. While the larger types generally approach the drogue from behind under the control of the VC10's captain, smaller aircraft operate under a slightly different method. Following a vector from ground control towards the tanker, the receiver aircraft acquires the VC10 visually and then switches to the VC10 captain who directs them to his port side. Fuel requirements are confirmed and/or checked and the receiver aircraft is directed to the pre-contact position behind the relevant hose. From behind, the receiver sees a series of lights on the pod (or centre HDU), not unlike road traffic signals. These cue him for the short sequence of moves he has to make to receive fuel. Red and amber initially glow as the flight engineer indicates to the pilot behind that fuel is available but not ready to flow. When the red is doused, the remaining amber light indicates that the receiver should move into the drogue contact position. When this is done and a firm contact is achieved, the hose is pushed in until a green light appears and fuel begins to flow. After delivery of the allotted amount of fuel, the procedure is reversed and the green light gives way to amber as the receiver drops back. Once clear, the red light appears until the immediate area is clear for the next receiver. If an emergency occurs during contact, the flight engineer can immediately order a break and replace the green with a red light and thus stop the flow of fuel.

Night refuelling is done with the aid of floodlights strategically positioned around the rear of the tanker aircraft to illuminate the wings and fuselage as

Tanker-to-tanker with the first of the nine conversions nearest the camera. Although the lighting system under the aircraft to aid refuelling is not apparent in this view, the dayglo markings around each of the refuelling points can be clearly seen. (BAe)

well as the HDUs. Triple refuelling is never undertaken due to the resulting close proximity of the three receivers, although both wing pods are routinely used together, all under the watchful eye of the CCTV camera.

Operating a tanker fleet is a complicated business. The aircraft are large and vulnerable and in a war could become high priority targets to any enemy. Therefore, the constant exercises in which the squadron participates are of paramount importance in preparing for operations in a conflict which everyone hopes will never happen. To the tanker crews, any pre-planned rendezvous with aircraft requiring fuel needs to be precise in timing and accurate in location; to linger high in a clear sky waiting for a thirsty customer who turns up late or off course is both dangerous and could see the loss of an expensive aircraft.

Peacetime tanking operations from the United Kingdom generally fall into three main categories. The first is support for tactical air training missions over the North Sea where the tanker transits to a pre-designated refuelling area and flies race track patterns with its drogues deployed. Receiver aircraft, known colloquially as 'trade', are vectored towards the aircraft by ground-based or airborne early warning (AEW) radar and are most likely to be Tornados, Harriers, Jaguars or Phantoms.

The second category is the support of overseas exercises or deployments which almost always require air-to-air refuelling at some stage. Typical missions are the regular armament practise camps in the Mediterranean where front-line squadrons do gun and missile firing as well as air combat manoeuvring against other NATO air force units. There are also squadron exchanges and Red Flag exercises in the USA which are best conducted with air-to-air refuelling support. For the longer overseas missions, the VC10 acts as 'mother' to the deploying aircraft, using its navigation fit (dual Smiths SFS6 flight director system, dual Sperry GM9 compass, dual HF and VHF/UHF communications, ADF, OMEGA and TACAN) to shepherd the 'offspring' to their destination. Spread out off the wingtips of the big aircraft, the smaller tactical aircraft can regularly top up their tanks and rely on someone else to take the additional strain out of the flight. Another advantage of the VC10 is the spare seating

which has been retained in the cabin forward of the fuel cells. This has accommodation for up to eighteen, allowing the aircraft to carry a relief crew and/or maintenance personnel. The front freight hold is also available for spares, and a Turboméca Artouste auxiliary power unit in the tail gives the aircraft complete independence from external support services. Thus, no matter how far from home base, the VC10 should still be able to start its own engines and systems.

If the receiver aircraft has the right kind of equipment, the VC10 can provide air-to-air refuelling to other countries' aircraft. An example of this occur-

In good company. A VC10 K.Mk.3 of No.101 Squadron plays host to two Harrier GR.Mk.3s already well connected. Two Tornado GR.Mk.1s await their turn.

red in June 1987 when six Italian Air Force Tornados on a Red Flag exercise in the USA made the transatlantic flight supported by No.101 Squadron RAF.

The third category involves operations with RAF fighters policing the UK Air Defence Region (ADR). This area covers the whole of the British Isles and the main threat, which has been present for some years and is likely to remain so for the foreseeable future, comes from Soviet long-range bombers. These aircraft, usually missile-carrying variants of the Tupolev Tu-95 Bear family, operate from their northern bases and fly practise missions down to probe the UK ADR and test for any weaknesses which could be exploited should the need arise. To keep these intruders at a respectable distance, RAF Phantoms and Tornado F3s intercept and escort

The underwing Flight Refuelling Mk.32/2800 lightweight pod unit which has digital electronic control and a ram-air turbine to drive the fuel pump which also powers the hose drum unit. (*BAe*)

Below: Pilot's eye view approaching the centre-line drogue of a VC10.

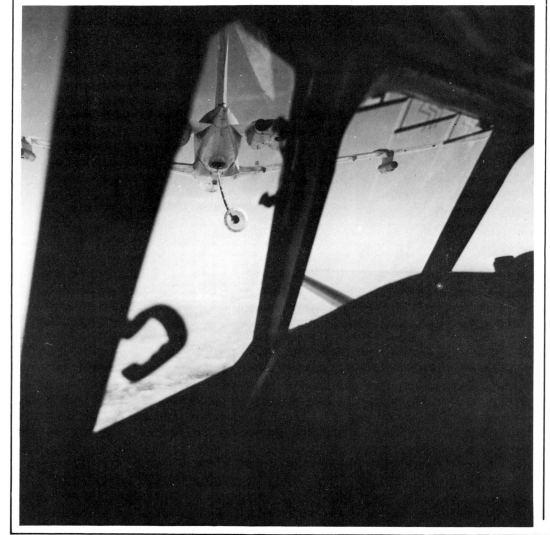

them until they depart from their area of responsibility. Lengthy shadowing is usually the order of the day, so tanker support is called for and fighter, Bear and VC10 often cruise together for some time before the Soviet aircraft heads for home. A tanker is always on alert ready to meet any threatening move by possible intruders and the heavily laden aircraft can be airborne within 15 minutes if necessary. With the pending retirement of the RAF's Victor tanker force in the early 1990s, the existing VC10 and TriStar tanker fleets will be unable to meet all the Service's refuelling requirements into the next century. However, such is the importance and demand for air-to-air refuelling that additional VC10 airframes, which have been in store since their sale by British Airways, will be converted to tanker standard in the next few years. In addition, some VC10 C.Mk.1s will be modified as emergency tankers.

Looking at the history of the VC10, it is ironic that the wheel has turned full circle. Vickers produced the Valiant bomber in a remarkably short time to meet a military requirement. This led to a civil variant design, the V1000, which never flew but which spawned the VC10 civil airliner. The Valiant bomber became a tanker and the VC10 became a military transport. Finally, the civil VC10 was modified, like the Valiant, as a tanker and on existing plans will be an essential element of the RAF front line well into the next century. If reproducing the design concept of another aircraft is an accolade, the final length of service of the military VC10 must be a tribute to the foresight and technical excellence of the original design and manufacturing team. □

RAF VC10 serials

VC10 C.Mk.1

XR806	*George Thompson VC*
XR807	*Thomas Gray VC*
XR808	*Kenneth Campbell VC*
XR810	*David Lord VC*
XV101	*Lanoe Hawker VC*
XV102	*Guy Gibson VC*
XV103	*Edward Mannock VC*
XV104	*James McCudden VC*
XV105	*Albert Ball VC*
XV106	*Thomas Mottershead VC*
XV107	*James Nicholson VC*
XV108	*William Rhodes-Moorhouse VC*
XV109	*Arthur Scarf VC*
XR809	became G-AXLR for RB.211 trials

VC10 K.Mk.2	VC10 K.Mk.3
ZA140	ZA147
ZA141	ZA148
ZA142	ZA149
ZA143	ZA150
ZA144	

VC10: A Niche in History

J R Finnimore

In 1952, the British Overseas Airways Corporation (BOAC) became the first airline in the world to operate a commercial jet, the Comet I. By 1956, the airline needed to take steps to secure its long-term future on North Atlantic routes on which the Britannia would be late coming into service.

In 1956, BOAC was undergoing changes in its top management structure and experiencing more than a little difficulty with its aircraft fleets, following the demise of the Comet I in 1954 and the continuing late delivery of the Britannia propeller-turbine aircraft.

Meanwhile, a major decision had to be taken on new non-stop transatlantic

All shipshape and Shannon fashion, Super VC10 G-ASGI takes off on a crew training flight.

jet equipment and, in November, the aviation minister Harold Watkinson, later Lord Watkinson, authorised the corporation to purchase fifteen Boeing 707s, making it clear, however, that further dollars for additional aircraft would not be forthcoming.

The outcome of this was that, in January 1958, BOAC signed a contract with Vickers for a total of thirty-five VC10s, which was far more than the corporation could envisage needing. Sir Basil Smallpeice, BOAC's resolute managing director of the period, gives a detailed account of events in his fascinating autobiography.* Suffice it to say that while BOAC would have been content to have a British type Boeing 707, built under licence, if it could not have the Seattle product, the airline's paymasters, the British Government, took a different view and decided to support the home aircraft industry with an all-domestic provision.

The arguments about VC10s, which prevailed for years until the type was in service, were essentially about the fleet mix, which eventually settled at seventeen Super VC10s and twelve Standards, and about the necessity for BOAC to commit itself to large numbers of British aircraft ahead of requirement, when the Boeing product could have been bought off the shelf as required.

BOAC had by that time more than its fill of aircraft made in small numbers and delivered late or very late. The launch customer of such aircraft has all the problems of the learning curve and no other operator with whom to share experience or to set up a pooling arrangement for route spares or maintenance support at outstations. By the autumn of 1964, with a small fleet of eight VC10s in service, the 707 had already been operating for 5½ years and over two hundred had been built, serving with fifteen airlines. A further six hundred and fifty already served the needs of the US Air Force as tankers and transports.

*Comets and Queens, Airlife Publishing, 1980.

G-ASGD as flown for the BOAC Super VC10 World Preview. (BAe)

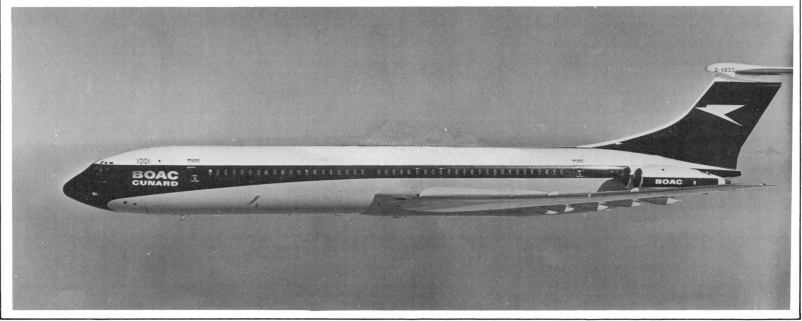

When the first VC10 made its maiden flight in June 1962, BOAC's Rolls-Royce Conway powered Boeing 707-436s had been in service for two years. Already, engineers in BOAC knew that the VC10's direct operating costs on entering service would be substantially higher than the 707's.

The challenge to Vickers to do better was met by a George Edwards aircraft of originality and flair. His VC10 had a stiff clean wing; rear-mounted Conways to give a low cabin noise level from the front first-class cabin to the rearmost economy seats; a lower over-the-threshold landing speed; and superior corrosion protection throughout the primary structure.

BOAC considered that these advantages were being won at the cost of accepting an untried aircraft with a heavy structural weight penalty, its powerful and fuel-thirsty engines leading to higher operating costs. Once the VC10s were in service, however, compensations were found as the aircraft achieved sustained high load factors well above the assumed break even figure. Passenger brand loyalty to the VC10 reached such proportions that it was not unknown for booked passengers arriving for departure and being told they were not on a VC10 to postpone their flight until a VC10 could be guaranteed. This affection for the aircraft was matched by flight deck crews, cabin crews and staffs wherever the VC10 flew, world-wide.

To BOAC it was of fundamental importance that the small fleet of VC10s matched the despatch reliability of the 707s. Not only was this achieved but the maintenance schedule development (increasing the period between maintenance, hence reducing non-flying time) achieved a similar pattern to the 707's in about a third of the time. This enabled BOAC to operate its total Big Jet fleet to the same interchangeable schedule (approximately 4,000 hours per annum).

Now increasingly rare on the ground, a British Triumph Spitfire sports car with proud owner J R Finnimore (author) against a Super VC10 backdrop.

When BOAC eventually sold its fleet of Standard VC10s, one aircraft was broken up for scrap. It revealed no corrosion and no fatigue distress, the basic structure being as good as new.

It had become habitual for the airworthiness authorities – the American Federal Aviation Administration (FAA) and the British Civil Aviation Authority (CAA) – to call for the mandatory major rework of many aircraft, to include wing spar replacement, wing skin and fastener rework, and fuselage skin replacement. There was no such rework on the VC10; there were running in problems, but nothing to detract from Vickers' dedication to structural integrity.

That said, the decision to change the fleet mix to seventeen Super VC10s was more than justified: the increased passenger capacity and improved operating economics really paid off. The Supers matched the 707s in most areas, and in terms of passenger preference and sustained high load factors, it bettered the 707. The original challenge to Vickers to 'do better' had been accepted.

The reliability of the VC10 prompted Dunlop to seek BOAC's co-operation in developing its carbon brake design by equipping one aircraft for operational proving. This exercise was very successful and Concorde was the first to benefit from the enormous weight saving over the steel assembly.

Another spin-off from the VC10 operation was that, under cover of its own Design Approval (CAA), BOAC could produce component salvage, repair and redesigns that facilitated the work being carried out in its own workshops (some of them copied by the parent manufacturer). BOAC's Design and Technical Departments dedicated to the VC10 were not prepared to await the pleasure of tardy suppliers who showed little interest in supporting the tiny production run.

When the VC10s were retired from service, it was an appropriate time to consider their contribution: they were late in the market-place; they were heavy; and they were expensive to buy. But they were not the lame ducks some had predicted: they were reliable; they made money; they had passenger appeal; they were safe; and they were loved by crew and staff. Their loss was almost a bereavement.

VC10: Two Operational Analyses

John Stroud

The operational statistics resulting from BOAC VC10 airline sectors flown between London and West Africa and from BOAC Super VC10 operations between London and New York in March 1964 and March 1965, respectively, are provided here, together with introductory remarks on the circumstances by the Editor.

VC10 Route Proving
– All in a Day's Work

In January 1964, Capt A P W Cane, BOAC's VC10 development pilot, with a small team of senior captains who had already been trained to fly the new aircraft, set out to complete a 1,000-hour flight development programme, before the new type received its full Certificate of Airworthiness (C of A) and entered revenue-earning passenger service.

The fifth production aircraft, G-ARVF, was selected to do most of the flying over the routes VC10s would initially operate in commercial service. One consequence of this decision was that many air correspondents, and even more of Vickers-Armstrongs (Aircraft) employees at Weybridge, Surrey, enjoyed day trips to West Africa. For many there was the opportunity to take off shoes and socks, only some 6 degrees north of the equator, and have a quick paddle in the South Atlantic before returning home. Because these flights were in a north-south, south-north direction, virtually along the Greenwich meridian, some quite interesting things happened. John Stroud remembers a conventionally early start from home to catch the 0800hrs VC10 proving flight from London Heathrow. By 1420hrs he was on the ground at Lagos, Nigeria, putting the hour's stop there to good use with a business contact. A further hour was spent on the ground at Kano on his return, leaving there at 1750hrs and arriving at London Heathrow, engines off, at 2335hrs, still in time after clearing customs to catch an underground train from Hounslow West for the journey home.

There is a little more to this tale than happy reminiscence, because in an era when we have grown accustomed to businessmen travelling to New York and back in the course of a working day on Concorde, using the 5 hour time benefit to advantage when going out westwards in the morning, we forget what was possible a quarter of a century ago. But the distance, you may say, must be very different. Not so. London to New York is some 3,450 statute miles; London to Lagos is some 3,150 statute miles. In its day, the VC10 created quite dramatic reductions in journey times on the West African routes compared with the propeller-turbine Britannia (introduced in April 1959) which itself took over from the piston-engined Boeing Stratocruiser.

The VC10 operating data reproduced in these tables were recorded on flights between London–Lagos–Kano–London on 14 March 1964. The units used are knots for speed, nautical miles for distance, kilograms for weight and feet for altitude; readers wishing to convert these factors may use the following:

Nautical miles to statute:	multiply by 1.1508
Knots to kilometres per hour:	multiply by 1.852
Kilograms to pounds:	multiply by 2.204622
Feet to metres:	multiply by 0.3048

Super VC10s
on the North Atlantic

When the Super VC10 (G-ASGD) first arrived in New York on 14 March 1965, under the command of Capt Tom Stoney of the British Overseas Airways Corporation (BOAC), but still in the ownership of the manufacturer, it was hailed in the American press as the 'world's largest jet airliner'. A series of demonstration flights were to be completed to satisfy the Port of New York Authority that the new aircraft could match the take-off and landing noise criteria. Also made were a series of Super VC10 World

Preview demonstrations at leading cities with appropriate courtesy flights. During these demonstrations, silky-smooth arrivals were frequently greeted by spontaneous applause from the non-revenue passengers; later, ordinary passengers did it too!

Four days earlier, the first BAC One-Eleven had also passed through New York (Newark Airport in New Jersey, in fact, rather than John F Kennedy International) on delivery to its first airline customer in the United States, Braniff International. Altogether, it had been quite a week for improving the morale of those Britons engaged in meeting the huge and competent US airline industry on its own ground.

How, then, did this new Super VC10 match the Boeing 707-320B, its nearest counterpart, then being operated by airlines such as Pan American, Air France, Lufthansa, TWA, Air-India and South African Airways? The 707-320B model with four Pratt & Whitney JT3D-3 turbofans, each of 18,000lb thrust, had been in service since June 1962, the month the Standard VC10 first flew. The two aircraft were similar in many respects. The 707-320B had a slightly longer cabin, permitting an extra two seat rows in the all-economy class, for up to one hundred and eighty-nine passengers. The wing spans of the two aircraft were the same within 5in and the gross wing areas the same within 78sq ft. Even fuel tank capacities were within 525 Imperial gallons. The forward underfloor freight holds were of exactly the same cubic capacity with the rear hold on the Super VC10 being slightly bigger than that on the Boeing.

Operationally, the two aircraft were different. The Super VC10 with 87,200lb static thrust from four Conway R.Co.43s, compared with 72,000lb from the 707-320B's powerplant, inevitably used more fuel but was much more sprightly. The Boeing had an aircraft-prepared-for-service (APS) weight considerably lower than the British product: in the order of 7-8 tons, although this varied from airline to airline.

What, then, were the benefits and the penalties, with a maximum power loading of 3.84lb per lb static thrust for the Super VC10 versus 4.54lb for the 707-320B? At maximum take-off weight, the Super VC10 grossed 8,000lb more than the Boeing and got off the ground to the 35ft mark in some 2,000ft less. The British aircraft was also capable of cruising a little faster, a matter of some 0.02 Mach (8-13 knots).

But perhaps the Achilles' heel, as far as BOAC was concerned, was that the Boeing could carry a reasonable commercial payload from London to San Francisco non-stop which the heavier Super VC10 could not. The British aircraft, with its capability of carrying a maximum payload of up to 50,400lb depending on internal arrangements for 4,720 miles versus the Boeing 707-320B's maximum 51,615lb for a still air 6,160 miles, was to prove a versatile aircraft virtually anywhere on the network, except on the non-stop sectors to the US West Coast which were becoming increasingly commercially significant, BOAC believed, because competing airlines could do them.

Throughout its working life, the Super VC10 did a superb job despite a headwind of prejudice which at times, before it entered service, reached gale force. So great was the basis for misunderstanding at the time that this

Standard VC10 G-ARVF visits Beirut, Lebanon, during the course of the route proving programme early in 1964.

Vickers-Armstrongs VC10 G-ARVF c/n 808

Rolls-Royce Conway R. Co.42
London–Lagos–Kano–London, BOAC proving flight VPF 1085
14 March 1964, Capts Peter Cane, Tom Stoney and Phillips

Sector	London Heathrow – Lagos	Lagos – Kano	Kano – London Heathrow
Stage length	2,737nm	452nm	2,477nm
Flight plan	5hr 45min	1hr 9min	5hr 35min
Block time	6hr 14min	1hr 23min	5hr 45min
Airborne time	5hr 59min	1hr 5min	5hr 34min
Take-off weight (brake release)	135,730kg	105,000kg	127,000kg
Runway	10R	19	07
Runway length	11,000ft	7,600ft	8,610ft
Runway elevation	80ft	132ft	1,563ft
Temperature at take-off	8 °C	34 °C	36 °C
Wind at take-off	110/8kt	−3kt	Zero
Runway length required	—	5,250ft	8,300ft
V1	128kt	116kt	130kt
VR	146kt	129kt	142kt
V2	159kt	145kt	155kt
Time to 100kt	28sec approx	18sec approx	25sec
Max take-off weight for conditions	141,500kg	135,000kg	130,000kg
Fuel	60,500kg	31,000kg	50,000kg
Endurance	9hr 5min	3hr 30min	8hr 0min
'Commercial' load	13,555kg	20,000kg	10,000kg
Cruise level	FL330↗360↘310↗350	FL330	FL360
Cruise level temperature	−60 °C/−70 °C/−40 °C	−38 °C	−45 °C
Cruise level wind	Zero	−20kt	−30kt
Average cruise TAS	—	480kt	485kt
Average Mach	0.86	0.86	0.86
Cruise technique	M: 0.86 ind	Constant altitude	Constant altitude
Est landing weight	95,160kg	94,000kg	93,000kg
Est fuel over destination	18,000kg	21,000kg	14,000kg
Actual fuel at end of landing run	19,000kg	21,000kg	13,800kg
Terminal weather	240/10kt, +33 °C	Light & variable +39 °C	210/15kt, 7 °C, rain
Alternates	Accra, Kano, Niamey	Lagos	Prestwick
Average cruise consumption	1,575kg/eng/hr	1,475kg/eng/hr	1,475kg/eng/hr

London Heathrow take-off run 35sec (wet runway)
London Heathrow VAT 120kt ind, 116kt true

Fuel figures London-Lagos

Start-up, taxi & to 1,000ft	1,000kg	
Fuel to destination	39,300kg	5hr 38min
Diversion	7,790kg	1hr 5min
Contingency & hold at 100kg/min	5,000kg	50min
Fuel required	52,090kg	7hr 13min
Excess fuel	7,410kg	1hr 32min
Load sheet fuel	60,500kg	9hr 5min
A Fuel to destination	39,300kg	
B Circuit & land	1,000kg	
C Water	270kg	
Actual take-off weight	135,730kg	
Burn off A+B+C	40,570kg	
Est landing weight	95,160kg	
Alternates 1 Accra	218nm	
2 Kano	452nm	
3 Niamey		
Actual fuel burn	41,500kg	
Fuel remaining after landing	19,000kg	
Actual landing weight	94,230kg	

Vickers-Armstrongs Super VC10 G-ASGD c/n 854

Rolls-Royce Conway R. Co.42
London–New York, 14 March 1965, Capt T Stoney
New York–London, 16 March 1965, Capts Robert Knights and Dexter Field
BOAC proving flights

Sector	Heathrow – John F Kennedy	John F Kennedy – Heathrow
Stage length	3,110nm	3,137nm
Flight plan	6hr 42min*	6hr 6min*
Block time	7hr 8min	6hr 37min
Airborne time	6hr 49min	6hr 12min
Take-off weight (brake release)	148,520kg	147,664kg
Runway	28L	31L
Runway length	10,500ft	14,572ft
Runway elevation	80ft	12ft
Temperature at take-off	8 °C	3 °C
Wind at take-off	220/7kt	270/15kt gusting 20kt
Runway length required	—	—
V1	130kt	135kt
VR	160kt	160kt
V2	167kt	167kt
Max take-off weight for conditions	151,953kg	151,953kg
Fuel at brake release	68,000kg	67,310kg
Endurance	9hr 51min	9hr 45min
'Commercial' load	9,272kg	8,793kg
Cruise level	FL310	FL330
Cruise level temperature	−60 °C	−50 °C
Cruise level wind	Zero	+32kt
Average cruise TAS	—	490kt
Average Mach	0.84 ind, 0.825 true	0.86 ind
Cruise technique	0.84 ind & constant FL310	0.86 ind & constant FL330
Est landing weight	99,548kg	101,694kg
Est fuel over destination	20,300kg	22,400kg
Actual fuel at end of landing run	19,100kg	21,480kg
Terminal weather	300/6kt, 8 miles	230/10kt, 12km
	Clear 2 °C	2/8, 4,000ft
Alternates	Washington–Dulles	Prestwick
Average cruise consumption	1,690kg/eng/hr	1,810kg/eng/hr
Max sector payload	18,153kg	Max

* On course time.

Copyright: *John Stroud*

operational analysis by John Stroud of two Atlantic flights was felt to be too sensitive for publication because others might choose to misconstrue the figures on fuel burned.

In the early 1970s, the wide-bodied Boeing 747 was coming into widespread use, and seats on offer were handsomely exceeding available passengers, as had happened for a little while during the first phase of the intercontinental jet age. At this point in October 1973, at the time of the Arab-Israeli Yom Kippur war, the member states of the Oil Producing and Exporting Countries (OPEC) decided to use oil as an economic weapon. Almost overnight, the price of aviation turbine fuel, JP4, increased from the range 9-13 US cents per gallon to 36 cents per gallon for new contracts. Fuel which had accounted for 10 per cent of aircraft direct operating costs would, within months, leap to 30 per cent of costs.

From 1974, therefore, with the airline industry plunged into gloomy reality, the outlook for the long-range narrow-bodied jets Boeing 707, Douglas DC-8 and Vickers Super VC10 on primary routes worsened. Wide-bodied jets already in use on these routes needed to be well filled to derive the benefit from their much lower seat-mile costs.

The earlier equipment would find a useful role on existing routes of lower traffic density and on new routes where, in both cases, aircraft-mile costs were important. Already, a looming obstacle for the narrow-bodied jets was anti-noise legislation for the 1980s. These, then, became the hurdles for the later 1970s. It is certain that the Super VC10 suffered most in its sales prospects by being the last of the narrow-bodied jets, of that overall period, to come to the market-place, and by carrying what had become the 'legacy of Empire': good airfield performance at the world's major airports, almost all at or near sea level, and many of them in the temperate zones. □

VC10: Engineering Pedigree of a Thoroughbred

Dr Norman Barfield

Having begun to design commercial airliners only in the closing stages of the Second World War, the Vickers-Armstrongs (Aircraft) team had a lot of ground to cover if they were to compete successfully against established US competition. Between 1945 and 1957, a range of design exercises resulted in a series of propeller-turbine and pure-jet powered aircraft for both commercial and military use, and all this relevant experience was directed into the VC10 concept.

Superb, paradoxical, enigmatic. These and other adjectives have been used over the past 30 years to describe the Vickers VC10 and Super VC10 intercontinental rear-engined jetliner family. The VC10 was Britain's largest (and many would argue, the world's best) passenger airliner contender when the airline industry leaped into the age of jet air transport in the late 1950s.

Technically, the concept and development of the VC10 demanded a courageous combination of vision, leadership, capability, experience and tenacity in the face of quite extraordinary vacillation in customer requirements and national political policy. Fortunately, Vickers possessed just such a matching combination (largely through the perceptive and dynamic leadership of George Edwards), bringing together in the VC10 the turbine-powered aircraft design, operational experience and technology that was then unequalled anywhere in the world.

In order to review the engineering of the VC10 it is important to look at the interlocking facts and project studies from which the aircraft and programme stemmed and matured, then to examine the major areas of advancement. By comparing Vickers' standards of engineering integrity and safety with both contemporary and subsequent standards and requirements, it is clear that the VC10 had a tremendous influence, and out of all proportion to its own programme scale.

Genesis of an Innovator

Having produced the 'interim' Vickers Viking civil version of the wartime Wellington bomber in 1945, Britain's first postwar airliner, and moved on in quick succession to the propeller-turbine Viscount for short-haul airline operation and then to the Valiant jet aircraft, the first of Britain's strategic V bomber force, the team at Weybridge, Surrey, consequently entered the 1950s in a strong position.

The Viking had provided invaluable exposure to numerous home and overseas customers for the first time and at the threshold of an expansive postwar commercial aviation business. The Viscount then brought a wholly new concept of passenger comfort thanks to the vibration-free Rolls-Royce Dart propeller-turbine engines and the introduction of controlled cabin pressurisation and air-conditioning. The Valiant incorporated compound swept wings to match its high-speed/high altitude mission requirement, together with an all-electric systems concept which overcame the problem of commutator brush wear in dry air at high altitudes.

With this reservoir of technical and business experience, secured in less than a decade, Vickers had a head start in fulfilling its ambition to become first in the field with a truly intercontinental jet airliner and, in particular, to compete for the prestigious and potentially lucrative North Atlantic operation.

Whereas de Havilland had already bravely but unsuccessfully tried to break into the jetliner field with the Comet I, the US airline business continued to rely on development of its well-proven piston-engined types: the Douglas DC-7, Lockheed Constellation and Boeing Stratocruiser.

Ambitious Vickers 1000

Given the still embryo perception of Vickers' objective by the British Overseas Airways Corporation (BOAC), and the still largely wartime-orientated (and American-equipped) Royal Air Force (RAF) transport capability, Vickers nevertheless came close to fulfilling its ambition with the Type 1000 project. This was a large intercontinental jetliner directly derived from the Valiant and conceived for both military and commercial transport duties. The V1000 had moved well into hardware with a starting order from the RAF before its premature cancellation in 1955, and this experience contributed to the conception of the VC10.

With its impressive performance in

By the summer of 1957, just over a hundred Valiants had been produced. All were designed to transfer or receive fuel in flight by the Flight Refuelling probe and drogue method, jointly developed with Vickers-Armstrongs (Aircraft) for the Valiant. (*Vickers*)

producing the Valiant, and in record time, it was natural that Vickers and the Ministry of Supply (MoS) should together consider a strategic transport development as a logical long-range successor to the Comet. Initial interest stemmed from an RAF requirement for a replacement for the Handley Page Hastings, which could provide logistic support for the V bombers throughout the world and, with a similar wing planform, act as a radio countermeasures (RCM) decoy with the same radar signature as the Valiant. All this meant that it would be essential for the aircraft to have similar speed and range characteristics and be able to operate from short runways.

The MoS thus requested Vickers to submit appropriate designs and BOAC, which also saw a requirement for a substantially bigger aircraft than the Comet, became involved in the studies from the outset. In order to achieve some kind of economic parity with piston-engined and propeller-turbine types, the aircraft would have to have a better engine than the Avon. Rolls-Royce, with whom the Vickers team had a close relationship in the turbine field, proffered the Conway dual-flow (then designated 'by-pass') engine which had first run in 1951 and by 1952 was developing a rating of around 13,000lb thrust (although its size was limited by the wing depth of the Handley Page Victor V bomber).

The military V1000 was regarded as an ideal stepping stone to the civil version, designated VC7, which could operate not only BOAC's Medium Range Empire (MRE) routes but also over the North Atlantic. For this reason, a common fuselage diameter was established to allow airline-type six-abreast seating and a generous wing area to meet both airfield and fuel/range requirements.

Work on the definitive V1000 started in October 1952 to meet MoS Specification C123D, which resulted from the British Government's decision that the country's global military commitments in peacetime would be most economically and efficiently met by air-transporting men and supplies from the United Kingdom to a limited number of overseas bases, and strategic reserve forces to local trouble spots in an emergency. This policy had been strongly influenced by events such as the Berlin Airlift of 1948 and the Korean War then in progress. The specification called for a Service load of one hundred and twenty equipped troops, or an equivalent freight payload including service vehicles. In similar vein to the Valiant,

the primary requirement of RAF Transport Command was an aircraft with exceptional performance from widely geographically dispersed airfields with limited runways of only 6,000ft in length in ambient temperatures up to ISA +30 °C.

The actual order for a prototype and a structural test specimen was placed in January 1953 and the initial production order for six aircraft for the RAF in 1954. Although relying heavily on the Valiant, the development programme for the V1000 was more detailed than for any transport previously procured by the RAF and it took the Vickers team into several completely new areas.

Prototype construction began in 1953. Though much larger than the Valiant, the V1000 had a similar compound swept wing planform, which was low-mounted, a span of about 140ft, four engines buried in the root sections, with the same moderate aspect ratio, generous wing area philosophy of the Valiant. Wings and tail surfaces incorporated curved Küchemann type tip shapes. The overall length was 146ft, height 38ft 6in and the gross weight about 225,000lb. Significantly, the Conway was the engine with which Rolls-Royce had pioneered the by-pass principle which was later so greatly exploited in the turbofan engine.

While much other valuable experience came directly from the Valiant, the structural problems with the Comet were not known at that time. However, with more than a decade of pressure cabin design experience – from the Wellington V and VI dating from as early as 1938 to the Viscount and Valiant – the Vickers team took great pains with this part of the design, and with the engine bay, by using the newly

Valiant and V1000 Models. (*Vickers*)

defined 'fail-safe' principle which has since become commonplace.

The primary airframe structural design of the V1000 was characterised by massive integrally machined members (first deployed by Vickers in the cruciform spine member of the Valiant fuselage to support the centre wing and the bomb bay) which were incorporated in the multi-spar wing and the supporting hoop frame in the fuselage. In particular, the two inner spars, through which the engines passed, were machined from massive solid slabs of metal, for which the raw material suppliers were considerably taxed in providing billets of homogenous specification and quality on such a large scale.

The V1000 was also the true ancestor of the cross-sectioned shape of the passenger cabins of the first generation range of jets with six-abreast seating.

The flying controls included an all-moving variable-incidence tailplane (also first introduced on the Valiant) and fully power-operated flying control surfaces, each divided into segments and operated by individual self-contained electro-hydraulic power units, thus maintaining the fail-safe principle (as in the airframe structure); consequently, a split AC electrical system was employed.

Political Indecision and Cancellation

Although there were problems as the V1000 design progressed in balancing airframe weight growth and the corresponding increased engine thrust, by 1955, when the first example was well advanced in hardware, powerful political forces were at work, principally in the Treasury, on the heavy expenditure involved in such a large-scale programme. This coincided with an Air Staff dictum that a jet transport was no longer

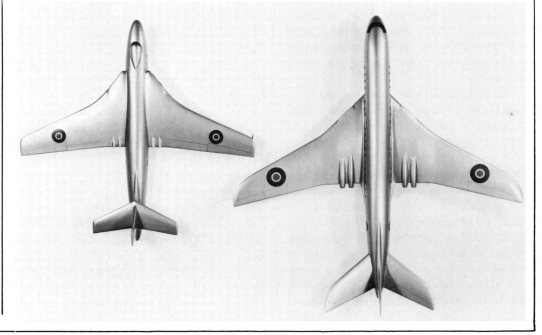

V1000. Photographic impression. (*Vickers*)

There was also governmental pressure for the RAF to support the propeller-turbine Bristol Britannia, the prototype of which had first flown in August 1952, three years after BOAC's original order, and with a joint MoS/BOAC investment of £20 million at stake in the face of a lengthening schedule to get the aircraft into service.
stake in the face of a lengthening schedule to get the aircraft into service.

The final blow for the V1000 was the urgent need to give production work to Short Brothers and Harland in Belfast – consistently Northern Ireland's biggest employer – which had been badly hit by the abandonment of the Supermarine Swift jet fighter and the Comet, for both of which the company had tooled up to build in quantity under contract from Vickers Supermarine and de Havilland, respectively.

When Whitehall decided to place a substantial order for military Britannia 253s to be built by Shorts to prevent large-scale unemployment in Northern Ireland, the RAF contract for the V1000 was cancelled and BOAC simultaneously dropped its interest in the VC7 civil derivative.

The political backtracking that followed is well recorded. On the civil part, although Trans-Canada Air Lines had also shown considerable enthusiasm for the VC7, the BOAC withdrawal, ostensibly on the grounds that the weight increase had penalised the aircraft's performance to such an extent that it would no longer be suitable for non-stop North Atlantic operation (even though a modest thrust increase from the Conway engine would have sufficed to restore the balance), totally sealed the V1000's fate. In reality, BOAC saw this as a neat way out of having to sponsor another British aircraft in addition to the developed Comet 4 and the Britannia.

The V1000 was duly cancelled by the MoS (the government agency and chief guarantor of the project) in a public announcement on 11 November 1955, when the first airframe was 80 per cent complete at the Vickers flight test aerodrome, Wisley, Surrey, and only months away from first flight in mid-1956.

The profound and extremely damaging effect that this cancellation had on the British aircraft industry, and so clearly and repeatedly stated by Sir George Edwards, cannot be overemphasised. Nearly £4 million of public money had been spent and, while RAF Transport Command would have received its aircraft in 1959, the civil VC7 could have been in non-stop Atlantic service in 1959/60.

American Movements

Although the Boeing 367-80 experimental prototype for a military jet tanker/transport had flown in July 1954, and the US Air Force had contracted for a fleet of the tanker version only 6 weeks later, the commercial jet transport 707 that eventually resulted was faced with direct competition from Douglas's projected DC-8.

The great Boeing-Douglas jet transport 'sales race' then took off with a vengeance, with world pacemaker Pan American (which had shown more than a passing interest in the V1000) placing a major launching order for the Boeing 707 in October 1955. Spurred by United Air Lines favouring the DC-8, Pan American also ordered the Douglas off the drawing board a month later. This enabled both types to go into production immediately, although it was not until 1958 that both companies had initial aircraft ready for service. Pan American

with the 707 was then just pipped at the post by BOAC with the revamped Comet 4 in opening up a jet service on the North Atlantic in October 1958, although capable of only one-stop operation. The long-range 707 did not begin non-stop service until August 1959 and the DC-8 two years later.

Less than a year after the V1000 cancellation on 24 October 1956, the British Ministry of Transport and Civil Aviation (MTCA) announced government approval for BOAC to buy the Boeing 707, powered by the Conway (and hence enjoying the prestige of the Rolls-Royce name tag). This also gave the lie to the alleged thrust inadequacy for the VC7. Whereas the government had gone through the political motions of asking whether the VC7 programme could have been resuscitated for BOAC, the V1000 prototype, jigs and tools had already been scrapped.

Advent of the VC10 design

The timing of the VC10 was such that, between the time the V1000 was abandoned and the VC10 design was frozen in early 1957, the Vanguard high-capacity short-haul propeller-turbine transport had been developed at Weybridge, from which the VC10 was to derive substantial benefit. In particular, the integral structural machining technique was greatly exploited for its multi-spar wet wing and introduced into series production. Numerous other valuable airline engineering improvements were introduced, notably a comprehensive structural corrosion treatment process which had devolved from operational tests with the Viscount, an aircraft by then in demand worldwide. Paradoxically, the series of project design studies from which the intercontinental VC10 stemmed began in May 1956 with a design for a much bigger jet replacement for the short-haul Viscount.

Whereas British European Airways (BEA), for which Vickers had already supplied the Viking and Viscount, had specified the 130-seat Vickers Vanguard to succeed the Viscount, it soon became clear that the airline would face stiff jet competition on its regional routes, just as BOAC would on its longer routes. Having thus introduced the Vanguard into full design development at the Weybridge design office, Vickers again focused its attention on the short-/medium-haul section of the market as an outlet for its commercial jet aspirations. Known at Weybridge as the 'Little Jet', and considered for the busy Australian domestic market, the design studies were based on the philosophy of building a cheap and widely applicable pure-jet transport with a first cost of not more than £500,000. However, little progress was possible at that point in the short-haul domain because of the lack of an appropriately economic engine.

Significantly, it was during this work that the Vickers team first became seriously involved in the rear-engined layout for commercial aircraft, for it soon became clear that the thinner wings required for the higher cruising speeds and the relatively large diameter of the emerging family of turbofan engines would no longer be compatible with the buried engine installation. It was therefore only a short and logical step to investigate the similarly close-coupled but external rear-engined configuration. First realised in civil aircraft with the French Sud-Aviation Caravelle, this novel layout had, in fact, been produced by Vickers in 1951 for a short-range pilotless expendable bomber project known as the SP2, for which the engines were to be mounted on the tips of the aft-disposed fins.

During the next few months, the Vickers project studies graduated into considerably larger aircraft, still in the short- to medium-range sector, by taking the Vanguard fuselage and mounting three Rolls-Royce Avon axial-flow turbojets (as used in the Valiant, Comet and Caravelle) in the rear fuselage, thus becoming known as the Vanjet. Variations on this theme were evaluated using adaptations of the wing shapes of both Vickers' contemporary high-speed military aircraft, the Valiant bomber, and the Supermarine Scimitar naval fighter.

After a brief period during which it was thought possible to adapt such a design to meet both the short-haul needs of BEA and the long-haul requirements of BOAC, by 1957 Avro, Bristol and de Havilland were all competing for the BEA order, which was ultimately won by de Havilland's all-new Rolls-Royce Spey-powered Trident.

Tough BOAC Specification

Thereafter, Vickers concentrated exclusively on the emerging BOAC requirement for the intercontinental class of airliner and consequently the design grew still further in both size and power. The studies then crystallised around the airline's initial specification in March 1957 for a Comet and Britannia replacement. The specification called for a payload of around 35,000lb to be carried over 2,500 mile sectors on the airline's African and Eastern routes to Australia. From the implicit combination of high ambient temperatures, high altitude and relatively short runways, plus difficult route sectors involving strong headwinds (35 knots), there emerged the basic design parameters which dictated

and defined the resulting Vickers response. These were the airports at Kano (Nigeria), Nairobi (Kenya) and Johannesburg (South Africa) and the sectors from Singapore to Karachi (Pakistan) and from Kano to London.

This was probably the toughest specification on which a commercial contract had ever been based, embracing some of the world's most arduous operating conditions. Moreover, both Vickers and BOAC recognised that the aircraft would have to be fully competitive in every aspect with the new American jets which were likely to set the trend of international air travel.

Rear-Engined Layout

By late 1956, it was apparent that the rear-engined configuration was clearly superior in providing the exceptional performance so essential for BOAC's particular route requirements. With airfield performance of overriding importance, the unclutterd wing became a vital part in achieving this primary design objective. Although this layout was accompanied by a higher basic airframe weight (due to the absence of bending relief afforded by wing-mounted engines), substantially better lifting characteristics could be obtained from this clean wing arrangement.

The advantages claimed for rear-mounted engines were:
- Greatly improved airfield performance, giving higher payload uplift, hence better operating economy.
- Greater versatility of operation for an airline such as BOAC, with its worldwide route network embracing virtually every kind of operational problem.
- Improved control characteristics and lower approach speeds.

Back into Brooklands, landing south to north (the only direction possible), comes E B Trubshaw with a Standard VC10 for modification to full production standard, after use in the test programme. (BAe)

- A new standard of passenger comfort derived from the drastic reduction in cabin noise level and vibration.

Other virtues which accrued from this installation were:
- Reduced fire hazards in a wheels-up landing because the engines were well isolated from the main fuel tanks.
- Excellent ditching characteristics.
- Structural surfaces less likely to incur fatigue failure from the effects of jet efflux.
- Reduced risk of damage to engines by the ingress of runway debris.

Although the rear-engined arrangement was accompanied by a higher basic weight, these substantial advantages not only convinced Vickers and BOAC that it was the best layout for such a demanding operational requirement, but this also evidently became the view of many other manufacturers and operators. This was especially the case in the short- to medium-haul domain where the benefits of both airfield performance and passenger comfort were equally appreciable – witness the de Havilland Trident, Boeing 727, Tupolev Tu-134/154, BAC One-Eleven, Douglas DC-9, Fokker F28/100 and the host of small business jets. However, only on the Soviet Ilyushin Il-62 was the idea adopted in the intercontinental class airliner.

In meeting BOAC's basic MRE operational requirements, it was then found that relatively small changes would give the aircraft a transatlantic capability as well, without compromising either the airfield performance or the overall fleet operating economy. These changes were incorporated in the definitive VC10 design during 1957 and BOAC's basic order was announced in January 1958.

New wing design concept

Having settled on the aerodynamically beneficial clean wing arrangement, the opportunity was also seized to provide an advanced wing section and shape.

The VC10 thus became the first aircraft to use the supercritical peaky wing design theory. This was the work of distinguished aerodynamicist H H Pearcey of the National Physical Laboratory (NPL), Teddington, and Barry Haines of the Aircraft Research Association (ARA) wind tunnel establishment, Bedford. Their aerofoil design concept gave a two-dimensional transition from supersonic to subsonic flow, without creating a strong shock wave, as with more conventional aerofoil sections, which provided benefits in both drag and buffet reduction.

However, considerable work had to be done to convert this two-dimensional benefit into three-dimensional reality in the actual working wing of the VC10. A complementary theory providing a reasonably accurate determination of pressure distribution over a definitive wing had been established by Dr Maria Weber of the Royal Aircraft Establishment (RAE), Farnborough. But the reversal of her work to provide data to design the VC10 wing came from the combined efforts of the Vickers Weybridge aerodynamics team and the RAE, for which the early Pegasus analogue computer was used.

To aid the fulfilment of the stringent airfield requirements, the clean wing was also able to incorporate highly efficient leading-edge slats together with powerful, large area, 65 per cent span Fowler flaps in the wing trailing edge.

Nevertheless, so advanced was the overall aerodynamic concept of the VC10 that it was not completely right first time and a 4 per cent wing leading edge extension, together with some reshaping, was needed after early flight testing. New work was also done to determine engine interference and installation drag effects, which notably resulted in changing the incidence of the engine nacelles and the fitment of a beaver tail fairing between the rear nacelles.

Passengers' Choice

Vickers' and BOAC's progress in developing the VC10 design was motivated by the objective of providing improved service to the passenger in a single aircraft type together with all the benefits of an overall operation with one fleet, which could be used throughout a worldwide route network and not just on selected routes. This philosophy not only characterised the basic layout of the aircraft, but also led to the completely new concept of interior design.

To exploit to the full the first step in improved passenger comfort, a major programme was then initiated, in collaboration with the team of American

Super VC10 G-ASGB at London Heathrow with spare engine pod. (*British Airways*)

specialist industrial designers (Charles Butler Associates of New York), to combine the best in contemporary decor with a new form of prefabricated laminated plastic fixed furnishing in place of traditional soft upholstery and craft carpentry, and to optimise the spacious payload compartment. Additionally, the passenger cabin appointments were complemented by carefully planned stowages for every item of equipment. All above- and below-floor door sizes exceeded the best of contemporary standards, making for easier and quicker handling. The main cabin passenger and service access doors incorporated an ingenious outward-opening plug mechanism with a parallel swinging linkage so that the doors were not fully to wind, as with a conventional outward-opening hinged design.

A 30 ton Freon vapour-cycle cabin refrigeration system was provided (in the wing root leading edge) for both ground and flight use in tropical conditions to eliminate the physical depression of passengers. This facility had not previously been available in intercontinental operation, although Vickers had experience of such a device fitted to the Capital Airlines Viscounts operating the densely travelled eastern seaboard states of the USA.

As well as the take-off, approach and landing speed improvements afforded by the clean wing design, the VC10 also had superior high-speed performance. This applied to both the maximum cruising speed and the minimum cost cruising speed; worthwhile reduction in block times were thus achieved. With a

20 per cent higher power-to-weight ratio than contemporary jets, a higher climb rate was also available. A higher cabin working differential pressure (9 lb per sq in) meant a wider band of cruising flight altitudes without impairing comfort. All these factors provided added benefits for the passenger.

Advanced Airframe and Systems

The principal and capital equipment for large area sculpture machining, first introduced for Vanguard airframe construction in the mid-1950s and which thereafter became a way of life at Weybridge, were greatly exploited in the VC10. Over 55 per cent of the weight of the airframe structure was machined from solid slabs of metal, including large sections of the fuselage skin, as well as the wings. This resulted in smooth and largely uninterrupted stress patterns and a greatly reduced parts count.

With universally applied fail-safe multi-load path principles, meticulous choice of materials and uniformly low working stress levels throughout the airframe (11,000 lb per sq in compared with 17,000 lb per sq in in the contemporary big jets), the VC10 thus began service with a commendably long-life structure: a minimum crack-free service life of 30,000 flying hours or 10,000 landings (equivalent to 10 years' operation). Special provisions were also made for the entire airframe to be drained of moisture and completely protected to eliminate corrosion, a highly effective innovation in the long-life objective.

Providing the aerodynamic and structural facilities for operation from secondary and tropical airports would

Uncowled, Numbers 1 & 2 Rolls-Royce R.Co.43s in a Super VC10. (*Rolls-Royce*)

have been to little avail if the undercarriage was not matched to low runway bearing classifications. Consequently, large low-pressure tyres and fully duplicated main wheel hydraulic braking systems were fitted to the Vickers-designed and -built 4-wheel bogie main units. Together with the low touchdown speeds, the undercarriage won high praise both from pilots and passengers.

Indeed, perhaps the finest tribute ever paid to a transport aircraft undercarriage was that repeatedly made during the initial operation of the VC10 and typified by the *Daily American* (Rome): 'The landing was such a masterpiece of perfection that the travel-hardened passengers burst into applause'.

To extend the fail-safe philosophy to the design of the systems and the use

of fully power-operated flying controls, all the primary aircraft systems were based on the split-system philosophy. This was developed logically from the Valiant and V1000. Safety being of primary importance, each system was split into two independent halves, each with its own engine-driven power supply source. Both half-systems were arranged to operate simultaneously and each designed to act as an emergency standby for the other, since all relative services could be operated from either of the half-systems, albeit at lower operating rates. This principle eliminated the possibility of an 'emergency only' system which itself may be 'rusty' and/or inoperative when needed. The split-system concept was applied to everything vital to the safety of the aircraft – hydraulics, electrics, air-conditioning, flying controls and landing aids – all systems being in constant use.

It was also targeted towards an autoland capability (for which it was eminently suited) and the aircraft eventually became certificated to Category I. Although this facility was not used in service, it provided valuable experience for the BAC One-Eleven domestic and regional twin jet and most other British and foreign airliners that followed.

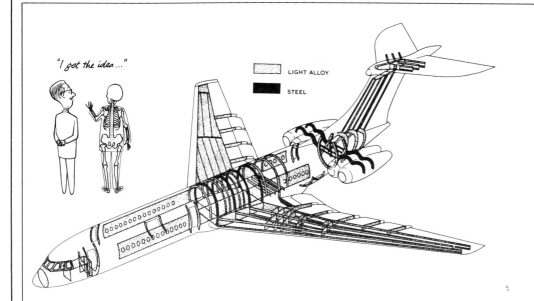

Over 50 per cent of the structure, by weight, was machined from solid metal.

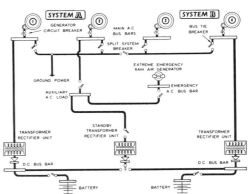

The split-system philosophy – electrics. System A is driven by engines 1 and 3. System B is driven by engines 2 and 4.

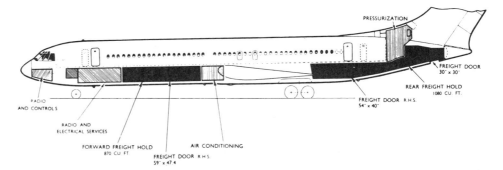

Super VC10. Arrangement of underfloor systems bays and freight holds.

(*All drawings BAe*)

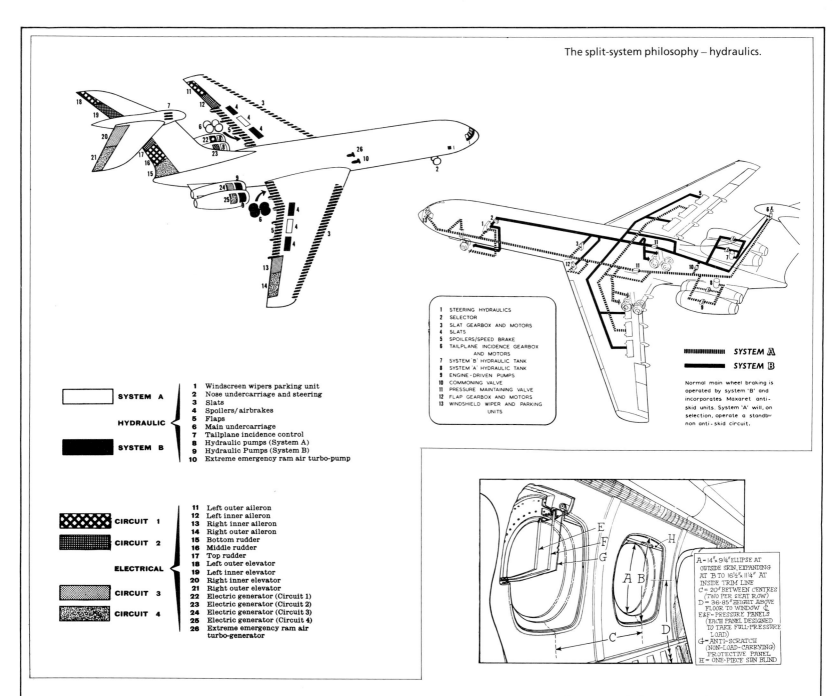

The split-system philosophy – hydraulics.

1 STEERING HYDRAULICS
2 SELECTOR
3 SLAT GEARBOX AND MOTORS
4 SLATS
5 SPOILERS/SPEED BRAKE
6 TAILPLANE INCIDENCE GEARBOX
     AND MOTORS
7 SYSTEM 'B' HYDRAULIC TANK
8 SYSTEM 'A' HYDRAULIC TANK
9 ENGINE-DRIVEN PUMPS
10 COMMONING VALVE
11 PRESSURE MAINTAINING VALVE
12 FLAP GEARBOX AND MOTORS
13 WINDSHIELD WIPER AND PARKING
     UNITS

SYSTEM A

SYSTEM B

Normal main wheel braking is operated by system 'B' and incorporates Maxaret anti-skid units. System 'A' will, on selection, operate a standby non anti-skid circuit.

SYSTEM A

HYDRAULIC

SYSTEM B

1 Windscreen wipers parking unit
2 Nose undercarriage and steering
3 Slats
4 Spoilers/airbrakes
5 Flaps
6 Main undercarriage
7 Tailplane incidence control
8 Hydraulic pumps (System A)
9 Hydraulic Pumps (System B)
10 Extreme emergency ram air turbo-pump

CIRCUIT 1

CIRCUIT 2

ELECTRICAL

CIRCUIT 3

CIRCUIT 4

11 Left outer aileron
12 Left inner aileron
13 Right inner aileron
14 Right outer aileron
15 Bottom rudder
16 Middle rudder
17 Top rudder
18 Left outer elevator
19 Left inner elevator
20 Right inner elevator
21 Right outer elevator
22 Electric generator (Circuit 1)
23 Electric generator (Circuit 2)
24 Electric generator (Circuit 3)
25 Electric generator (Circuit 4)
26 Extreme emergency ram air turbo-generator

A – 14″ × 9¼″ ELLIPSE AT OUTSIDE SKIN, EXPANDING AT B TO 16½″ × 11¼″ AT INSIDE TRIM LINE
C = 20″ BETWEEN CENTRES (TWO PER SEAT ROW)
D = 36.85″ HEIGHT ABOVE FLOOR TO WINDOW ₵
E & F = PRESSURE PANELS (EACH PANEL DESIGNED TO TAKE FULL-PRESSURE LOAD)
G = ANTI-SCRATCH (NON-LOAD-CARRYING) PROTECTIVE PANEL
H = ONE-PIECE SUN BLIND

Arrangement of passenger windows.

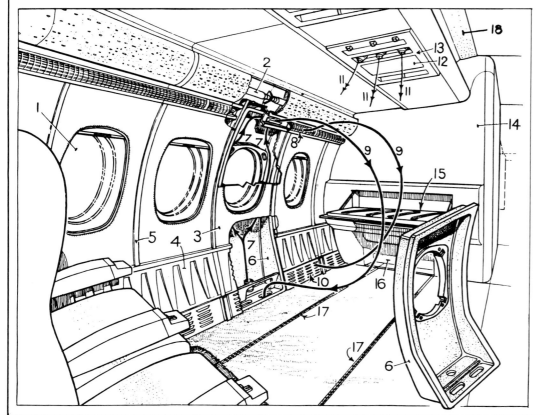

Engineering passenger wellbeing. Cabin hard trims and conditioned airflow distribution.
Key: 1 14 by 9¼in elliptical windows at 20in pitch with single piece blinds. 2 fluorescent cornice lighting. 3 rigid moulded plastic trim panels easily removable and replaceable for maintenance and access to air-conditioning ducts and airframe structure. 4 fluted plastic dado panel. 5 sidewall trim panel retaining strip. 6 pre-moulded conditioned air duct fits behind trim. 7, 8, 9, 10 transverse air distribution path. 11 individual cool air outlets. 12 emergency 'drop-out' oxygen masks. 13 therapeutic oxygen connection. 14 non-structural bulkhead. 15 'carrycot' shelf. 16 foot well. 17 seat rails. 18 public address speakers behind perforated trim.

Taking up the principle established with the V1000, the primary flying control surfaces of the VC10 were also split into segments, each driven by a separate electro-hydraulic power unit: ailerons – four, elevators – four, rudders – three, making a total of eleven identical power units to which the power supplies were also cross-coupled to complete the overall fail-safe provisions. The power supplies for the six wing-mounted, hydraulically operated spoiler/speed brake sections (three per side) were similarly arranged. However, to provide adequate control in the unlikely event of a four-engine flame-out, drop-down electrical and hydraulic ram-air driven turbines (Elrat and Hyrat) were fitted, the first to be installed in a commercial airliner. These units were actually proved once in anger when a fuel management error on a scheduled flight caused a complete loss of engine power at 35,000ft. As intended, the aircraft was flown under control from ram-air turbine power only until main engine power was restored at 23,000ft.

A clinical approach was adopted in the choice of system equipment. Tried and tested components, particularly those with a large background of experience on contemporary jets, were incorporated whenever possible. To obtain the greatest benefit from both the British and American equipment industries, advantage was taken of reciprocal design and production licence agreements between companies with complementary interests on both sides of the Atlantic.

To convey the benefits of engineering integrity and fail-safe design concepts to the flight crew, a spacious flight deck with superb visibility was logically and meticulously laid out. The aim was to reduce the workload and eliminate crew fatigue with the same thoroughness as was devoted to the airframe and systems design. The fully power-operated flying control system (hence substantially fly-by-wire) was publicly acknowledged as the best in the world by the engineer of a rival American company.

The complete engineering programme for the VC10 family was supported by a major test schedule, both on the ground and in the air. Full-scale replicas of the structure and systems underwent exhaustive tests and fatigue and automatic cycling to extend component reliability and overhaul life. This resulted in the most modern concepts of planned maintenance being applied, which led not only to lower introductory and maintenance costs but, significantly, also resulted in excellent down-the-line reliability.

Conception of the Super VC10

As the VC10 development programme proceeded, there were fundamental movements in the global jet transport scene. The characteristics of the first generation of American intercontinental jets necessitated major runway extensions in many parts of the world. It became clear that, although the Standard VC10 would be able to show superiority on BOAC's African and Eastern routes, there was the potential for a larger development for the airline's trunk routes, and in particular the North Atlantic operation where airfield length was not limiting and where larger payloads would be available.

This decision was made easier to implement by the intrinsic nature of the basic aerodynamic and powerplant layout. Because the engines were close-mounted, no significant changes would be necessary to the empennage configuration through increased design problems of asymmetric thrust in the engine failure case. The added payload capacity could be provided by the straightforward addition of fuselage length in the appropriate positions ahead and behind the wing to maintain longitudinal balance.

The amount of extended fuselage capacity was conditioned by the engine power development prospects. For a period, it was planned to extend the fuselage by 28ft and to fit a 24,000lb thrust development of the Conway in order to optimise the aircraft to carry up to two hundred and twelve passengers solely between London and the gateway cities of the eastern seaboard of North America, making full use of the long runways and plentiful termini facilities at each end of the route.

The complete practicability of this design (named Super VC10) proves that the basic concept of the VC10 family was able to take full advantage of improvements in engine development without prohibitive geometric, structural or aerodynamic changes.

In the event, a more modest fuselage extension was adopted to provide an aircraft that would specialise mainly in the operation of BOAC's North Atlantic and transcontinental routes while retaining complete versatility through all its other routes as the need arose. Thus, the definitive Super VC10 was developed with a 13ft fuselage extension to increase passenger capacity from the one hundred and thirty-five economy class of the Standard VC10 to one hundred and sixty-three economy seats. Lower seat-mile operating costs were a natural corollary to this still substantial increase in size.

Military Developments

From its evolutionary background, it was natural that the VC10 should be considered by the RAF as a multi-mission strategic transport (in which concept Vickers was already well versed with its Valetta of a decade earlier). This stemmed principally from the fact that the same kind of short, high altitude and tropically located airports which had so characterised the whole VC10 conception for BOAC were again specified by the RAF in its own strategic jet transport reappraisal made towards the end of the 1950s. This study resulted in Air Staff Requirement (ASR) 378 which was translated into Ministry of Aviation (MoA) Specification C239 in 1960 and built round the VC10, which had by that time reached production status.

Known in the RAF as the VC10 C.Mk.1, and at Weybridge as the Vickers Type 1106, the military adaptation of the commercial airliner was begun in September 1961. This followed the government's decision to increase the size and versatility of the RAF's strategic jet transport force with an initial order for five aircraft to begin service in 1966. With its greatly increased size and range and superlative airfield performance, the VC10 would be complementary to the RAF's long-range Britannia 253s and Comet 4Cs already in service and the Shorts Belfast propeller-turbine strategic freighter to come. The VC10 was to operate over similar routes to the Britannia, but with a greater versatility of loading and at much faster speeds. Six further VC10s were ordered a year later and the total was increased to fourteen in July 1964 when the government made over three former BOAC Super VC10 production line positions to RAF Transport Command. All fourteen aircraft were, in fact, built to the same standard.

The detailed specification of the military VC10 called for worldwide inter-theatre operation in the diverse roles of troop transport, freighter and aero-medical evacuation. A flight refuelling facility was also specified, initially as a receiver aircraft with a possible later adaptation to the tanker role.

Basic design features were to meet British Civil Airworthiness Requirements (BCAR) and Air Registration Board (ARB) conditions and the aircraft was to be delivered with a passenger transport category Certificate of Airworthiness (C of A). These requirements were to be supplemented by the traditional military AP 970 where no relevant BCAR existed. In fact, the whole RAF VC10 programme was very much similar to that for a commercial operator.

Adopting an aircraft and engine that were already established and fully proven in worldwide service was a great comfort to the RAF. However, the VC10 did bring the RAF a number of major innovations in addition to being considerably larger than any of its previous aircraft in any Command. It was the first RAF rear-engined T-tail aircraft, and the first with turbofan engines, outward-opening plug doors, an advanced automatic flight control system with auto-flare/autoland capability and a Skydrol 500A synthetic oil hydraulic system.

On the other hand, there was already considerable experience available in the RAF on large aircraft in such important areas as fully powered flying controls, high-frequency AC electrical systems and constant-speed drives. The handling characteristics of the V bombers and the Comet, with engines mounted close to the fuselage, also provided a directly relevant preface to the VC10 operation. The main cabin side-loading freight door was essentially similar to that of the RAF Britannia 253, forerunner of the widespread use of combination passenger/freight transport.

The extensive use of integral machining techniques in the VC10 airframe structure, as noted earlier, was extended

to the freight floor of the RAF aircraft. This floor was made up of replaceable machined light alloy panels incorporating around four hundred tie-down lashing points on a 20in grid throughout the length and breadth of the cabin, the whole floor being carried on integrally machined transverse beams at each fuselage frame station along the length of the cabin. The floor was designed to carry a local ultimate loading of 1,000lb on each square foot and the support structure to accept a typical maximum vehicle axle load, without spreaders, of 5,500lb centrally placed. The lashing points were designed to restrain an ultimate load of 10,000lb in any direction. Further freight lashing points were provided in two rows along the side walls.

Full provision was also made for the carriage of up to eight pre-loaded freight pallets of 108 by 88in (the well known commercial standard).

Standard RAF practice since the introduction of the Britannia in 1960 has been the transport of mixed loads of personnel and freight in the same aircraft and the same procedure has since been adopted by many of the world's airlines. There were clearly great advantages in moving troops and their equipment together in the same aircraft so that the complete force arrived at its destination equipped and ready for immediate action. In this way, five VC10s could airlift an infantry force of five hundred and thirty fully equipped troops,

eleven general service heavy utility cars and their trailers, and 14½ tons of freight over a distance of about 4,000 miles in roughly 8½ hours. Additional freight and equipment could be carried in the underfloor holds.

The RAF VC10 thus became a worthy, faster and more useful complementary type to the RAF Britannia. Both shared the civil pedigree, which ensured a long airframe life, with design for military service when needed. In an economy drive to reduce overseas commitments, the British Government ordered the phasing out of the Britannias long before the end of their useful lives, leaving the VC10s to carry on the strategic fetching and carrying duties.

Commercial experience with the BOAC VC10s and Super VC10s eventually led to a doubling of the aircraft's initially designed life of 30,000 flying hours/10,000 landings to 60,000 flying hours/20,000 landings. A much greater residual life was now available in the ex-airline aircraft, making it a thoroughly practical decision to convert nine of them into full air-to-air refuelling (AAR) tanker/receiver aircraft for the RAF.

The article on page 7 deals with the RAF aspects of VC10 operation; the observations made here confirm the engineering pedigree which is making that operation possible well over 30 years after the start of the initial design by a company which had been making military transports for the RAF since the Vickers Vernons of the early 1920s.

Hands off as BAC test pilots Eddie McNamara and Bill Cairns monitor the final 20ft before automatic touchdown, with the Elliott system, of Super VC10 G-ASGG at RAE Bedford. (*Elliott-Automation*)

What Might Have Been

Several other notable and imaginative project design studies were made in an attempt to exploit the development potential of the VC10 in both the commercial and military fields, but all were frustrated by prevailing political climates and/or the radically changing world air transport scene and, ultimately, the small overall production run of the VC10.

In 1961, proposals were made for a 4,000 mile range military VC10 mounting four of the American Skybolt 1,000 mile range two-stage solid fuel air-launched ballistic missiles (ALBM). Although the British Government's brief flirtation with Skybolt as a primary nuclear deterrent weapon resulted in the cancellation of the United Kingdom's own Blue Streak intercontinental ballistic missile (ICBM), Skybolt itself was abandoned by the Americans due to technical problems in both propulsion and guidance, and in 1962 the government finally adopted the undersea-launched American Polaris system and commissioned a major programme for the construction of the supporting submarines.

Following several projected fuselage stretches well beyond the definitive Super VC10, Vickers proposed a 265-seat double-deck version with three Rolls-Royce RB.178 engines (predecessor of the later RB.211) which BOAC publicly dubbed the Superb. The upper deck was to have been identical to the basic VC10 and the lower deck multi-arc

East African Airways Super VC10 5X-UVA models the 140 by 84in cargo door incorporated to customer specification in both Standard and Super VC10s. (BAe)

form, the side radius being identical to the upper section to accommodate the standard VC10 windows and furnishing panels. However, this project was not proceeded with, partly on the grounds that emergency exit and ditching requirements were deemed by the authorities to be unsatisfactory at that time, but mainly because BOAC opted to continue its American buying policy by, later, ordering the Boeing 747.

Almost from the beginning of design work on the VC10, it was clear that its high volumetric capacity, combined with its advanced wing design, high power loading, modest approach speed and superb airfield performance, could certainly provide an excellent basis for a high work capacity economic freight transport.

Although a wave of contemporary opinion was emphasising the need to consider the project as a complete freight transport system (using containers, loading closely integrated with freight terminal, road and rail trucking support systems) there were differing views on the merits of side, and end, loading. Both methods were extensively evaluated so that the considerable experience gained in side loading could be perpetuated, if required, or to assess the particular suitability of the VC10 design to nose loading, either by a hinged nose section or clam shell loading doors with a raised flight deck position for access to the full fuselage cross-section.

In the event the emergence of the American Lockheed C-5/Boeing 747 generation precluded the development of this latter approach and the preference was to retain side loading as a logical first step from existing designs and the experience gained with the conversion of numerous piston-engined airliners and military transports. Consequently, the operation of combination passenger/freighter aircraft became the

The flight engineer's station on the VC10 concentrates all the aircraft systems information and control, laid out diagrammatically, in one place. A duplicate set of throttles is provided. (BAe)

VC10 flight deck. Captain and first officer's positions with outboard entry to their seats, allowing maximum dual use of a generous centre pedestal housing throttles, trimmers, radio controls and high-pressure cocks aft. To the right is part of the flight engineer's station. (*BAe*)

widespread preference, and when Freddie Laker's British United Airways led the way by specifying a large side-loading door, as noted earlier, all subsequent production aircraft (for Ghana Airways, East African Airways and the RAF) incorporated this feature.

Other more exotic VC10 development possibilities included twin and even triple fuselage designs, with four and six tail-mounted engines, respectively. One ex-RAF aircraft was also used as the flying test-bed for the Rolls-Royce RB.211 large-fan engine (on one side in place of two Conways) and the possibility of the twin RB.211-powered VC10 was a natural corollary to this (with the split-systems concept being particularly attractive).

Licence production of the complete Super VC10 in China was also examined but could not be brought to fruition within a timescale that was compatible with active programme continuity in the United Kingdom.

Retrospect and Prospect

Like its military contemporary, the TSR.2, in the intervening 27 years the VC10 also seems to have acquired its own mythology. But the qualifying facts do expose and explain just what a monumental task it was, both physically and politically, to bring it to reality with the degree of technical, if not commercial, success it achieved and sustained, especially in its airfield performance and passenger appeal – all essentially from the Weybridge team and factory. Of the multitude of aircraft projects and programmes sponsored by the British industry during the Second World War, the VC10 was (and still is) the largest aircraft type to be put into series production in the United Kingdom and the

largest single-company sponsored type in Europe (excluding the Airbus consortium).

The VC10 was also one of the most paradoxical aircraft. On the one hand, it was customer-specified and built for the British national airline BOAC to a greater extent than ever before or since for a major international operator. Yet it was grossly frustrated by the interplay of the extreme vacillations of the management, the politics of the airline itself and national political factors far beyond Vickers' ability to influence, let alone control – hence the numerically small production out-turn. All this was also a measure of the inordinate risk and vulnerability of a single company, even as large as Vickers then was, and the inexorability of international collaboration on such large-scale projects in the future. On the other hand, so well was the VC10 engineered that the combination of its operational utility and longevity has kept it in the forefront of diplomatic prestige throughout the world for nearly 30 years.

In retrospect, meeting the tough technical and operational specifications for the BOAC MRE routes meant that Vickers was deluded into providing a better performing aircraft off the ground than the market eventually needed. However, this provided an excellent basis for a premier North Atlantic aircraft as well. Unfortunately, the commercial considerations imposed by the prime customer would not permit Vickers to realise an entirely feasible full-

stretch aircraft. This, coupled with the difficult – indeed almost impossible – economic situation on production numbers, meant that the aircraft genuinely could not reach its full potential. Nevertheless, the all-round superb engineering of the aircraft stood it in good stead in the limited applications to which it was put.

Most notably, its exceptional airfield performance proved to be an excellent safety feature vis-à-vis the first generation American jets that were designed exclusively to operate from primary runways at the major cities which were invariably at or near sea level. The split dual-systems arrangement was both imaginative and reassuring, providing adequate and safe performance in the air from only half of one system, yet without the complexities and weight of completely isolated active and passive duplication – especially in the demonstrated despatch reliability 'far down the line'.

Although engineering-wise the VC10 could fairly be regarded as a generation ahead of its competitors, the subsequent runway lengthening programme throughout the world effectively reduced its lead to only half that margin. However, large-scale structural machining, the heavier and stiffer wing and the 'all over, inside and out' painting and protection from the detail manufacturing stage all provided long-life advantages which are of continuing benefit to the RAF and hence to the British economy more than 30 years later.

The ultimate Super VC10 achieved the distinction of being the biggest airliner on the Blue Riband North Atlantic, as well as the best, at a time when the traditional sea-going holders of that accolade were finally abandoned – but not for long. Now, the majority of the huge mass traffic volume knows nothing other than wide-gauge room-size cabins and no longer have the option.

Whatever the tribulations, the VC10 has demonstrated beyond doubt that, as the ultimate product of the world-renowned five-decade Vickers-Weybridge lineage, it was as proud and commendable a symbol of British ingenuity and capability in the eyes of both the engineering and the aeronautical world as the first Vickers transatlantic aircraft, the Vimy, had been 50 years earlier. Today, the Anglo-French supersonic Concorde (to which the Weybridge team made the largest single contribution in design, manufacture and test of any and all of the eight factories involved on both sides of the English Channel) will be partner to the RAF VC10 into the early years of the twenty-first century. Achievement indeed.

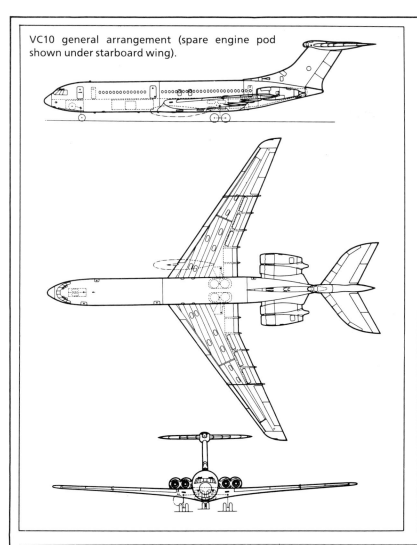

VC10 general arrangement (spare engine pod shown under starboard wing).

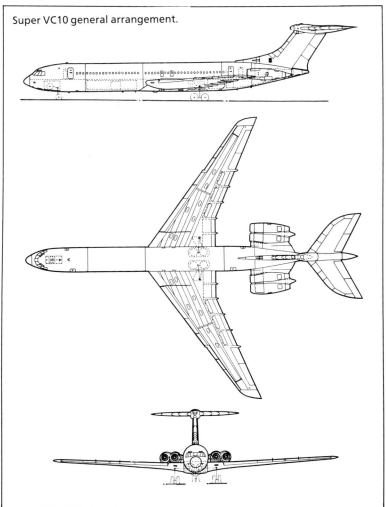

Super VC10 general arrangement.

Leading Particulars of the VC10 Family

Model	Standard VC10	Super VC10	VC10 C.Mk.1	VC10 K.Mk.2	VC10 K.Mk.3
DIMENSIONS:					
Overall length (excl refuelling probe)	158ft 8in	178ft 8in	158ft 8in	158ft 8in	171ft 8in
Span	146ft 2in	146ft 2in	146ft 2in	146ft 2in	146ft 2in
Height	39ft 6in	39ft 6in	39ft 6in	39ft 6in	39ft 6in
Gross wing area	2,932sq ft	2,932sq ft	2,932sq ft	2,932sq ft	2,932sq ft
Aspect ratio	7.5	7.5	7.5	7.5	7.5
Sweepback at ¼ chord	32.5°	32.5°	32.5°	32.5°	32.5°
Tailplane area	638sq ft	638sq ft	638sq ft	638sq ft	638sq ft
Wheel base	165ft 10½in	72ft 1½in	65ft 10½in	65ft 10½in	72ft 1½in
Wheel track	21ft 5in	21ft 5in	21ft 5in	21ft 5in	21ft 5in
CABIN DETAILS:					
Length (excl flight deck)	85ft 6in	105ft	85ft 6in	85ft 6in	105ft
Max width	11ft 6in	11ft 6in	11ft 6in	11ft 6in	11ft 6in
Max height	7ft 5in	7ft 5in	7ft 5in	8ft 6in	8ft 6in
Usable volume	5,620cu ft	7,850cu ft	5,620cu ft	5,620cu ft	7,850cu ft
Underfloor hold volume	1,412cu ft	1,942cu ft	1,412cu ft	610cu ft	750cu ft
ACCOMMODATION:					
Passengers/troops/fuel	Up to 151 seats	Up to 174 seats	150 troops	Additional fuel up to MTOW	Additional fuel up to MTOW
POWERPLANT:					
Four Rolls-Royce Conway 42 turbofans	Mk 540	Mk 550	301	Mk 550B	Mk 550B
	20,400lb thrust	21,800lb thrust	21,800lb thrust	21,800lb thrust	21,800lb thrust
WEIGHTS AND LOADINGS:					
Max take-off weight (MTOW)	312,000lb	335,000lb	323,000lb	312,000lb	335,000lb
Max landing weight (MLW)	216,000lb	237,000lb	235,000lb	216,000lb	237,000lb
Max zero fuel weight (MZFW)	188,400lb	215,000lb	205,000lb	188,400lb	215,000lb
Typical operating weight empty (OWE)	147,000lb	158,600lb	146,000lb		
Max payload	39,800lb	50,400lb	59,000lb		
Total basic fuel capacity	17,925 Imp galls	19,365 Imp galls	19,365 Imp galls	17,925 Imp galls	19,365 Imp galls
PERFORMANCE:					
Typical cruise speed	550mph	550mph	550mph	550mph	550mph
Take-off distance to 35ft (ISA SL MTOW)	8,300ft	8,300ft	8,400ft		
Landing distance (ISA SL MLW)	6,400ft	7,000ft	6,500ft		
Range with max payload (no reserves)	5,040 miles	4,720 miles	3,900 miles		

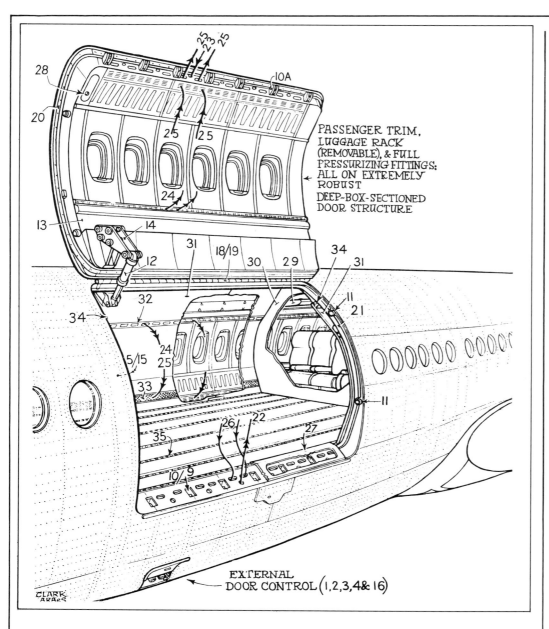

PASSENGER TRIM, LUGGAGE RACK (REMOVABLE), & FULL PRESSURIZING FITTINGS: ALL ON EXTREMELY ROBUST DEEP-BOX-SECTIONED DOOR STRUCTURE

EXTERNAL DOOR CONTROL (1, 2, 3, 4 & 16)

Combined passenger freighter version incorporates large door while retaining full passenger services.

Key: External door control: selector **1** selects electric pump **2** or emergency hand pump **3**; external operating switch **4** or internal switch **5** then operates latch-jack (at **9**) to operate latches **10** (hook latches **10A**) and pins **11**, and operates two-stage door jack **12** to open door to horizontal position for side loading. For vertical (crane) loading, support the door **13** with strut supplied, disconnect jack **12** and retract it, rearrange linkage **14** and couple-up jack again to expand it and so give further door-opening to vertical position (as shown). Hand control **15** (inside door) and **16** (outside) master-control the latches **10** and override **4** and **5** before take-off and release after touchdown. Door **13** has flexible hinge-pin **18** threading hinge-bushes **19** (anti-bind during fuselage racking). Pressure-seal **20** (on door). Noise-seal **21** on door frame. Cabin air supply **22** up into door via **23**, out into cabin at **24**, back at **25** and out via **26**. Debris guard **27** to air-ports **22**, **26** when door is open. At **28** is container door with sight hole to transparent bottle of silicagel crystals for window demisting. With freight, hat-racks (as **29**) are removed from freight compartment (in sections). End wall **30** positioned and wall trim protection guards **31** erected and clipped up grilles **32** and **33** in **31** allow circulation of cabin air **24**, **25**. Spotlights **34** light up door sill. Eight floor rails **35** run length of cabin. All luggage racks are taken down for freighting but air ducting remains except across door where it is built into that section of luggage rack spanning the door and so is removed with it.

Operational variations, VC10 and Super VC10.

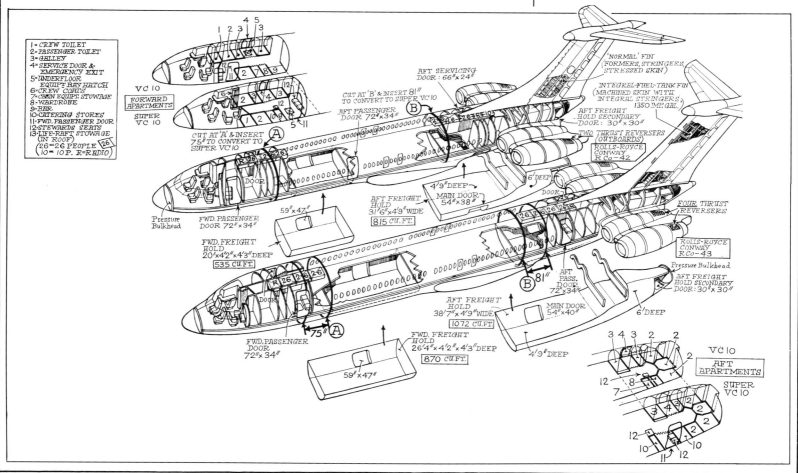

1 = CREW TOILET
2 = PASSENGER TOILET
3 = GALLEY
4 = SERVICE DOOR & EMERGENCY EXIT
5 = UNDERFLOOR EQUIP'T BAY HATCH
6 = CREW COATS
7 = CABIN EQUIP'T STOWAGE
8 = WARDROBE
9 = BAR
10 = CATERING STORES
11 = FWD. PASSENGER DOOR
12 = STEWARDS SEATS
13 = LIFE-RAFT STOWAGE (IN ROOF)
(26 = 26 PEOPLE 10 = 10 P. R = RADIO)

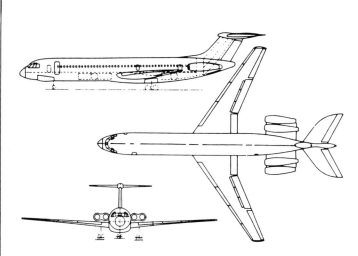

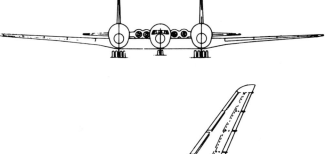

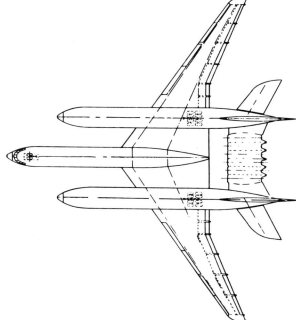

VC11. A project for a scaled-down VC10 for short- to medium-haul operation.

A notional possibility from spring 1964 shows a study using three Standard VC10 fuselages and Super VC10 outer wings, with power supplied by six Rolls-Royce Medways — one possible way to carry four hundred and fifty passengers largely from existing jigging and tooling.

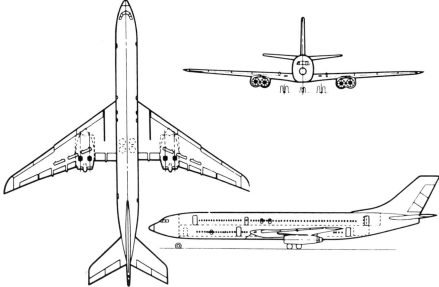

Another study, radically different for Vickers, shows a 300-seater with twin decks, powered by four underwing, paired Rolls-Royce Conways.

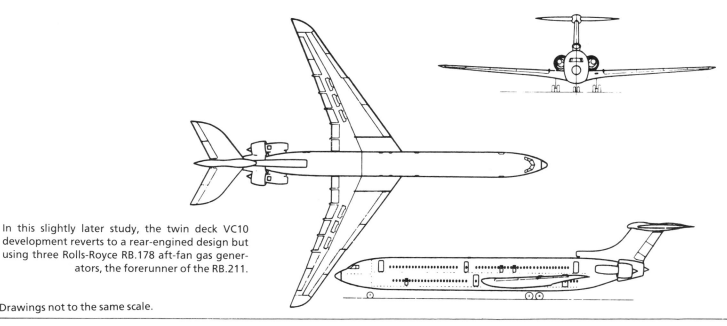

In this slightly later study, the twin deck VC10 development reverts to a rear-engined design but using three Rolls-Royce RB.178 aft-fan gas generators, the forerunner of the RB.211.

Drawings not to the same scale.

Postscript

The following extract from a feature on the Super VC10, written by the author for the leading aviation technical journal *Aircraft Engineering*, April 1965, encapsulates the VC10's contribution to the air transport business:

Transport aircraft have traditionally been procured on grounds much broader than the purely technical. The effect of passenger attraction on an otherwise technical evaluation of competing types can prove decisively in favour of the aircraft being able to derive higher load factors over long service life due to new passenger-attracting features. This cannot be reflected in conventional technical evaluation methods and operating cost formulae.

The development cost to provide the all-round high standards achieved in the Super VC10, with modest production, has remitted a slightly higher unit cost than for earlier jets. However, in the overall economic concept this is more than balanced by the better value for money to the operator and the greater passenger appeal. Comparisons with other jets are thus incongruous unless full account is taken of these cardinal factors.

Consumer preference is an imprecise concept of airline promotion but the extraordinary preference and ecstatic reactions from wide-ranging Standard VC10 operations are proving the sound-

Factory-fresh at Wisley, the thirteenth VC10 C.Mk.1 for the RAF, XV108. The point on the slightly extended tail fuselage is the exhaust for the Bristol Siddeley Artouste auxiliary power unit. (*BAe*)

ness of its original concept – and augurs well for the Super VC10. It is providing an impact as marked and fundamental as the ubiquitous Viscount.

Though unfortunately not permitted to exploit this fundamental axiom to the extent that it deserved, nevertheless the VC10 and Super VC10 have demonstrated it as clearly as any other type in modern times. □

What's in a Name?

Vickers Commercial Transports

Electing to focus on the newly emerging commercial aircraft market at the end of the Second World War, Vickers began to use the (Douglas-style) VC (Vickers Commercial) designation for its civil airliner designs.

The first of these were the VC1 and VC2 in the late 1940s, which subsequently became the Viking (Britain's first postwar airliner) and the Viscount (the world's first turbine-powered airliner), respectively. Names rather than numerical designations were in vogue (and, as it turned out, many hundreds of Douglas DC-3s were now in commercial service throughout the world).

Apart from these two, the only other VC-designated type to go into production was the VC10 two decades later, by which time the general fashion had reverted to numerical designation.

The designation VC3 was given to a proposed civil development of the Varsity, the ultimate (military trainer) derivation of the Viking family.

The VC4 covered several studies for Vickers' initial series of jet transport designs for a North Atlantic aircraft from 1946.

The VC5 was a long-range derivative of the Valiant V-bomber, intended for a possible British Overseas Airways Corporation (BOAC) requirement in 1951.

The VC6 was a generally similar design to the VC5 but was a short-range version intended for British European Airways (BEA) in 1952.

These last two designs were later incorporated in the VC7 which culminated as the designation of the commercial derivative of the V1000 strategic military jet transport project in 1952.

In the mid-1950s, when Douglas's DC-8 was well established and committed to production as an intercontinental jetliner and it was already notionally using DC-9 for its supporting jet transport studies (but well before the definitive DC-9 series), Vickers opted to leapfrog the VC8 and 9 to become ahead of the field with VC10 for its own intercontinental jetliner contender.

Meanwhile, the large regional propeller turbine Vanguard airliner, having evolved directly from the Viscount and in close association with BEA and Trans-Canada Air Lines (TCA), thus took its name early on and outside the VC series – and while the V1000/VC7/Vanjet/VC10 study continuum was still in progress.

The final study in the Vickers VC commercial transport series was the VC11 between 1959 and 1961. Essentially a scaled-down VC10 for short- to medium-range operation, and in the same general category as what was eventually realised as the de Havilland Trident and Boeing 727, the VC11 project was keenly pursued for BEA and TCA, following their strong alliances with Vickers and the Viscount and Vanguard.

With the formation of the British Aircraft Corporation (BAC) in 1960, the imaginative Hunting 107 project came into the fold and this was developed by the Weybridge team into the BAC One-Eleven domestic and regional twin jet airliner – the designation being a combination of BAC's first new commercial airliner design and the retention of a vestige of the VC11 which, although not proceeded with, had already engendered goodwill in the market-place and a certain interest from the British Government. □

VC10 & Super VC10 in Service with BOAC & British Airways

R A R Wilson

The author recaptures here the flavour of marketing British Overseas Airways Corporation (BOAC) seats in a new airliner, a quarter of a century ago. The VC10 was produced in two sizes: the Standard, of which there were twelve, for the technically demanding and sometimes commercially difficult routes to Africa and Asia, was in service from April 1964; one year later the longer-bodied Super VC10, of which BOAC took seventeen, began to fly the important routes to North America. Throughout the airline's diverse network, travel by BOAC VC10 was marketed to a degree quite unknown before.

From the outset, the VC10 had been designed for BOAC's Eastern and African routes, so it was predictable that the first scheduled services would be run on an out-and-back basis to West African destinations. The first scheduled service operated between London and Lagos,

The choice: tailor-made Super VC10s or Boeing 707-436s almost off the shelf. Both BOAC aircraft used the pioneering Rolls-Royce Conway by-pass jet engine, the Boeing having four of the earlier and less powerful Mk.508s, each of 17,500lb thrust. (*British Airways*)

Nigeria, on 29 April 1964. However, in the 6¼ years since the contract had been signed with the manufacturers, much had changed. The 'politics' of that change were extremely complex and caused much anguish and misunderstanding at the time. The task now, a quarter of a century later, is to try to see and appreciate the wood without getting caught up in every patch of brambles along the way.

The VC10 was the last of the big

aircraft designed for BOAC and its predecessor, Imperial Airways, to serve specifically what were known as the Empire Routes. At the time of the VC10's predecessor, the propeller-turbine Bristol Britannia of the 1950s, the design capability existed to transform the Medium Range Empire (MRE) aircraft, by fuselage stretching and power-plant development, into an airliner with intercontinental range. Because the Britannia was late into service, its period of dominance on the North Atlantic could be measured almost in months before it was overtaken commercially (and literally) by the first generation intercontinental jets, the early Boeing 707s and, a little later, the first Douglas DC-8s and the short-lived (on the Atlantic) BOAC de Havilland Comet 4s.

The VC10, which had similarly started out as more suitable for the Empire routes, was also transformable into an Atlantic Blue Riband airliner. As

a result, the initial order for thirty-five VC10s became modified to just twelve Standard VC10s with all subsequent deliveries to be Super VC10s. By rescheduling both the development and production programmes, this was achievable. But then a more serious problem arose. In selecting the British aircraft, BOAC had committed itself, it believed, to too much 'big jet' capacity.

The final outcome, after contract negotiation and penalty payments, and following much public debate in Parliament and the press while many in the aviation fraternity and elsewhere watched aghast, was a BOAC fleet of twelve Standard VC10s and seventeen Super VC10s.

We take up the story in the summer of 1963 by which time, in a series of special flights, BOAC board members and senior management, including route managers from around the world, had flown in the VC10 which was then just a year into its flight test programme. This gave the executives their own feel for what was coming. As the airline's managing director Mr Basil Smallpeice (later Sir Basil) expressed it, in an article to staff, the VC10 would provide lower noise in the cabin, a good pressurisation system, a landing speed slower than a Boeing 707 and much greater airfield flexibility.

No effort was to be spared to market the VC10 effectively to prospective passengers, resulting in an unprecedented advertising campaign which was to contrast strongly with the arguments over capacity, which would soon occupy almost equally prominent space in the news columns of many of the same newspapers and magazines.

Long before the new airliner flew, research revealed the advantage of retaining the designation VC10, but prefixing it with the airline's name to make it the BOAC VC10. Beginning in November 1963, *The Times* produced a full supplement on the new aircraft, a vehicle for BOAC's promotional efforts. The advertising made the most of the new shape of the VC10, and its quietness in the passenger cabin, under the slogan 'Swift, silent, serene', and its slower landing speed was compared with that of the American jets.

As the advertising gained momentum towards the introduction of the Super VC10 in April 1965, one year after the Standard, the slogan had become 'Triumphantly swift, silent, serene – the BOAC VC10'. Accompanying the slogan was an accomplished and eye-catching Super VC10 in-flight picture by

Frank Wootton, British aviation and landscape artist of distinction. In nearly fifty languages, on 100ft hoardings, on the banks of the Suez Canal, on 'spectaculars' in New York's Time Square, in newspapers and magazines, in folders and brochures, on posters and in window displays, the message was drummed home: 'Something new and exciting in air travel – by BOAC'.

By October 1963, all twelve Standard VC10s had been manufactured, although none was yet at acceptance standard and a number were still heavily encrusted with test instrumentation. By 22 April the following year, the VC10 had its full Certificate of Airworthiness (C of A) and services could begin. By then, VC10s had flown 4,230 hours, 1,800 of them on crew training, and 1,000 hours of aircraft route proving had also been completed. Intensive flying by senior BOAC captains had taken VC10s to the Middle East and West Africa and across the Atlantic to Canada. BOAC commercial staffs had also been 'educated' by means of visual and other presentations to ensure the smooth introduction of the VC10 scheduled service.

Meanwhile, flight training of line crews had been proceeding apace at Shannon, Ireland, where, by April 1964, seventy-five captains, one hundred and twenty-seven first officers and seventy-two flight engineers had completed conversion courses. Route training had been based on the Standard VC10 initially taking over the existing Comet 4 operations; it had been limited during the remainder of 1964 to West Africa, the Gulf, East, Central and South Africa, South America, India, Pakistan and Aden during the summer, and continued with services to Hong Kong and Tokyo

by the end of the year – paving the way for the introduction of the Super VC10 the following year. Initial plans were to include nine BOAC/Nigerian Airways VC10 services a week to West Africa; to Central and South Africa six services a week; the Gulf four; East Africa four; and Far East destinations three. Thus was the VC10 proved, over the routes for which it had been specifically designed, for operation into and out of limiting airports, hot or high or both with long sectors in between.

First Year in Service

On 29 April 1964, the first commercial scheduled service began from London to Lagos. In G-ARVJ were: in command, Capt A S M Rendall (flight manager VC10), senior first officer W Harnett; first officer W Wallace; navigating officer W P Robinson (flight navigating superintendent); engineering officer G Sears; flight steward W A Hemming; flight stewardess Tess Curtain; stewards C Stubbings, F Clauson and A Critchell; and stewardesses Josephine Dixon and Susan Graham. The 3,100 mile sector was flown in 5hr 50min.

Expansion – at the cost of the Comet 4 fleet – continued rapidly throughout the remainder of 1964. Services to East Africa began on 6 June, the Gulf on 22 June, Nairobi (Kenya)–Salisbury(Rhodesia)–Johannesburg on 21 July, Lusaka, Zambia, being included later in the month. A regular London–Bahrain non-stop flight began on 16 December and a weekly London–Bombay on 19 December. In 1965, the VC10 was introduced on the London–Cairo–Bahrain–Bombay–Singapore route (returning via Bombay, Teheran and Cairo) on 5 January; 16 March saw the VC10 taking over from the Comet 4 on the

London–Aden route, with a time saving of 1hr 30min for the flight.

By the end of April 1965, the Standard VC10 had completed a year of service carrying over 216,000 passengers, flown 12 million miles, 25,000 revenue hours, and had an average daily utilisation of 8½ hours per aircraft. Average cost per flying hour was £475 compared with £371 for the Boeing 707. However, in its initial year, the attractiveness of the aircraft to its passengers resulted in passenger load factors averaging 80 per cent on the West African routes and a surprising 98 per cent on Central and South African routes where traffic was turned away because of the demand.

Super VC10 and the Integrated Fleet

On 1 February 1965, the manufacturers made available to BOAC a Super VC10 and on 2 March a second, on loan, to make possible an intensive training programme with complete integration and interchangeability between crews of Standard and Super VC10s.

Meanwhile, BOAC was preparing fresh advertising promotions for the Super VC10 in the sophisticated and highly competitive North Atlantic travel market. Although all VC10s shared exactly the same clean architectural lines to the cabin and the same advanced air-conditioning, on the ground as well as in the air, for the Super there were to be changes in decor planned by Robin Day: for example, a striking black and white print of seventeenth-century London on the cabin bulkheads. The promotional copy read that 'Ceilings and walls of smooth pastel shades and a cherry red carpet blend brightly with charcoal grey seat covers. Scatter cushions of Venetian scarlet, coffee, white and pale blue' added to a colourful aircraft interior. 'Only a bed could be better', said a passenger of the new first-class seats which, with the help of Marlene Dietrich, had achieved worldwide publicity. Extolling passenger appeal, the airline advertised that 'it will not take the passenger long to discern what the VC10 has to offer. The revelation comes the moment one steps inside the warm, softly lit cabin and soon one notices with surprise that with the merest whisper of engine power the Super VC10 is rolling forward'. There was no doubt about the favourable reception by passengers, and BOAC began to receive an increasing number of booking enquiries specifically linked to VC10 travel.

Hastening out of Heathrow – a Super VC10. (*British Airways*)

As the first Super VC10 commerical scheduled service under the command of Capt Norman Todd taxied out at London Heathrow on 1 April 1965, there was an unforeseen delay of 57 minutes as the aircraft passed over an iron bar, bursting two tyres. The sector from Heathrow to New York took 6hr 36min, the aircraft continuing to San Francisco. Initially, the New York daily flight from London was extended once a week to San Francisco and six times a week to Bermuda. Progress with the takeover from the Comet 4 routes and from some Boeing routes continued, routes being thrice weekly London–New York–Nassau (Bahamas) – Montego Bay – Kingston (Jamaica) from 30 April 1965; from 8 August the Standard VC10 replaced Comets on the London–Colombo (Sri Lanka) route and on 26 October the Kingston service was extended to Lima (Peru). The end of November 1965 saw the last Comet 4 service. Utilisation on the VC10s was increasing and by the winter of 1965/66 it had exceeded 10 hours per aircraft per day.

During the first 6 months of Super VC10 operation some of the fleet bore the markings BOAC-Cunard. This company, formed in 1962, achieved little but it did, for a time, eliminate potential competition on the North Atlantic routes from Cunard-Eagle. It was dissolved by mutual consent in September 1966.

With full integration of the Standard and Super VC10 fleets, crew members reached a total of over 900 (300 pilots, 120 flight engineers, 260 stewards and 250 stewardesses). By the end of 1966, twelve Standard and eight Super VC10s had been delivered. During the following year route deployments were made as deliveries increased until the final fleet total of twelve Standard and seventeen Super VC10s was reached in

May 1969. On 12 December 1965 a once weekly service by Super VC10 began via New York to Freeport (Bahamas). On 28 February 1966 the Super VC10 replaced the Boeing 707 on the London–Montreal–Chicago service and on 1 April on the daily Manchester–Prestwick–New York route. Meanwhile, on 2 April the Standard VC10 began a once weekly service to Dubai, UAE, and a twice weekly service to Perth, Western Australia.

The second anniversary of the Standard VC10 was 29 April 1966, by which time the fleet had flown nearly 30 million miles and more than 65,000 revenue hours with an average passenger load factor of 59 per cent.

However, earlier that month an event of more wide-ranging significance had occurred when Juan Trippe, president of Pan American with some 30 years' continuous leadership of this then world-leading airline, announced orders for twenty-five Boeing 747s, all planned to be in service by May 1970. The potential scale of intercontinental air transport was about to change again, this time by whole orders of magnitude. The first long-range jet services which had become properly under way in 1960 had provided big increases in individual aircraft productivity, mainly through increased speed. The next stage, a decade later, would provide an enormous geometric increase in size for each aircraft. For example, instead of sixteen first-class seats and one hundred and twenty-three economy seats aboard a BOAC Super VC10, the Pan American 747, from 1970, would offer three hundred and seventy-eight seats in mixed class in a fuselage capable of taking four hundred and ninety all-economy. It was inevitable and proper that BOAC would have to follow suit by ordering, initially, six Boeing 747s.

Meanwhile, the expansion of VC10

services went ahead, on 1 November 1966, the new aircraft beginning a London – Frankfurt – Zürich – Nairobi – Blantyre (Malawi)–Johannesburg service. The one-millionth VC10 passenger was flown on 9 July 1967. Route changes continued: 16 December 1967 saw the introduction of a Super VC10 service four times a week Manchester–Prestwick –New York–Antigua–Barbados and on 2 May 1968 a non-stop VC10 service began from London to Jeddah.

When the VC10 had been in service for just over four years, the first BOAC fully automatic landing in commercial service was recorded on 16 May 1968. On the following 21 September, a non-stop London–Barbados service was started, becoming the longest VC10 commercial sector – 4,420 statute miles, the previous longest being Nairobi–London, 3,970 miles. The year 1969 saw the introduction of a Super VC10 service westbound from London to Sydney via New York, Los Angeles, Honolulu and Fiji on 26 October.

While it is all too easy and possibly misleading to snatch isolated figures from old company reports, it is worthwhile to look at the period April 1968 to April 1970 spanning the Super VC10's fourth and fifth years of service. BOAC's chairman, Sir Charles Hardie, presenting the report and accounts in August 1970 for the year ending 31 March 1970, had a generally good story to tell, both in respect of that year and the current operations. There was then sufficient confidence in the performance of the VC10s and Supers to increase their amortisation periods, to nil residual values, from 8½ years to 12 years, indicating Super VC10 services at least until 1976. At this point, the best economic performance from the Super VC10s was still to come, as reference to the Super VC10 utilisation tables on page 38 shows.

In the year 1970/71, sixteen Super VC10s flew 70,347 revenue hours, that is, 4,397 hours per aircraft per year, or just over 12 hours per aircraft for each and every day of the year. At that time, nearly 20 years ago, this said much for the efficiency of BOAC's scheduling, for that use represented entirely scheduled service revenue-earning capacity, and much for the manufacturers' ability to produce an aircraft capable of meeting those demands. For, on multi-sector VC10/Super VC10 routes, there was no back-up support from other operators; the aircraft had to be capable, within its basic design, of carrying its own defects.

VC10 utilisations on the normally shorter commercial sectors flown by that fleet – as opposed to the Supers – were lower and the cost per capacity ton-mile

was inevitably higher. However, BOAC financial statistics do not record how well the VC10/Super VC10s performed in their best years, either in their own right or in comparison with other types operated by the corporation. The tragedy of the VC10 story is that by the time both operator and manufacturer had become good at making big British aircraft earn their keep efficiently, the national decision had been taken to stop making them.

Meanwhile, in 1971 Super VC10 service was still expanding. From 29 April a new Belfast (Aldergrove) to New York service via Prestwick was introduced, and a daily London to Philadelphia and Washington service began on 2 July.

The early 1970s were to bring many changes, not only through the introduction of the Boeing 747s. In December 1970, the government announced the routes which would be taken away from BOAC to enforce the recommendations of the Edwards Committee on the Second Force airline. Initially, this was

Setting down at Shannon – a Super VC10. (*British Airways*)

that after 31 March 1971, BOAC would cease to serve Lagos and Kano (Nigeria) and Accra (Ghana); this, at the time, ended Imperial Airways' and BOAC's efforts in the field of West African civil aviation. The intention was that the routes would be taken over by Caledonian/BUA from London Gatwick. Again, from 1 July 1971 BOAC was required to cease its activities between the United Kingdom and Libya and in addition Caledonian/BUA were to be permitted to carry first-class traffic between London and East Africa. All the routes and areas affected were those of the VC10. BOAC estimated the loss in revenue at £10 million – but no compensation was offered. Nevertheless, the first service to the Seychelles from London began on 3 July 1971.

By 1972, full utilisation of the joint fleet was complete and some services had begun to be replaced by the wide-bodied Boeing 747. On the African and Eastern routes of BOAC some changes were made, but there were also additions to the network. From 5 January, VC10s made a once weekly stop at Addis Ababa on the Mauritius route –

The major portion of the economy-class cabin (one hundred and twenty-three seats) in a BOAC Super VC10. (*BAe*)

Nairobi being the other transit stop. The success of the Seychelles service led to an imaginative plan to extend the services to the East and West. The Indian Ocean service began on 18 February when the service London–Entebbe (Uganda)–Nairobi–Seychelles–Colombo–Hong Kong–Tokyo began twice weekly with Standard VC10s. On 6 January 1973, VC10s were used to close the gap between South Africa and the Seychelles and open up markets, in partnership with South African Airways, to Hong Kong and Japan. This route, which was the last VC10 new route, was from Johannesburg to Seychelles, Colombo and Hong Kong, with a connection to Tokyo. But the writing was on the wall when the Super VC10 London–Chicago service was replaced by the 747 on 25 May 1972.

Thereafter, the VC10s were gradually withdrawn from service with a final operation on 31 March 1981. The type had been in BOAC service for 17 years – difficult years which saw not only the development of high capacity long-range jet aircraft but also the start of many political changes, nationalist movements and disruptions to normal flight communications.

Air Piracy

In relation to numbers operating, BOAC's VC10s suffered more than their share of acts of air piracy. Super VC10 G-ASGN, under the command of Capt Cyril Goulborn, was blown up on the ground at the 'Revolutionary Airstrip' in Jordan on 9 September 1970, having been seized in the air and forced to make

an intermediate landing at Beirut, Lebanon. Fortunately, the passengers (including thirty-four school children) were unharmed. Three and a half years later Super VC10 G-ASGO, en route from Beirut to London, was seized in the air and diverted to Amsterdam where passengers and crew were released but the aircraft was blown up.

The third hijacking had a more satisfactory outcome. On 21 November 1974, Super VC10 G-ASGR, with twenty-three passengers on board, was seized at Dubai and forced to fly to Tripoli to refuel, landing at Tunis-Carthage Airport the following day. In command was Capt James Futcher with first officer Michael Wood and engineer officer Frank Sharples completing the flight crew. Eventually, the hijackers laid down their weapons, the episode leading to the award of the Air League Founders

Standard VC10

C/n	Registration	Delivered	Disposal
804	G-ARVA	8 Dec 1964	To Nigeria Airways 29 Sep 1969, written off 20 Nov 1969
805	G-ARVB	6 Feb 1965	Scrapped Oct 1976
806	G-ARVC	1 Dec 1964	To Gulf Air 9 Feb 1974
807	G-ARVE	1 Oct 1964	Scrapped Oct 1976
808	G-ARVF	4 Sep 1964	Withdrawn and flown to Germany for preservation
809	G-ARVG	12 Jun 1964	To Gulf Air Jun 1974
810	G-ARVH	2 Jul 1964	Scrapped Oct 1976
811	G-ARVI	22 Apr 1964	To Gulf Air Mar 1974
812	G-ARVJ	23 Apr 1964	Withdrawn 1980 – sold to RAF Sep 1982
813	G-ARVK	2 May 1964	To Gulf Air Jun 1975
814	G-ARVL	16 Jun 1964	To Gulf Air Apr 1974
815	G-ARVM	22 Jul 1964	Withdrawn and flown to Aerospace Museum, Cosford, for preservation in British Airways collection Oct 1979

Fleet, Hours and Utilisation

Year	Fleet in-service numbers	Revenue hours flown	Annual utilisation hours
1964/65	12	21,828	1,819
1965/66	12	39,867	3,322
1966/67	12	45,871	3,823
1967/68	12	45,199	3,767
1968/69	12	42,148	3,512
1969/70	11	44,247	4,022
1970/71	11	43,462	3,951
1971/72	11	40,838	3,713
1972/73	11	39,820	3,620
1973/74	9	31,401	3,489
1974/75	6	12,887	2,148
1975/76	4	1,837	830
		409,405	

Medal jointly to the flight crew, for 'the most meritorious achievement in the whole field of British aviation in 1974'. The citation stated, 'This award is in recognition of the bravery and fortitude of these officers. Magnificently supported by the cabin staff of the airliner, they displayed exemplary calm, dignity and consumate tact in their dealings with armed and hostile men during an ordeal lasting 85 hours during which one passenger was shot dead. A number of deadlines passed before everyone in the aircraft was eventually released and the hijackers laid down their weapons on 25 November. For the last 5 hours the flight deck crew were themselves hostages in a tense negotiation for the freedom of the hijackers ... Capt Futcher and his staff did their utmost to ensure the safety and sustain the morale of the passengers held captive.'

VVIP Role

Her Majesty The Queen accompanied by the Duke of Edinburgh first made use of a VC10 Special Flight during the state visit to Ethiopia in May 1965. Subsequently, between February 1966 and March 1974 the Queen used Super VC10s for state and other visits on seven major occasions. At many other times members of the Royal Family flew by BOAC VC10s on their scheduled services.

End of Service

While BOAC did not overstate the low noise level inside the VC10, the noise outside, as anyone under a VC10 flight path around Heathrow could testify, was a different matter. International noise legislation alone, in time, would have forced the withdrawal of the VC10s from commerical service. With such a small number of airframes, re-engining, a technical possibility, was in no way a practical option.

Indeed, it was ironic that the by-pass principal, of which the Rolls-Royce Conway was pioneer, leading to significant reductions in external noise in much later designs, was fitted in its most powerful and noisiest form (the Rolls-Royce Conway R.Co.43 of 22,500lb static thrust each in the Super VC10). Later, the Rolls-Royce RB.211, the first of a pre-eminent family of 'quiet' commercial engines, was flight tested in an ex-RAF VC10.

The combination of higher power and high lift which the VC10 provided (operational virtues to meet the specification as runways were lengthened to accommodate American jets in new, young countries) made the VC10 an expensive solution. But for passenger confidence and comfort and pilot confidence, whatever the situation, the VC10 set entirely new standards in its generation.

Sunday evening, 29 March 1981, saw the arrival at Heathrow of G-ASGF operating the last British Airways commerical Super VC10 flight, from Dar-es-Salaam, Tanzania. It was time to check the records but, first, starting at lunchtime the following day, there was a special 2½ hour flight around the United Kingdom. All one hundred and thirty-seven seats had been sold out, one passenger coming over especially from the USA. Another, E E 'Ernie' Marshall, former chief project engineer at Weybridge, had known the VC10 'when it was a blank sheet of paper'. During its 17 years of service, British Airways' fleet of twelve Standard and seventeen Super VC10s had carried at least 13 million passengers and made 250,000 landings. Altogether, the fleet had accumulated 1,207,106 flight hours and flew an estimated 560 million miles.

BOAC VC10s can still be seen, displayed statically. Standard G-ARVM made its last flight from Heathrow to join the British Airways collection housed in the Aerospace Museum at RAF Cosford on 22 October 1979, and Super VC10 G-ASGC went to the Imperial War Museum's display at Duxford, Cambridgeshire, on 15 April 1980. Significantly, ex-British Airways Super VC10s fly on, playing a major role as aerial tankers in the defence of the United Kingdom – a less glamorous task but scarcely a less vital one. □

Super VC10			
C/n	Registration	Delivered	Withdrawn from use and sold
851	G-ASGA	31 Dec 1965	Apr 1980 to RAF
852	G-ASGB	30 Apr 1965	Apr 1980 to RAF, scrapped Apr 1987
853	G-ASGC	1 Apr 1965	Oct 1979 – to Duxford for preservation Apr 1980
854	G-ASGD	1 Apr 1965	May 1980 to RAF
855	G-ASGE	27 Mar 1965	May 1980 to RAF
856	G-ASGF	2 Apr 1965	Apr 1981 to RAF
857	G-ASGG	21 Jun 1967	Apr 1981 to RAF
858	G-ASGH	4 Nov 1965	Apr 1980 to RAF
859	G-ASGI	12 Feb 1966	Apr 1980 to RAF, scrapped 1987
860	G-ASGJ	7 Mar 1967	Apr 1980 to RAF, scrapped 1987
861	G-ASGK	27 Oct 1967	Mar 1980 to RAF
862	G-ASGL	25 Jan 1968	Apr 1981 to RAF
863	G-ASGM	9 Apr 1968	Jul 1980 to RAF
864	G-ASGN	7 May 1968	Hijacked and destroyed Dawson Field Sep 1970
865	G-ASGO	27 Sep 1968	Destroyed by fire at Amsterdam by hijackers 3 Mar 1974
866	G-ASGP	6 Dec 1968	Apr 1981 to RAF
867	G-ASGR	31 May 1969	May 1981 to RAF

Fleet, Hours and Utilisation

Year	Fleet in-service numbers	Revenue hours flown	Annual utilisation hours
1965/66	8	16,584	2,073
1966/67	9	30,245	3,472
1967/68	12	38,894	3,241
1968/69	16	50,192	3,137
1969/70	17	64,390	3,788
1970/71	16	70,347	4,397
1971/72	16	69,001	4,312
1972/73	16	66,992	4,187
1973/74	15	60,990	4,066
1974/75	15	57,999	3,867
1975/76	15	59,237	3,949
1976/77	15	57,732	3,849
1977/78	15	53,586	3,572
1978/79	15	48,708	3,247
1979/80	15	43,629	2,909
1980/81	4	9,175	2,249
		797,701	

VC10 and Super VC10s also served with a number of other airlines including British United Airways, East African Airways, Ghana Airways and Laker Airways. Those operators, and others, will be summarised in an article by Dr Norman Barfield appearing in the second issue of the Review.

Wingtip Technology

Maurice Allward

In long-distance aircraft operations, particularly, the search for greater efficiency in relation to fuel burned is unending. One key area for improvement concerns the aerodynamic efficiency of the wing. Here, attention is focused on wingtips and the improvements that can be obtained through reduction in vortex drag.

Improvements in aircraft efficiency rarely come in giant leaps in these days of advanced wind tunnels and high-speed computers. More often they are the result of relatively small steps involving one or more of the many technologies, such as better fuel management, lighter structures, computerised systems and smaller crews.

Even the quantum improvements rarely bring the rewards originally promised. One example of this is composite materials. Basically, many non-metallic fibres, such as Aramid and Boron, are up to six times as strong as aluminium alloy. This basic strength, however, cannot be used in practice because of the cumulative effect of a number of modifying factors, such as notch sensitivity, lay-up, the environment and variability properties. These combine to reduce the practical strength of composites to between one and two times the alloy value. This, of course, still provides a useful saving in weight, but nothing like that originally anticipated.

Another example of a potentially large improvement is the propfan, or unducted fan engine. When originally conceived in the days of soaring fuel prices, these new engines were claimed to have the potential of reducing fuel costs by 30 per cent or even 40 per cent. However, the result of steadying fuel prices, improvements in 'conventional' engines and technical problems that have arisen during the development of the new engines will probably halve this figure.

Vortex Drag

A major factor influencing the efficiency of an aircraft is aerodynamics, and improvements in this area tend to develop in small increments. The fundamental aim is to minimise drag. All parts of an aeroplane create drag, especially the wings doing their job of keeping the machine airborne. A wing develops lift by creating a region of relatively high pressure below and a region of lower pressure above the wing. During normal flight the air flows over most of the wing quite smoothly along the line of flight. At the tips, however, the difference in pressure causes the higher pressure air under the wing to flow around the end of the wing to the upper surface. As the air swirls around the tip, it rolls up to form a vortex which is left trailing behind the wing. The energy required to create this vortex gives rise to an extra drag force on the wing known as trailing vortex drag.

The existence of wingtip drag has been known since the early days of aviation. The drag can be reduced by increasing the span of the wing, which is the conventional method used. However, increasing the wing span increases the bending at the wing root. Consequently, it has to be stronger – hence heavier – which can counterbalance the benefit gained from the reduction in vortex drag.

Aerodynamicists have sought other means of reducing vortex drag, without incurring penalties. In fact, the problem has been the subject of detailed study for nearly 80 years. Early attempts, and some contemporary ones, achieved the desired result by refining the design of the wingtip itself. Special devices date from at least 1921, when a patent was filed covering an arrangement of finger-like surfaces at the end of the wing which moved in a complex manner to increase lift and reduce drag.

End Plates

At that time, however, and for many years, such devices were far from cost-effective. Active interest thus lay dormant until the late 1940s when the United States sought a solution by fitting simple end plates to the wingtips. A large number were tested but none was entirely satisfactory. In order to stop the air flowing around the wingtip, the plates had to be quite big, thus developing a great deal of drag on their own account which, again, equalled the reduction in vortex drag so that there was no overall benefit.

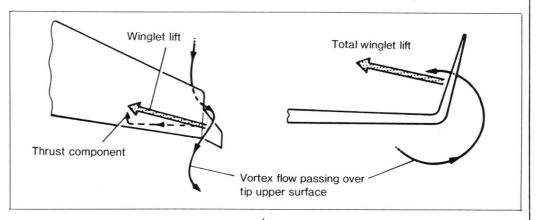

How winglets work. Note the forward thrust component and the total winglet lift. This lift bends the wing less than a comparable wing extension.

Below: Learjet 31 winglets are particularly beneficial because the jet routinely operates above 45,000ft.

Winglets

Interest languished for some 20 years, but was rekindled by the oil crisis which saw aviation fuel costs soar. In 1973, the devices sprouted again in the United States under the umbrella of the National Aeronautics and Space Administration (NASA). This time the devices emerged as winglets. Their aerodynamic theory was developed by Richard Whitcomb, an aerodynamicist who had earlier leapt to fame for his work on area ruling. (In the 1980s, theories have proliferated as manufacturers strive to make their aircraft as efficient and competitive as possible.)

Left: Cessna 414AW fitted with winglets by RAM Aircraft Corporation which increase the payload by 300lb with no loss of climb or cruise performance.

Winglets are relatively small, near vertical surfaces attached to the tips of a wing. On some configurations an additional, smaller winglet mounted forward below the wingtip is necessary. Although some winglets have the superficial appearance of a wingtip extension bent upwards, they offer a significantly greater reduction in drag coefficient and in cruise conditions than a basic tip extension, for the same amount of wing bending at the root.

A winglet operates in the circulation field around the wingtip. As explained, because of the difference in pressure above and below the wing, the airflow at the tip tends to move along the wing's lower surface, around the tip, and inboard along the wing's upper surface. Thus a winglet tends to produce a large side force, even at low angles of attack. Because the side force vectors are approximately perpendicular to the local flow, the side force produced by a winglet has a forward, that is a thrust, component, which reduces the aircraft trailing vortex drag. This is similar to the principle that enables sailing boats

to travel upwind by tacking.

For a winglet to be most effective, the side force must be produced as efficiently as possible. This means that what appears to be a simple surface is, in fact, of very advanced aerodynamic design. The side force produced by a winglet, and hence the thrust generated, depends on the strength of the circulation around the wingtip. Since the circulation strength is a factor of the lift loads near the wingtip, winglets are more effective on aircraft requiring high lift, such as high-altitude executive jets and heavy commercial transports.

One rarely gets a benefit in the world of aviation for nothing. The penalty looming in the background where wingtip devices are concerned is the increased wing bending caused by the additional force on the winglet. However, as the additional force vector points inwards from the tip, it has only a small moment arm about the root and produces a much smaller additional bending load than, say, an increase in the wing span.

As might be expected, winglets

offer the greatest benefit to long-range aircraft and have thus been fitted to second generation airliners and some business jets, such as the Gulfstream III and IV and the Challenger 601-3A. Winglets are not applicable to all business jets: many new ones have high-aspect ratio wings with super critical sections and these cannot benefit from such devices.

Learjet was the first manufacturer to use a winglet on a production aircraft, NASA-style surfaces being fitted initially to the Models 28 and 29 in 1979 and today on the new Learjet 31 and its heavier and bigger companion, the Learjet 55C. The winglets are particularly effective on the two new Learjets, which can routinely operate above 45,000ft and can fly at altitudes up to 51,000ft. At such heights the drag reduction afforded by the winglets is more pronounced.

Zip-Tip winglets are an interesting feature, developed by Colemill Enterprises. This company, based in Nashville, Tennessee, specialises in performance improvement conversions of light single- and twin-engined aircraft. Zip-Tips fitted to the Piper Panther Navajo are claimed to improve stability at low speeds, down to the stalling speed, and increase the cruising speed by 4 to 9 knots. The converted version is, therefore, able to match the cruising speed of the standard Panther Navajo at engine settings which offer greater fuel economy.

Another company engaged on such conversions is RAM Aircraft Corporation of Waco, Texas. This company has fitted NASA-type winglets to a Cessna 421C (then designated 421 CW). The winglet installation first involved removing the standard wingtip and replacing

Boeing 747-400. The initial motivation for the winglets on this airliner was the need to limit the wing span which takes up nearly all the available space at airport loading bays.

it with a wing extension and winglet assembly of slightly greater span. Landing lights, position lights and strobe lights are all installed in the new winglet assembly. The RAM winglets account for an increase of 50lb in payload and a 25 per cent increase in the maximum rate of climb. The cruising speed is increased by 10 knots and the aircraft is more stable. The winglet assembly is constructed of glass fibre and stainless steel; the outer layer of the winglet is an aluminium cloth for lightning protection.

Similar RAM winglets fitted to a Cessna 414A Chancellor reduce drag to the extent that an extra 300lb in payload can be carried with no loss in climb or cruise performance. Alternatively, with the normal payload, cruise speeds increase and the time to climb to 18,000ft is reduced by 25 per cent. An additional benefit is that aircraft stability is improved as the winglets dampen out Dutch roll tendencies.

The biggest winglet developed so far is the 6ft canted surface fitted to the Boeing 747-400. Together with a wing extension piece, these can provide either a fuel burn improvement of 2 per cent where additional range is not required, or an increase in range of 1 per cent, about 100 miles or so, where this is required. The initial motivation for the -400 winglets was the need to limit the span of the new airliner, which is over 17ft greater than earlier versions and takes up nearly all available space at airport loading bays.

Not all winglets are intended primarily to reduce drag. On some unconventional aircraft, such as the Gyroflug Speed Canard, the winglets basically fulfil the task of fins and rudders. Consequently, they are much larger than they would be if their main function was to reduce vortex drag. On the Speed Canard the winglets reduce the total resistance during climb and thus improve climb performance, but during cruising they increase the drag slightly.

Wing Fences

Winglet design is a demanding task. Even when providing the optimum reduction in drag, these miniature wings can stall in the same way as a main wing. This tends to happen when least wanted, for example, during take-off and landing when high lift performance is required. The winglet stalls abruptly, preventing further drag reduction and significantly altering the handling characteristics at a critical period.

Main wing stalling can be delayed by the installation of slats on the leading edge, but this solution is not practical for a winglet. The problem was how to overcome the winglet stalling without adding a complicated device on the winglet itself.

British Aerospace (BAe) aerodynamicists at Hatfield, Hertfordshire, considered the problem and proposed a solution which includes what can be called the delta wing factor. The result is a winglet, or fence as BAe prefers, with a general, sharply swept shape similar to the wing of the supersonic airliner. This planform, when combined with a special leading edge, results in a drag-reducing device with gentle qualities over the

The McDonnell Douglas MD-11 winglet installation includes a smaller surface below and ahead of the upper surface to prevent the main winglet stalling in high lift conditions.

The large wingtip surfaces on the Gyroflug Speed Canard are primarily fins and rudders for directional stability, not winglets.

whole flight envelope. In conditions when some other tip surfaces would stall, the Hatfield fence continues to generate a well ordered vortex flow, which although causing some drag, does not result in sudden changes in aircraft handling characteristics. As on the full-sized airliner, it changes smoothly from one type of flow to the other. Another advantage of the Hatfield fence is that this gentle flow pattern is maintained if conditions cause ice to form on the leading edge. This is a major benefit, as it would not be practical to de-ice the fence in ways used for the leading edges of main flight surfaces. Yet another advantage is that it minimises the increase in wing bending, reducing the wing root strengthening required.

The first airliner to enter service with the BAe delta wing fence was an Airbus A300-600 delivered to Thai Airways International in autumn 1986. Later that year, an A310-300 delivered to Swissair had fences specially developed to maximise their effect on the wing of this airliner.

On the A300-600, the tip fences provide a fuel saving of 1 per cent at cruising speed, and on the A310-300, 1.5 per cent. A saving of 1 per cent may not seem a great deal but it reduces the fuel burn on an Airbus at cruising speed by as much as 12 gallons an hour. Hence, over a typical year the fuel saved is more than 36,000 gallons – sufficient to fill the tanks of 3,600 family-sized motor cars. The financial savings on even a small fleet of airliners amounts to many hundreds of thousands of pounds a year.

The first few aircraft of the next Airbus airliner, the A320, did not have

tip fences because of the refined aerodynamics of the wingtip itself. However, on aircraft No. 22, the first A320-200, BAe-type fences have been fitted to reduce the fuel burn in the cruise by a useful 1.5 per cent.

The next two members of the Air-

Airbus Industrie A300-600 wingtip fences reduce fuel consumption by 1 per cent at cruising speed. The fuel saved each year by each aircraft is about 36,000 gallons. (*Airbus Industrie*)

bus family, the much larger A330 and A340, may have conventional-looking upturned winglets, because these are designed with slender wings with relatively small wingtip chords which would allow only a diminutive tip fence.

Wingtip Sails

Yet another approach to the vortex problem is the use of wingtip sails. These are a number of small lifting surfaces fitted to wings in such a way that they use the disturbed local airflow around a wingtip to produce lift. Like the single winglets, they do this in a manner similar to the sail of a yacht when it is sailing close to the wind, hence their name.

Much work on wingtip sails has been undertaken by Professor John Spillman at the Cranfield Institute of Technology, Bedford. The work has covered the addition of sails both to existing wings and to new wings designed with sails as part of the normal outer wing. Spillman has observed that birds make use of their tip feathers (acting as sails) to derive benefit from the disturbed airflow around the ends of their flapping wings. This reminds us of the classic observation of birds by Wilbur Wright which gave him the vital clue to the solution of roll control, so important in the story of the development of the powered Flyer of 1903. Wilbur recorded his bird studies thus: 'My observation of the flight of buzzards led me to believe that they regain their lateral balance when partly overturned by a gust of wind, by a torsion of the tips of the wings'.

Spillman laments that it took him some years to realise that a spectacular photograph of a marsh harrier about to land showed magnificently how the

Airbus Industrie prototype A310 with development wingtip fences. Production fences have a swept trailing edge. (*Airbus Industrie*)

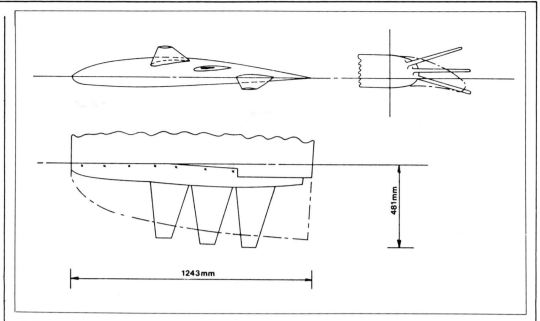

primary wing feathers were acting as wingtip sails. He explains, 'they are well separated, like a wide-spaced venetian blind, set at a marked nose-down attitude, yet clearly they are carrying a large positive lift judging from the extent to which they flex upwards. Obviously, the flow directions in their vicinity are quite different from those elsewhere.'

First one and then three sails on each wingtip were tested in a wind tunnel. The results were sufficiently encouraging to persuade the National Research Development Corporation to provide a small grant for further tunnel tests and for real flight tests. Three sails, each 23in long and 7in wide, were fitted to each of the wingtip tanks of the Institute's Paris twin-jet aircraft. The sails were arranged in a spiral around the tank, in the opposite direction to the

Wingtip fences fitted to the A320-200.

Airbus Industrie A310-300 with delta wing-type fences developed by British Aerospace, Hatfield. The fences reduce fuel consumption by 1.5 per cent at cruising speed. (*Airbus Industrie*)

airflow, so that a rear sail was not in the wake of a forward one. Flight results were even better than those predicted by the tunnel tests: zero-lift drag was not increased significantly and the maximum lift-drag ratio of the aircraft was increased from 12.5 to 15.8.

Flight tunnel tests showed that similar lift-dependant drag reductions could be achieved by fitting sails straight on to rounded wingtips. To prove the tunnel tests, a set of sails was designed to replace the standard wingtips of a Cessna Centurion, chosen partly because at the time it was one of the biggest single-engined aircraft in quantity production. Flight tests showed that the sails reduced the stalling speed, marginally improved some of the handling characteristics and reduced the fuel consumption between 3.5 per cent and 5 per cent relative to the standard Centurion.

Financial savings to a Centurion operator would be minimal, but would increase proportionally to the weight of the aircraft, with significant savings resulting from sails being fitted to large airliners with a high annual utilisation.

Future Prospects

The vortex produced by the wingtip of a large aircraft results in a swirling mass of disturbed air which can persist for many minutes. This disturbed airflow can make landing for a following aircraft a hazardous process and has, in fact, caused some small aircraft to overturn completely. The persistence of the vortex drag patterns is one of the reasons for the stipulated interval required between aircraft landing at busy airports.

Winglets tend to straighten the airflow and thus slightly reduce the strength of the wingtip vortex. However, the basic purpose of current winglets is to increase aerodynamic efficiency, not to improve flight safety. It would obviously be beneficial if, in addition to reducing drag, wingtip devices could reduce or change the pattern of the vortex. This could provide, for example,

better control of the spreading and dispersal of insecticides and fertilisers behind agricultural aircraft, more flexible ground manoeuvring, and reduced spacing between airliners on the approach and landing. It could also increase airport productivity, a useful bonus these days when the construction of a new airport in a developed country is almost impossible.

Ideas to reduce the vortex range from the suggestion for a ring-like aerofoil or duct at each wingtip to mounting engines at the wingtips, and using the fan exhaust to break up the vortex

A 'windmill' located at the wingtip could be used either to reduce drag or to generate power.

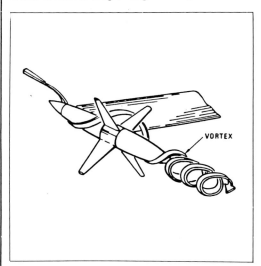

structure. One of the most ingenious suggestions is to place a small windmill in the tip vortex. A 4-bladed propeller is mounted at the trailing edge of the wingtip. When the propeller is stopped and the blades are feathered, the windmill works like a Spillman sail and reduces drag. If the windmill is allowed to rotate, the drag reduction is less, but power could by extracted from an electrical generator in the hub. Flight tests have been made with such windmills on a light aircraft with apparent success.

Lockheed, in a major study of unconventional aircraft design concepts, included a transonic biplane. This actually stems from work done in 1934 which showed that the smallest possible induced drag for a given span and height would be obtained by what the technicians call a 'closed rectangular lifting system'. More simply, this means a biplane with fins connecting the wings. Theoretical studies suggest that if the vertical separation between the wings is one-third of the wing span, the trailing vortex drag can be reduced by 40 per cent compared with a monoplane wing. To enable this aircraft to fly at high subsonic speeds, the wings are swept in opposite directions to form a diamond planform. In eliminating the wingtips completely, this would appear to solve the vortex drag problem effectively. □

Boxplane. Lockheed's concept for its 'closed rectangular lifting system': a biplane with fins connecting the wingtips. Studies indicate that this design could reduce drag by 40 per cent over a conventional monoplane of the same aspect ratio.

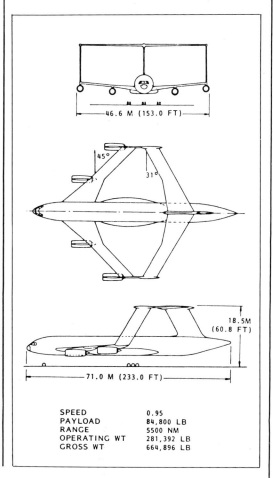

SPEED	0.95
PAYLOAD	84,800 LB
RANGE	5500 NM
OPERATING WT	281,392 LB
GROSS WT	664,896 LB

READERS' FORUM

How do you propose to tackle reader feedback to the *Review*? A fair question. It is our ambition by editorial encouragement to stimulate a forum of readers' views with the objective of increasing overall knowledge about particular subjects covered in the *Review*.

This almost puts an onus on the reader who believes he or she knows a little more than the contributor about some aspect of a historic subject to respond and let us all in on the secret before it is lost. We are sufficiently

realistic to know that occasionally we shall end up with a forum of just one reader's letter. However, even that response will be treated in the forum spirit.

The more succinctly points can be made, obviously the greater chance they have of being quoted fully. We shall not respond to blanket recriminations about past policies but we shall welcome observations from experience on the lines of, 'this did not work at the time because ...' or 'I was there and ...' or even 'on good evidence ...' and publish them.

While readers in Western Europe will be able to respond quickly, and we hope they will, we welcome information that will inevitably take longer to reach us from overseas, where the *Review* has been in the surface mail delivery system.

There are a hundred and one or more ways of responding positively, so how about making a start now? On the VC10, for instance, the main subject of this issue, we have neglected direct pilot comment and should very much welcome it. □

Istres–Damascus–Paris Air Race

Dr Roberto Gentilli

In 1936, the Paris-based Fédération Aéronautique Internationale (FAI) decided that a transatlantic air race should be held the following year, the tenth anniversary of Charles Lindbergh's solo New York to Paris flight. However, the contest was cancelled, being considered too hazardous, and a race from Istres to Damascus and back to Paris substituted. Dr Gentilli records the zeal with which the then Italian Government entered the event and considers its outcome within the political climate of the day.

Among the many events that were popular in the interwar years and which kept people from worrying too much about dictatorships and depression, two outstanding ones were Charles Lindbergh's solo flight and the transatlantic Blue Riband. The fascination of the *Lone Eagle* survives to this day, while the Blue Riband has faded from memory together with the ocean liners that competed for it. But in 1933, when the Italian liner *Rex* won the award for the fastest crossing of the Atlantic, Mussolini's fascist regime could boast one more accomplishment for Italy. The government was enjoying a high reputation abroad and the enthusiastic allegiance of most of the population, thanks in part to the exploits of Italian aviation.

In 1933, twenty-four Savoia Marchetti S.55X flying-boats had crossed the Atlantic, led by Italo Balbo. In 1934, Agello's Macchi MC.72 floatplane had established a world speed record which was not beaten until 1939, and Donati's Caproni set the absolute height record. These and many other flights were enthusiastically reported in the Italian press and on radio. Popular novels and films confirmed that the pilot was the most admired man in Italy. The Regia Aeronautica was the best example of the 'new' Italy which Mussolini boasted he was creating.

At the end of 1936, the FAI decided to create an aerial Blue Riband, an air race for landplanes from New York to Paris to commemorate the tenth anniversary of Charles Lindbergh's flight, and entrusted its organisation to the Aéro-Club de France. The Italian Air Ministry consequently decided to make a concerted effort to secure an Italian winner. On 15 April 1937, the Nucleo Addestramento Nastro Azzurro (NANA – the Blue Riband Training Unit) was established at Guidonia airfield, Italy's experimental and test cen-

tre, near Rome. Command of the unit was given to Lt-Col Attilio Biseo, a pilot from the seaplanes squadrons who had risen to become flight adjutant to the air minister, Mussolini himself, and to the under secretary and chief of staff, Gen Giuseppe Valle. Biseo had flown in all of Balbo's transatlantic cruises and in 1933 his S.55X I-BISE had led the way across the Atlantic. He was reputed to be Italy's best long-distance pilot.

Biseo organised an impressive team, drawing its strength from the 12° Stormo, an élite unit based at Guidonia that was made up of veterans from Balbo's Atlantic cruises and fighter pilots, and charged with the introduction into service of the new SIAI Marchetti S.79 bomber. Designed as a fast, eight-passenger airliner, the S.79 had first flown in October 1934. Its military conversion had appeared in 1936, but special procedures were deemed necessary to bridge the gap between 120mph biplanes and 270mph cantilever monoplanes – bombers, that is – which were much faster than any fighter then in service. The S.79 had been operational in Spain, with the Aviaciòn Legionaria supporting Franco in the Civil War since February with expert 12° Stormo crews, and was fast becoming appreciated by the air force staff. It was clear

that it could be a good transatlantic racer, and SIAI received an order to adapt five airframes for this special event.

The S.79 'Corsa' (racer – it was not designated S.79 C) was stripped of the ventral bomb aimer's gondola and the dorsal defensive position, or 'hump', and given the streamlined profile of its airliner prototype. Its fuselage had two new fuel tanks, which meant moving the entrance door from the port side, at the rear, to a new location just behind the cockpit. Some of the wing fuel tanks were enlarged, and the racer's total fuel capacity was 7,000 litres compared with 3,320 litres for the standard bomber. Other special installations included a fuel mixture analyser for each engine, cylinder temperature gauges, auxiliary oil tanks and special oil radiators, new radios and radio direction-finders, Smiths autopilots and fixed tailplanes (which were of variable incidence in the bomber) with trim tabs on the elevators. The finish was highly polished, with burnished NACA cowlings, faired tailplane bracing wires and a smooth, glossy red paint scheme with white and green trim, representing the national flag colours.

The S.79 racers had construction numbers 51 to 55 and serial numbers MM.355, 356, 357, 358 and 359, respectively. The first one flew on 4 March 1937, the last one on 22 May, and during the spring they were taken on charge by the Regia Aeronautica. At first they had flown with standard SIAI Marchetti two-position P.W. 0/2 propellers, but these were replaced by electrical controllable-pitch constant-speed Ratier propellers from France, recognisable by their more pointed spinners.

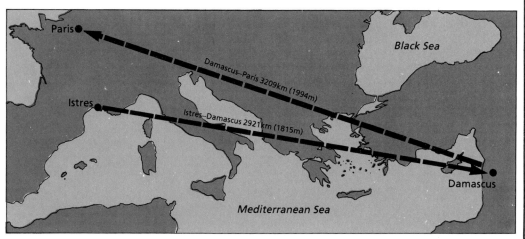

The highly specialised Guidonia unit planned its Atlantic race very carefully, making several hundred flying hours of navigation and engine testing. The engines could be boosted, allowing take-offs at a maximum overload weight of 14,000kg, corresponding to a wing loading of 227kg per sq m, with a 650m take-off run. It was determined that the S.79 could cross the ocean at an average speed of 384km/h in 16hr 4min. However, SIAI was not the only company involved in this alluring event. Rival Fiat had prepared two special versions of its BR.20 bomber, the BR.20As, serialled MM.21241 and 21242. The BR.20A was also a racing version of the standard twin-engined bomber, with all defensive positions and bomb bay removed, pilot's cockpit moved forward about 20in and fuel capacity increased from 3,620 to 7,700 litres. Finally, the Reggiane company of the Caproni industrial group built its own racer, the Ca.405 Procellaria (Petrel). It was a big mixed construction (wooden wings and fuselage, tubular metal nose) twin-engined monoplane designed by Giovanni Pegna, a gifted but unlucky Italian engineer who had recently left the Piaggio company. The Ca.405 used the same wing, with double camber flaps stretching along the whole trailing edge (their outward portions also acting as ailerons), as was used on the unsuccessful Piaggio P.32 bomber, coupled with a new slender fuselage with twin fins and rudders. Powered by two 850hp Isotta-Fraschini Asso XI R.C.40 engines, the Procellaria (serialled MM.375) first flew on 19 May 1937, and was taken to Guidonia. A second machine, MM.376, was on order and was entered in the race, but does not seem to have been completed.

Over two hundred crews had expressed their intention to participate, and twenty-two had actually registered by 1 May. Departure was scheduled for a day from 1 to 31 August, and the arrival was to be at Le Bourget, Paris, before 1200hrs on 3 September. The flight time to qualify had to be shorter than the 33½hrs of Lindbergh's *Spirit of St Louis*, and the prize for the winner was 1½ million francs. However, the American authorities were not at all receptive to the idea of a transatlantic race, and offered the French no co-operation. By the end of May, it was obvious that the race would not be held, and the Aéro-Club de France, which had already received the entrance fee of 5,000 francs for each contestant, had to find an alternative solution.

Europeans could acidly comment that the American authorities had sunk the Blue Riband race because the USA had no aircraft that could compete, but this is probably unfair. A more likely reason is simply that the race was considered too dangerous. 1937 was regarded in all countries as the year of the 'Atlantic eve', transatlantic air services were about to start (actually, two Short Empire Flying-boats of Imperial Airways and Pan American's Sikorsky S-42 were already making experimental crossings) and the idea of racers ditched all across the ocean, in a sort of replay of the Dole Hawaii Race tragedy 10 years before, was not appealing to the American airline industry. One can add, incidentally, that Charles Lindbergh himself never played any part in the 1937 competition, or 'Lindbergh Derby', as the US press called it.

The solution proved to be a race from Istres, near Marseilles, on the south coast of France, to Damascus, capital of French-administered Syria, and then back to Paris. Instead of one 5,850km flight, the race would now comprise two legs of 2,921 and 3,269km, respectively, with the possibility of mid-stop refuelling. The rules were quite basic: any multi-engined landplane provided with a radio, that was considered fit to fly by its national aeronautical authority and that could be at Istres by 1500hrs on 18 August was acceptable. The race would be judged on the effective time from take-off to the overflight of a white marking line at Le Bourget, making a quick refuelling in Damascus a necessity. Prizes included 1.5 millon francs to the winner, plus 50 per cent of the entrance fees, 1 million francs and 25 per cent of the fees to the second, and 500,000 francs and the remaining 25 per cent to the third.

Classic views of the S.79 racer as it first appeared at the factory, with bright red, white and green livery but with no race numbers. It is still fitted with SIAI P.W.0/2 propellers, soon to be replaced by Ratier units.

An S.79 racer on the factory aerodrome.

Contestants

The replacement race had a final roster of eighteen competitors: five from France, two from Romania, one from the USA, one from the United Kingdom and nine from Italy. France's best hope was the Amiot 370, a record aeroplane developed from a new metal twin-engined bomber, featuring an elegant unstepped nose. The beautiful Amiot was to be piloted by Maurice Rossi and Paul Codos, but it had first flown only on 25 July 1937, and there was little time to eradicate its teething troubles, particularly the cooling of its Hispano-Suiza engines. Codos and Rossi were famous for their long-distance flight from New York to Syria on 5 August 1933, with the Blériot-Zappata *Joseph le Brix* which set the new world record at 9,063km. The same crew flew from Paris to New York in 1934, and attempted an Istres to Argentina flight in 1935. These two great pilots, however, would not miss the race. Codos, teamed up with Maurice Arnoux, test pilot with the Caudron company and winner of the 1934 Coupe Deutsch de la Meurthe, flew another prototype, entered at the last moment, the Breguet 470 Fulgur F-APDY named *Raoul Ribière*, a modern-

looking twin-engined airliner powered by two 800hp Gnome-Rhône GR.14 Kirs engines, which had first flown on 5 March 1936 and which, late in 1936, made a proving flight to Indochina. It was claimed to be the fastest airliner in the world, and received race number F-1.

Maurice Rossi entered a Caudron Typhon named *Louis Blériot*, F-4. He was the only pilot to fly alone, on this sleek racer, a real thoroughbred that possibly owed some of its inspiration to the de Havilland Comet. The Amiot 370 and Rossi, however, were to cross the path of the Savoia trimotors again, because in 1938 and 1939 this elegant prototype set some speed-with-load records (2,000km with 2,000kg, 5,000km with 1,000kg, speed over 10,000km) in a long record-setting rivalry with the Italians.

A French entrant that did not show up was the Air Couzinet 10, a twin-engined low-wing monoplane that was to be flown by Dubourdieu, Macaigne and Cavaillès but that had flown too recently, on 3 August, to be ready. The third French aeroplane to be present on the starting line was the Farman 2231 F-APUZ, a magnificent high-wing four-engined mailplane that had first flown in

Biseo (centre, with sunglasses) is welcomed by officers and ground crewmen after having set the world speed record over 1,000km with 2,000kg load. The aircraft was the Piaggio P.XI-powered S.79 and co-pilot was Bruno Mussolini.

Air Race Aircraft

	Breguet 470 Fulgur	Farman 2231	Bloch 160	Caudron 640 Typhon	SIAI Marchetti S.79	Fiat BR.20A	de Havilland D.H.88 Comet
Span	20.5m	33.6m	27.4m	14.8m	21.2m	21.56m	13.41m
Length	14.3m	22m	26.5m	11m	15.6m	16.5m	8.84m
Wing area	58sq m	132.6sq m	105sq m	28sq m	61.7sq m	74sq m	19.74sq m
Empty weight	3,986kg	9,790kg	9,400kg	2,000kg	6,650kg	6,400kg	1,288kg
Loaded weight	8,600kg	22,600kg	14,500kg	3,600kg	12,500kg	11,400kg	2,413kg
Engines, number of	2	4	4	2	3	2	2
Engines, type	Gnome-Rhône 14 Kirs	Hispano-Suiza 12 Xirs	Hispano-Suiza 12 Xirs	Renault	Alfa Romeo 126	Fiat A.80	de Havilland Gipsy Six
Power (each)	800hp	720hp	720hp	220hp	900hp	1,000hp	205hp
Speed	385km/h	345km/h	355km/h	370km/h	450km/h	435km/h	381km/h
at altitude	1,600m	2,000m	2,100m	—	4,000m	3,900m	—
Serial/ registration	F-APDY	F-APUZ	F-ARFA	F-AODR	MM.355 MM.356 MM.357 MM.358 MM.359 MM.21117	MM.21241 MM.21242	G-ACSS

Above: An S.79 racer with Ratier propellers, showing the entry hatches. (*SMA*)

Below: A line-up of the Italian racers before the start: the two Fiat BR.20As I-8 and I-10 and the S.79.

Instrument panel of the S.79. This is not a racer but c/n 57, the Piaggio-powered one, showing a non-standard Nistri panel, a predecessor of today's head-up display.

July with test pilot Coupet. This big aeroplane, with an elegant high-aspect ratio wing, had excellent lifting capacities and it was planned that it would fly over Damascus, without landing to refuel. Its crew included Guillaumet, Lanata, Le Duff and Vauthier. Guillaumet, one of the greatest pilots of French civil aviation, died on 27 November 1940, shot down by fighters over the Mediterranean, in the Farman *Le Verrier*. The Farman (later SNCAC, after the nationalisation of the French aviation industry) F-APUZ set the speed record on 1 October 1937 over a circuit of 1,000km with a 10 ton load. It was then named *Chef pilote L Guerrero* after an Air Force commander lost in a crash, and was flown by Paul Codos, inspector general of Air France, on a record-setting flight to Chile, ending its career as a patrol aircraft in the Second World War.

The Bloch 160, F-3, was another nice airliner with four Hispano-Suiza 12 Xirs, but of more modern design than the Farman. The Bloch had been designed for colonial use for Air Afrique, as a 12-passenger airliner, and although not successful, it set a record for speed with 5,000kg loads over 1,000km in October, piloted by Curvale, and it sired the Bloch 161 Languedoc airliner that was much used after the war. Its crew for the race included pilots Col François and Laurent and crewmen Beucher, Joly and Faure. It was named *Lieutenant Genin*.

Romania had entered two contestants, the Caudron Typhon YR-ADD *Dacia* of Prince Cantacuzene, one of the many flight-loving Romanian aristocrats of considerable wealth who gained notoriety in the 1930s, and the Bellanca 28-92 YR-AHA of Alexander Papana. The Bellanca was an odd-looking record aeroplane designed for the Lindbergh Derby, powered by a nose-mounted 420hp Ranger and two 250hp Menasco engines, claiming a promising top speed of 285mph and a cruising range of 4,000 miles. Both aeroplanes had the qualities needed to be serious contestants, but did not arrive at Istres.

If one racer embodied sporting qualities at their best, it was the British entry, the D.H.88 Comet G-ACSS of A E Clouston and Flt Lt George Nelson. This was in fact the *Grosvenor House* which had won the 1934 Mac. Robertson race from England to Australia. It had been taken over for tests by the Royal Air Force as K5084, had been damaged and sold as scrap. Recovered and restored, it was called *The Orphan*, but

The Caproni-Reggiane Ca.405 Procellaria at the Guidonia test aerodrome. In the background is an S.79 racer. Here, the Ca.405 is not painted with the white rectangle for its race number.

luckily this historic aeroplane was not foresaken, and it is now the only survivor of the race, owned by the Shuttleworth Trust.

Finally, there was an American competitor, James Mattern, a pilot of high reputation who, in 1932, had attempted a flight around the world, with Griffin, on the Lockheed Vega *Century of Progress*, getting as far as Borissov in the USSR, where he force-landed on a swamp. In 1933, he tried again with the same Lockheed Vega, flying alone from New York to Norway, but once more he had to terminate his flight in the USSR. It is not known which aeroplane he intended to fly, but he, also, did not arrive at Istres.

Green Mice from Italy

Italy had originally entered five S.79s, two BR.20As and two Ca.405s. The NANA had been changed to the Nucleo Addestramento Voli Ampio Raggio (NAVAR – Long-Range Flight Training Unit) and its personnel, waiting to know what would happen to the Blue Riband, spent the spring of 1937 working on a highly specialised schedule for setting new world records – not with a single aircraft, but with the five S.79s in formation – and studying mass fast flights across the Atlantic, which were later realised with the Rome to Rio de Janeiro flight in 1938. Their aeroplanes were emblazoned with the insignia of the 12° Stormo, 'I sorci verdi', three green mice, referring to an Italian manner of speech (to show somebody the 'green mice' meant to show incredible things, to cause trouble).

When it was confirmed that the Istres–Damascus–Paris flight was replacing the Blue Riband, the organisation of the Regia Aeronautica was put to work to support the Italian competitors.

Fiat BR.20A I-8 of pilots Rolandi and Bonino being serviced while the pilots look on. (*Emiliani*)

They would fly over the whole range of Italian airports, weather reporting stations and radio direction beacons, across the peninsula and over the Italian-held island of Rhodes. The Air Ministry spared no expense to ensure an Italian victory. It had ordered nine special aeroplanes from three different companies, plus all ancillary equipment, and organised strong support groups in Istres and Damascus: fifty men and twelve train wagons of fuel and equipment to France, seventy men and nine tank trucks equipped for fast refuelling of the racers in Syria.

Moreover, the SIAI S.79 prototype, I-MAGO, was used for reconnaissance flights along the route. Designed as a fast airliner and a possible contestant in the London to Melbourne race, it had first flown on 8 October 1934, too late for that event, but now its path again crossed that of the Comet. Originally powered by three Piaggio P.IX radials, it was re-engined with Alfa Romeo 125s and on 24 September 1935, it had set several world speed records over 1,000km and 2,000km with loads up to 2,000kg. Finally, it received three Alfa Romeo 126s, the standard powerplant of the S.79 bombers and racers, a particularly successful and reliable powerplant that was a development of the Bristol Pegasus, for which a licence had been

After the race, the Ca.405 was modified with a lower fuselage shaped like a flying-boat keel to facilitate emergency landings on water. The aircraft was written off on 19 June 1939.

acquired by Alfa Romeo in 1931. During these scouting flights I-MAGO flew from Rome to Le Bourget in 3 hours, impressing the French with its 400km/h average.

As the Procellaria, assigned racing number I–9, could not compete, the Air Ministry ordered the preparation of one more S.79, standard bomber MM.21117, in order to have at least an eighth entrant. This aircraft had its ventral gondola removed but the dorsal hump was left, and it had the standard SIAI propellers. More purposeful than the sleek racers, this S.79 bomber had the same high-gloss racing colours.

The men were as remarkable as their aircraft. The pilots at Biseo's orders were among the best and most famous Italian airmen. Lt Bruno Mussolini was the dictator's son. He and his brother Vittorio had served, with much fanfare, in a bomber unit during the Ethiopian war, but while Vittorio was not much of a pilot (he wrote a book on his experiences, *Voli sulle ambe*, that is particularly distasteful), the shy and earnest Bruno was very good, respected and appreciated for his abilities and not for his lineage. In July, he flew with Biseo on the Piaggio P.XI-powered S.79 (c/n 57), setting another world record, and later flew with the 12° Stormo in Spain. Later, he piloted one of the 'Green Mice' to Brazil, then held a management position with the airline Ala Littoria, and then with LATI, when a commercial service was established to South America. During the war he joined a bomber squadron and died in the crash of a Piaggio 108 bomber in 1941.

The other pilots were Lt-Col Ranieri Cupini, Majors Angelo Tondi, Antonio Lippi, Umberto Fiori, Giuseppe Gaeta, Enrico Rolandi, Umberto Rovis and Captains Amedeo Paradisi, Stefano Gori Castellani, Giovanni Lucchini, Luigi Questa, Guido Bonino, Luigi Klinger and Stefano Trimboli. This was really a Who's Who of Italian aviation. Cupini, Questa, Bonino, Lippi, Rovis, Fiori, Klinger and Trimboli had been among the pilots with Biseo on Balbo's mass flight to Chicago in 1933. Cupini was later director of the Air Force Academy and Inspector of Schools. Angelo Tondi was later chief test pilot at Guidonia, holder of several records (including the strange one of being the only Axis pilot to shoot down an enemy, using a captured Lockheed P-38). Moscatelli, among other feats, was later the only Axis pilot to fly to Japan and back during the war, with a SIAI Marchetti SM.75. Gori Castellani was commander of Squadron 274 'Bruno Mussolini', the only unit equipped with the P.108 four-engined

bomber. Giuseppe Gaeta became commander of the Co-belligerent Air Force, and died after the war in a Beech C-45 crash. Luigi Questa was another remarkable pilot, who had flown the special S.55 I-SLAN in a scouting mission to Iceland before the 1933 *Trasvolata*. Maj Rolandi was an engineer, formerly technical director of the Gabardini factory and now Fiat test pilot, who had first flown the BR.20 bomber. Guido Bonino had won the Bibesco Cup of 1931. Umberto Klinger became no less than the director of Ala Littoria, and during the war he commanded the Air Transport Command. After the war he was president of Società Aerea Mediterranea airline and of Officine Aeronavali of Venice.

If the pilots were famous, the NCO crew were also remarkable men, the very backbone of the air force, most of them veterans of Balbo's seaplane flights. At least two must be remembered: wireless operator Davide Giulini, who held the record for the first man in the world to have flown an aircraft over the Atlantic four times (in 1931, twice in 1933 with Balbo, then in 1934 with the SIAI S.71 I-ABIV), and engineer Armando Palanoa. Palanoa was the wizard of aero engines, the man who had prepared all the powerplants of the Italian Schneider Trophy racers, and who had done all the fuel consumption calculations for this race.

Biseo and Mussolini flew S.79 MM.359 I-5, Tondi and Moscatelli MM.355 I-6, Lippi and Castellani MM.357 I-7, Rolandi and Bonino the BR.20A I-8, Gaeta and Questa the other Fiat I-10, Cupini and Paradisi S.79 MM.358 I-11, Rovis, Klinger and Trimboli S.79 MM.21117 I-12 and Fiori and Lucchini S.79 MM.356 I-13. All these

Lt Bruno Mussolini, son of the dictator, during the stop in Damascus of his S.79 I-5.

aircraft had radio call signs derived from their pilots' names: respectively, BIMU, TOMO, LICA, ROBO, GAQU, CUPA, ROTR and FILU.

The Race

Considering the whole affair, there was no real contest: there was a whole proud air force against some unproven company prototypes and a few sportsmen, or 'gentlemen drivers'. The whole point of the event was not technical achievement, because in 1937 few would have found much difficulty in flying about 2,000 miles over well-known routes in the summertime, but it was political. In 1937, Italy's fascist régime was at its zenith. It had conquered Ethiopia against the opposition of the world, and was now involved in the Spanish Civil War. Aviation was the core of fascist

Maj Fiori, pilot of S.79 I-13, talking to a French officer.

Take-off of S.79 I-5 of Biseo and Mussolini from Damascus.

propaganda and a powerful tool with which to intimidate the democracies. The rivalry with France was very high, French aviation minister Pierre Cot and people such as pilot Bossoutrot, a leftist Member of Parliament, were often attacked in the press, while Italian aviation successes, new records or victories over French aeroplanes supplied by the Front Populaire government to Republican Spain were polemically emphasised. The nationalisation of the French aircraft industry was ridiculed (not only in Italy, but also in Britain by people such as C G Grey). A victory in the Istres–Damascus–Paris race would show the superiority of fascism over the 'decaying' democracy of Paris. In France itself, the right wing was viciously opposed to the socialist government, accusing it of leaving the country defenceless, and one has to wonder whether many were regretting that the race was being held at all.

The Breguet 470T Fulgur F-APDY took part in the race as F-1. This aeroplane later flew for Republican Spain.

The Italians were accommodated in Aix-en-Provence, 50 miles from Istres, and forced to commute several times in the days preceding the race, days that were particularly hot. After long discussions, the departures were arranged according to the speed of the competitors in order to allow arrival and refuelling in Damascus in the early morning, when the air was coolest. The Italian team had planned a demonstration mass flight, with a formation landing at the arrival in Paris. In this way, the crew that had started last would be the winner. According to Mussolini, positions were drawn by lot, but it seems that the five S.79s had their order of departure set according to each pilot's seniority: first, the slower S.79 I-ROTR, then the two Fiats, Majors Fiori, Lippi, Tondi, Lt-Col Cupini and, finally, Lt-Col Biseo, ensuring the victory of the commander and Bruno Mussolini, which was obviously of political significance.

First to be waved off by the starter was Rossi's Caudron Typhon on Friday 20 August 1937, at 1725hrs, followed by the Bloch 160 F-3 of François and

Radiogram by Bruno Mussolini

From on board the S.79 Sky of Albania, 0100hrs (Italian time).

I am transmitting this via the onboard radio as we fly over the coast of Albania. Weather conditions have been good so far. There are some storm clouds here and there. We are informed of good conditions also above Greece, which we shall be flying over presently, and above the Mediterranean. We are in constant contact with Rome and with the Italian aeroplanes, so I am sending you these radiograms in the midst of interruptions.

We flew over Italy half an hour ago, at a height of 4,200m, in the Orbetello-Bari direction. The peninsula was starry with the beacons of the airfields' searchlights. The organisation planned by the air force to assist our flight has meant for us not just support, but also a greeting from the Fatherland.

We took off from Istres on the thirteenth at 2234hrs, after four French, the British and the other seven Italian aeroplanes. We had drawn lots for this schedule. Somebody said it was a good omen. Visibility has been good so far, particularly helped by the full moon that shines in a fascinating way over the landscape. Anxiety about the race disappeared, as usual, as soon as we took off. All our attention is now concentrated on the not-so-easy navigational calculations, on which our chances of victory mostly depend.

I shall now stop transmitting because the radio has to be used for more important matters. Goodbye until tomorrow morning.

Bruno Mussolini

Published by the *Popolo d'Italia* (Mussolini's newspaper), 21 August 1937.

Above: The Breguet 470T Fulgur F-APDY.

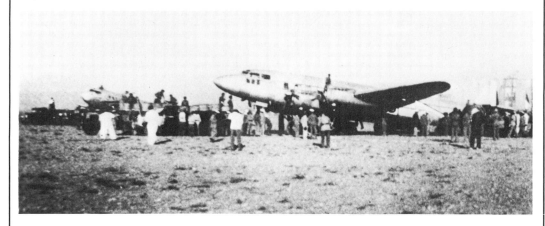

Below: The elegant Bloch 160 four-engined airliner and the Breguet Fulgur during their refuelling stop at Istres.

Left: The French Government-owned Farman 2231 F-APUZ, seen here at Paris Le Bourget, on another occasion, was a formidable load lifter but scarcely built for speed.

Laurent, then the Breguet Fulgur at 1901hrs, the Farman 2231 at 2012hrs, the Comet G-ACSS at 2020hrs and then the Italians (the S.79 I-12, the BR.20As and the five S.79 racers) at intervals of a few minutes.

The night flight over the islands of Corsica and Elba, across Italy, then over the Adriatic to Corfu, Athens, southern Turkey, Cyprus and Damascus was smooth and easy for all aeroplanes, as the night was clear, with a full moon, although there was a storm front to the east. The only contestant to have any trouble was Rossi: his Caudron was buffeted by the wind and its automatic pilot was disengaged. The lone flyer landed in Damascus at 0654hrs on 21 August; the first leg of his flight had lasted over 13 hours and Rossi decided to abandon the race. Another aeroplane, the S.79 I-7 of Lippi and Castellani, had to fly with its starboard propeller stuck at minimum pitch, forcing it to reduce speed. (Ratier propellers gave problems to most SIAI Marchettis, but they were retained for the 1938 flight to Brazil.)

The first aeroplane to land in Damascus was the Breguet Fulgur, which landed at 0340hrs after an uneventful flight. Ten minutes later it was joined by the Bloch 160, and the two French airliners were refuelled side by side, rather leisurely. The Farman 2231 flew at low level over Damascus, without landing, at 0543hrs. It saved some time by not landing, but its flight time had been almost 10½ hours, that is to say, an average speed of 279km/h, much too slow for any chance of victory. The de Havilland Comet landed at the Syrian

The sole British entrant, de Havilland Comet, G-ACSS, raced as G-16, was entered and flown by A E Clouston with Flt-Lt G Nelson as his navigator. Clouston, New Zealand born, had joined the RAF on a short service commission, afterwards becoming one of the first permanent civilian test pilots at the Royal Aircraft Establishment.

G-ACSS, after winning the 1934 England–Australia air race, and technically far in advance of its day, had been bought by the Air Ministry for tests. Clouston flew it at Farnborough. Later, the aircraft was written off during overload tests at Martlesham, the wreckage being sold as scrap. At that point Clouston, with financial backing, bought the remains and had them rebuilt, with two Gipsy Six Series II engines equipped with DH controllable-pitch propellers.

The rest is history – the story of Air Cmdr Clouston, who restarted his RAF career at the outbreak of the Second World War, winning a DSO and DFC and exceptionally an Air Force Cross and bar for his services to test flying, and of G-ACSS flying again today, is the very stuff of Britain's aviation heritage. [Editor]

I-12, the standard S.79, at Rome's Littorio Airport after the return from France. The hangar, built by the Compagnia Nazionale Aeronautica, was unusual because it had two storeys and a ramp from which the aircraft could roll out and take off.

town at 0456hrs, having flown at an average speed of 346km/h, a good performance that was enhanced by a very fast fuelling stop: after 22 minutes the two Englishmen were back in the air.

At this time the Italian team had already arrived, first Rovis and Klinger's I-12 at 0405hrs, then at very short intervals I-13, I-6, I-8, I-11, I-5 and finally I-7, slowed down by its propeller. Still, all eight Italian racers had been faster than the Comet, the best of their foreign rivals, and all the other S.79s had averaged speeds of between 426km/h (I-6 and I-11) and 403km/h. Biseo felt that victory was no longer in doubt, and the Italians, greeted by their fellow countrymen in Syria, had time to organise a new plan for the return leg. This included individual take-offs, navigation on a Rhumb-line track at a slightly reduced speed and maintaining radio contact in order eventually to formate in flight, landing at Le Bourget together.

However, something went wrong during that hectic time on the ground.

Some aircraft were only partially refuelled and the pilots, overtired because of their hard commuting in France and because of the night flight, did not realise they were taking off with half-empty tanks. The BR.20 of Rolandi touched a lamp post and ripped part of the fabric covering of the rudder, while the I-7 of Lippi and Castellani, the last one off, hit a ridge and snapped an undercarriage leg, crashing on its belly. The race was over for I-7. Eventually it was dismantled, the fuselage and en-

gines were shipped back to Italy and its wings were scrapped.

On the way back, all remaining aircraft fell foul of the storm that they had just skirted on their way to Syria. Very soon all radio communications

On 30 November 1937, an S.79 racer set the world speed record of over 1,000km with a 5,000kg load. That same day another S.79, with Piaggio P.XI engines, set another speed record for over 1,000km with 500, 1,000 and 2,000kg loads. The pilots Bacula and De Ambrosis are pictured here in front of the former.

Air Race Performance

Race no.	Aircraft	Serial/ registration	Crew	Start time	Av.speed to Damascus	Arrival time	Racing time	Average speed	Place	Remarks
F-1	Breguet 470 Fulgur	F-APDY	Codos, Arnoux, Agnus	1901hrs	334km/h	1603hrs	21hr 2min	294km/h	5	
F-2	Farman 2231	F-APUZ	Guillaumet, Lanata, Le Duff, Vauthier	1912hrs	279km/h	1317hrs 22 Aug	42hr 5min	147km/h	9	No landing in Damascus. Forced landing in Belgrade
F-3	Bloch 160	F-ARFA	François, Laurent, Beucher, Joly, Faure	1849hrs	322km/h	1727hrs	22hr 38min	273km/h	7	
F-4	Caudron 640	F-AODR	Rossi	1725hrs	215km/h	—	—	—	—	Withdrawn from race at Damascus.
I-5	SIAI S.79	MM.359	Biseo, Mussolini, Mori, Gadda, Bernazzani	2134hrs	423km/h	1537hrs	18hr 3min	342km/h	3	Forced stop at Cameri
I-6	SIAI S.79	MM.355	Tondi, Moscatelli, Reverberi, Giulini	2128hrs	426km/h	1902hrs	21hr 34min	286km/h	6	Forced stop at Venice-Lido
I-7	SIAI S.79	MM.357	Lippi, Castellani, Palanca, Suriano, Trezzini	2124hrs	350km/h	—	—	—	—	Written off in Damascus
I-8	Fiat BR.20A	MM.21241	Rolandi, Bonino, Rossi	2107hrs	400km/h	—	—	—	—	Forced landing in Cherso
I-10	Fiat BR.20A	MM.21242	Gaeta, Questa, Zoppi	2114hrs	378km/h	—	—	—	—	Forced landing near Venice
I-11	SIAI S.79	MM.358	Cupini, Paradisi, Ardu, Vaschetto	2131hrs	426km/h	1503hrs	17hr 32min	352km/h	1	
I-12	SIAI S.79	MM.21117	Rovis, Klinger, Trimboli, Forconi, Murolo	2101hrs	403km/h	1115hrs 22 Aug	38hr 14min	161km/h	8	Forced stop at Alture di Pola
I-13	SIAI S.79	MM.356	Fiori, Lucchini, Negro, Cubeddu	2120hrs	422km/h	1517hrs	17hr 57min	344km/h	2	
G-16	de Havilland D.H.88 Comet	G-ACSS	Clouston, Nelson	2020hrs	346km/h	1600hrs	19hr 40min	314km/h	4	

Each Italian crewman who took part was given a beautiful silver medallion, minted in 1920 for another purpose, as a souvenir.

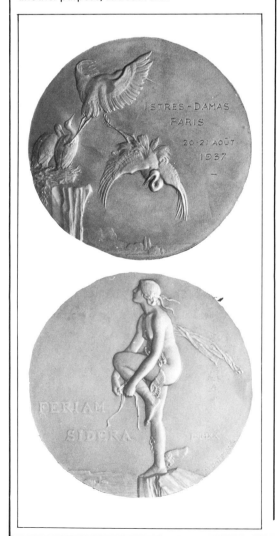

were lost and the pilots could neither get in touch with each other nor home on the radio beacons. The Farman, heading towards France, had to turn back and finally landed in Belgrade at 1805hrs, spending the night there. The Fulgur carried on, slowed down by the wind, as did the small and valiant Comet. Meanwhile, the Italians were having more than their share of troubles. Tondi and Moscatelli realised that they were short of fuel, and headed for Ronchi dei Legionari Airport, where a support base was organised, but had to land at Venice-Lido, where they got stuck in the mud. After refuelling, they took off again only 2 hours later. I-12 had to land at Alture di Pola airfield, in the Istria peninsula, at 1325hrs, but their magnetos had been flooded by the storm, and they had to wait overnight, not starting again until 0800hrs on Sunday 22 August.

The Fiat BR.20A had no better luck. Test pilot Rolandi, short on fuel, headed for a landing in the North Adriatic area but came into the worst part of the storm, so he force-landed on the island of Cherso (now Cres, Yugoslavia) and his aeroplane was written off. The racer of Gaeta and Questa was forced to shut down an engine that was vibrating badly, and with the remaining engine it tried to reach Venice-Lido Airport but could not, and force-landed on a nearby beach. The Fiat participa-

tion in the race ended with failure, which did not go down well with the Turin company. Fiat had strong hopes of its BR.20, the winner of the Italian Air Force 1934 competition for a medium bomber, only to see the popularity of the S.79, which was not even designed as a bomber to start with, growing in favour in the air force's higher circles, particularly with Gen Valle. That this wonderful aeroplane could amass records and successes did not mean, in Fiat's opinion, that it deserved to be ordered in ever-growing numbers for Italy's bomber squadrons instead of the steady and businesslike BR.20, which was not much slower anyway.

Biseo and Mussolini managed to get a glimpse of the other S.79s and tried to catch them up before crossing the Alps, but their centre propeller became stuck at minimum pitch. They chose the best possible landing place, Cameri, the SIAI Marchetti company aerodrome, where they landed at 1425hrs. The Ratier airscrew could not be repaired quickly, so it was hurriedly set at maximum pitch, the aeroplane was refuelled and Biseo and Mussolini took off 30 minutes later. Only two Italian racers, I-11 and I-13, flew the second leg without forced landings, but they also had their share of problems. The automatic pilots had been affected by turbulance, so that the pilots had to fly the trimotors manually (a particularly unstable aircraft) in a strong headwind. The oxygen system was not working properly and had to be turned off, and the aeroplanes had to climb to over 20,000ft in order to fly above the great mass of clouds stretching from the Aegean Sea all the way to France. Finally, one of Cupini's wing-engine propellers malfunctioned and it was stuck at cruising pitch. All radio communications were impossible, so no pilot knew where his team mates might be. Finally, the S.79 homed on Le Bourget's radio beacon and by early afternoon they had reached the outskirts of Paris.

No matter what was written at the time, the unspoken order was that it would be better if Biseo and Mussolini won and the French were treated to a formation landing. So Cupini's hapless S.79 I-11 started circuiting above Le Bourget, without crossing the finishing line, hoping that another Savoia Marchetti might appear. Suddenly, Cupini saw a small grey twin-engined aeroplane zooming by. Fearing that it could be *The Orphan*, he rushed in to land in order to obtain at least second place. It was not G-ACSS but a French touring aeroplane on its way, so Cupini and Paradisi found themselves winners of the race, having landed at 1503hrs. A quarter of an hour later, I-13 also arrived and landed, and at 1537hrs it was the turn of Biseo and

Mussolini who, because of landing at Cameri, gained only third place.

At 1600hrs, Clouston and Nelson's Comet landed. They had gained fourth place, but their performance had been a sporting feat. Without the assistance of an air force, and against newer competitors that had seven times their power, they had averaged a most impressive 314km/h, also showing remarkable regularity on both legs of their trip. Three minutes later, the first French competitor arrived, the F-1 Breguet Fulgur of Codos and Arnoux. Later, at 1727hrs, the Bloch 160 arrived and, finally, at 1902hrs, the S.79 I-6 of Tondi and Moscatelli who had been stuck for two hours in Venice. However, considering their starting time, the Italians were scored sixth and the Bloch of François seventh. The two strayed aeroplanes, the bomber I-12 and the large Farman, arrived the next day, around midday.

The crowds at Le Bourget must have been disappointed by the foreigners' victory but it was very dear to the Italian winners. Minister Pierre Cot, certainly not a friend of fascist Italy, was forced to award the prizes to the dictator's son and to the other Italians, possibly fearing what would happen the next day: the rightist French press would (most unfairly) claim that the Front Populaire and its nationalisation

had ruined the French aircraft industry. One must also consider that in July an air meeting had been held in Zürich in which the new German military types, the Messerschmitt Bf 109 and the Dornier Do 17, had humiliated the obsolete French Dewoitine 510 fighters (even the Do 17 bomber was faster than them). Obviously, air power was not judged on sporting events only, but an alarm had been sounded and, shortly after, the first French missions were sent to the United States in order to approach the American aviation firms to evaluate their possibilities.

All Italy rejoiced at this victory, which was boasted by the press. The 'Green Mice' became national heroes, and the elegant shape of the Savoia Marchetti S.79 became familiar to all, becoming a sort of nationally recognised symbol, like the Spitfire in the United Kingdom. The effect was emphasised five months later when three S.79s, again led by Biseo and Mussolini, flew to Brazil, paving the way for the transatlantic service inaugurated by LATI in 1939. Meanwhile, on 25 August the four surviving S.79 racers, the I-12 and the faithful I-MAGO, returned to Rome's Littorio Airport, where they were welcomed by a crowd and all the air force commanders, with chief of staff Gen Valle. Mussolini was there to speak to

Le Journal de Paris **SIXIEME EDITION** Le Journal de Paris

L'INTRANSIGEANT

APRÈS LA VICTOIRE ITALIENNE

Lundi 23 Août 1937 SIXIEME EDITION

L'Italie n'accomplit pas de MIRACLES : elle travaille

Elle a gagné la course Istres-Damas-Paris avec des **avions de bombardement de série,** dont le premier exemplaire a battu, il y a 23 mois, en un seul vol, six sensationnels records

ISTRES-DAMAS-PARIS Ce qu'il devait être et ce qu'il a été

La course telle que « L'INTRANSIGEANT » l'avait fait prévoir vendredi dernier :	La course telle qu'elle s'est déroulée hier :
"A 350 à l'heure et en 18 heures les 8 avions italiens gagneront-ils la course dont le départ sera donné ce soir ?"	A 352 à l'heure et en 17 h. 32 m. les avions italiens ont gagné la course Istres-Damas-Paris

« Ma course » par Bruno MUSSOLINI

C'est la foi qui fait le succès

Une épreuve : une leçon

Ce n'est pas un succès de l'aviation italienne, c'est un triomphe ! Les équipages italiens remportent avec éclat une double victoire :

CELLE DE LA VITESSE : 427 kilomètres à l'heure contre 336 ;

CELLE DE LA MASSE :

R. PEYRONNET DE TORRES

ISTRES-DAMAS : 2.931 KM.

1" équipage italien :	427 km. à l'heure
1" équipage français :	336 km. à l'heure
Équipage anglais :	335 km. à l'heure

the victors, among them his son.

The Italian victory had been fully deserved, and it was due to the careful preparation of the team (as opposed to the French improvisation), to the abundance of means, but most of all to the expertise of the pilots and crewmen, whose performance, particularly during the difficult return leg through the storm, was most admirable. The main

Below and bottom: Mussolini and Gen Guiseppe Valle, chief of staff and under secretary of aviation, welcome the victorious crews at Littorio Airport, Rome (now Aeroporto dell'Urbe). Bruno Mussolini is the pilot in white overalls without breast insignia.

propaganda point, which was much emphasised on both sides of the border, was that the S.79, which was merely the standard bomber of the Regia Aeronautica and had been in service for a long time and in large quantities, had nevertheless beaten the newest prototype from France.

The Istres–Damascus–Paris race had in reality been a political event. The Italians had 'shown the green mice' to the French when it was all a chauvinistic but harmless game. Unfortunately, it did not remain a game, and two years later Europe was at war. □

Bibliography

Cupini, R, '1937 – Istres–Damasco–Parigi', *Rivista Aeronautica*, Rome, Jul-Aug 1977.
Gentilli, R & Bignozzi, G, *Aeroplani SIAI 1915-1935*, Edizioni Aeronautiche Italiane SrL, Florence, 1982.
Gianvanni, P et al, *SIAI Momenti di Storia*, Ed AI SrL, Florence, 1977.
Noel, Jean, 'La course Istres–Damas–Paris', *Aviation Magazine 753, 1 May 1979.* 'Tre equipaggi italiani vittoriosi nella Istres–Damasco–Parigi del 1937', *IARB* (now *Aerofan*) magazine, Milan, Jul-Sep 1977.
Spaggiari, L & D'Agostino, C, 'SM-79 il gobbo maledetto', *Il Castello*, Milan, 1979.

Regional Airport: Liverpool

David Smith

Liverpool Airport   5321N   0253W

ICAO location indicator: EGGP
IATA location indicator: LPL

Visited on: 12 December 1988

Population resident within 50km:
3½ million
Runway: 09/27 2,286 x 46m
Operational: 24 hrs
Aids: ICAO Category II Approach and
Runway lighting
ILS Category II on Runway 27
NDB
Surveillance radar approaches down to
½ mile on Runway 09
Owned and operated by: Liverpool Airport
plc
Airport director: Christopher J Preece (acting
airport director at time of visit, Bryan
Trunkfield)
Passengers per year: 360,078
Principal users at date of visit: Manx
Airlines, British Midland Airways,
Ryanair, Aer Lingus
Movement totals: air transport 18,249,
other 59,471

Scheduled service destinations: (winter
1988–89): Isle of Man, Belfast City,
London Heathrow, Dublin
Customs: Yes

The one certain thing about regional airports is that each is unique. Only slightly less certain is that their future now looks better than at any time as more passengers and more cargo shippers become aware of the advantages of using local airports at both ends of their operation. Gradually, air operators are acquiring equipment better suited to the demands of serving many more traffic points which may have wide fluctuations in traffic demand. Each airport has its own special problems and its own marketing opportunities, as this recent visit to Liverpool shows.

Only 15 minutes by car or taxi from the city centre, Liverpool is one of the most easily accessible of United Kingdom airports. A rail and bus link ensures that public transport takes only about 10 minutes longer. The national motorway network is also close at hand, as well as mainline railway stations.

Existing in the shadow of Manchester International Airport, which is about 30 miles way, Liverpool Airport has had a long struggle for survival with threats of closure looming on several occasions. Its operators concede that it can only ever be complementary to Manchester, but it still has an important role to play, particularly in support of the current revival in the area's long-depressed fortunes.

The airport became a Public Limited

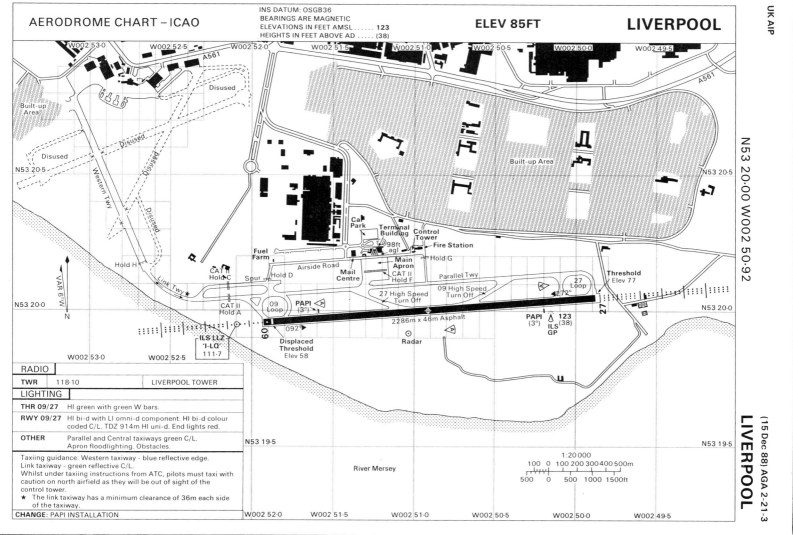

A TNT BAe 146, freight charter for the Ford Company in autumn 1986.

Company in March 1988 with backing from three local councils, and a Five Year Business Plan has been agreed to make it profitable. Drastic trimming in staff levels has already produced considerable economies without reducing the standard of service.

The airport's chief asset is the new terminal complex which opened in 1986. It was designed to cater, initially, for 400,000 passengers a year with provision for expansion, as necessary. It is a spacious, well-planned building, the one and a half storey design making passenger handling an efficient and speedy operation with special benefit to the elderly and disabled.

The main concourse incorporates a large check-in area, comfortable domestic and international departure lounges and a shop and information desk. A buffet bar is located on the first floor, overlooking the apron. The compact, purpose-built design is one of the airport's major advantages. The time taken to transit the terminal from car or public transport to the aircraft is as little as 15 minutes and, in reverse, as short as 5 minutes for a passenger with hand luggage only. An effective baggage handling facility systematically linked to passenger handling ensures rapid aircraft turn-rounds. The duty free shop, opened in 1985, is reputedly the cheapest in Britain.

A new management team has recently been appointed, the previous airport director, Rod Rufus, having left in September 1988 for a post in industry. Bryan Trunkfield, the airport's deputy director, was in charge during our visit and was able to give me an outline of current plans and answer a number of questions. The new director, Christopher J Preece, was formerly with British Aerospace (BAe).

Looking back at previous achievements, Mr Trunkfield observed that one of the airport's biggest coups was securing the 'Spokes for Speke' mail operation in 1979 (Speke is the airport's local name). Liverpool was chosen not only for its central position but also because of its proven weather record. It is said to be the second most fog-free airport in Britain after Prestwick.

The intention was to improve the first-class post between the more remote parts of the United Kingdom and is the reason why, theoretically, a letter posted the evening before in Exeter can be read at breakfast in Aberdeen next morning. The service has expanded to the point at which Liverpool is the centre of a hub linking Aberdeen, Bournemouth, Bristol, Cardiff, Exeter, Southend, East Midlands, London Stansted, Norwich, London Gatwick, Belfast, Edinburgh, Glasgow and Newcastle airports. About 60 tonnes of mail (6,000 bags) are handled five nights a week in the Post Office's own distribution depot next to the terminal building.

A considerable amount of regular income is generated by the Post Office operation and this has been augmented by the decision of TNT Express Europe to set up a northern hub at Liverpool from January 1989. The contract was won amid much competition from other locations, including Manchester and Leeds. TNT specialises in overnight parcel delivery, using a fleet of BAe 146QTs (the Whispering Jet).

Asked about TNT's policy, Mr Trunkfield explained, 'There will be five flights a night to Belfast, Prestwick and Cologne, the last being the centre of a network of European routes flown by TNT and its associates, The company

Congestion on the apron, viewed from the tower, during the strike at Manchester International in February 1987.

has indicated that it may increase the flights to fifteen a night, but this depends on initial results'.

The airport is responsible for supplying temporary accommodation for freight handling equipment, but TNT intends to build its own cargo building at a cost of about £1 million. Liverpool Airport's financial director, Rod Hill, believes that TNT's move will enhance the airport's credibility in the commercial sector and provide additional business. It may show users of Manchester Airport that Liverpool has a great deal to offer and is a viable alternative.

The other good news in 1988 was Ryanair's decision to transfer its north-west England centre from Manchester to Liverpool. For years, Aer Lingus has had a monopoly on the Liverpool–Dublin route and there was little incentive to reduce the relatively high fares. Then Ryanair introduced a daily BAC One-Eleven service at a 50 per cent lower fare. Aer Lingus was forced to upgrade some of its Shorts 360 services to BAC One-Eleven operations and reduce fares to a similar level.

This healthy competition has resulted in extremely high load factors for both carriers and plenty of business for the duty free shop, always a significant part of any airport's income. More important, people who would normally consider flying beyond their means have suddenly discovered that it can be both cheap and infinitely preferable to the tedious and uncomfortable sea journey.

Mr Trunkfield comments, 'The extra traffic generated by Ryanair and increased Aer Lingus flights is very welcome. Ryanair is looking into the possibility of a service to Knock, western Ireland, and to Amsterdam. This is not the first time that a Liverpool–Amsterdam route has been operated, the last being Genair's,

suspended in 1982. Ryanair believes a market exists, but we shall have to see what develops'.

Derek O'Brian, Ryanair's commercial director, has gone on record as saying that he is determined to continue the provision of a low-cost air bridge to the north of England, for which Liverpool is ideally placed. 'It is a compact and friendly airport, meeting the needs of the passenger and airline alike'.

Two other Irish Sea routes, which have been a traditional part of the Liverpool scene since the mid-1930s, are those to the Isle of Man and Belfast. However, by 1980 services to the island were barely profitable and British Midland Airways (BMA), which had taken them over from British Airways, decided to form a new airline. It was to be based on the Isle of Man and tailored to its special needs. Known as Manx Airlines and 75 per cent owned by BMA, it was to operate scheduled flights to London Heathrow and major provincial airports.

Manx Airlines began services into Liverpool in 1982 and soon became the airport's biggest operator, using the 36-seat Shorts 360 and, more recently, the Viscount, SAAB 340 and BAe 146. The demise of the Liverpool–Isle of Man

ferry service brought further expansion and bargain fares which included the highly successful 'Farecracker' service. Using Belfast City Airport, Manx Airlines was able to establish another profitable route.

Mr Trunkfield, quoting figures from Tim Stevens, commercial director of Manx Airlines, said, 'Services to the island are breaking all records. Carryings in summer 1987 were up no less than 22 per cent over 1986. During the summer of 1988 the figures were 23 per cent higher than those of 1987. On Saturday 20 August 1988, Manx Airlines carried a record number of passengers between Liverpool and the Isle of Man, a total of 1,277 on fourteen return flights'.

Manx Airlines also inherited the Liverpool–London Heathrow route from BMA but results were often disappointing. In 1988, aircraft technical problems wreaked havoc with its reliability, and many disgruntled business travellers obviously switched their allegiance to the British Airways Shuttle at Manchester – at least, until BMA took over the route in October 1988 with DC-9s and began to win them back.

The BMA Diamond Service was

A TNT BAe 146 bringing Vauxhall Motor Company freight from Nuremberg in November 1987.

British Airways Concorde at Liverpool in 1986, just before the opening of the new terminal.

Liverpool's third major success in 1988. The airport director takes up the story: 'BMA has announced that it is investing £350,000 in promoting the service and improving handling facilities in Liverpool. A new departure lounge especially for London Heathrow passengers has been established at the far end of the terminal building on the ground floor, and the whole operation is very much geared to the business passenger'.

The new service is an undoubtedly valuable connection, with check-in time reduced to just 10 minutes. The special lounge has a direct luggage feed to the aircraft, which is doubly important where speed of handling is a necessary selling point for the airport. The flight takes 45 minutes and a full meal service is offered. In addition, passengers can check in for onward British Airways flights and thus avoid the usual congestion at London Heathrow.

A field in which Liverpool is not making significant progress is that of inclusive tours. Manchester has effectively cornered the market and local people have no option but to travel the extra distance. However, a demand to fly from the local airport does exist, as is shown by the popularity of the few inclusive tours which do operate through Liverpool.

The holiday destinations announced for 1989 include Pula and Split, Yugoslavia, and a number of weekend departures to Jersey. There are continuing rumours of a large tour operator moving to Liverpool but Mr Trunkfield was unable to comment on this. He did agree, however, that congestion at Manchester might bring inclusive tour business to Liverpool when the companies saw what the smaller airport could offer. Some serious marketing is required and this will be set in motion by the new management team.

Other Activities

Liverpool's weather record is generally much better than Manchester's and it makes a convenient alternative for diversions. When an industrial dispute closed Manchester for two weeks in 1987, Liverpool handled more than a third of

A typical midnight mail aircraft scene with several de Havilland Canada Twin Otters and an Embraer Bandeirante.

British Eagle BAC One-Eleven G-ATVH outside the old Speke terminal, summer 1967. (*B Trunkfield*)

the former's regular traffic. The limiting factor, unfortunately, is the amount of apron space, the old airport no longer being available for parking.

Mr Trunkfield does not anticipate any increase in apron size in the foreseeable future, although this does cause problems when mail flights coincide with Ford Motor Company freight charters or Manchester diversions. The Ford Plant at Halewood is only a mile from the airport and freighters up to DC-8 and Belfast size are frequent visitors, bringing in parts from other Ford factories in Europe. Fluctuating demands may jeopardise the production lines and supplies are often flown in at short notice to maintain the output.

A further shortcoming is the lack of hangarage on the south airfield, all locally based aircraft being housed on

A BEA Douglas C-47 and DH Dragon Rapides at Liverpool in 1947.

the original airport site which requires lengthy taxiing to the runway. The land is to be developed and sooner or later everthing will be transferred to the new airport. As Mr Trunkfield points out, who is going to pay for the new accommodation? The airport does not have the necessary capital and the local operators are unwilling or unable to put up the money. The space is available, however, but careful planning will be required to avoid piecemeal development along the airside road.

There are three airport-based flying schools: Cheshire Air Training School, Liverpool Flying School and Air Nova, as well as a number of privately owned aircraft. In addition, because of intensive commercial traffic at the larger airport, much of the circuit training needed by Manchester-based flying

schools has to be done at Liverpool. It seems likely that some private aircraft currently operating from Manchester will move to Liverpool as the former expands.

There is no air taxi operator at Liverpool, but Dollar Air Services, whose headquarters is at Coventry Airport, maintains a subsidiary at Liverpool. This helicopter firm is kept busy with contract work and frequent charters on behalf of the news media, as well as ad hoc work. A limited amount of training is also undertaken.

A steadily increasing source of income is airline training. A variety of companies, including British Airways, Air Europe and Air 2000, find the airport ideal for circuit training, most of which can be flown over the River Mersey, which is almost 3 miles wide at this point. The Instrument Landing System (ILS) is available for instrument work, and it also attracts RAF aircraft,

Handley Page Halifax production, probably in 1942, from the Rootes Securities shadow factory built adjoining the airport, before the Second World War. (*Via P Summerton*)

ranging from Hercules and VC10s to Jetstreams and Hawks.

A further factor in the airport's future development is the adjacent Skypark site, which has meant clearance of the former Rootes shadow factory, used by Dunlop until about 1980. A very large area of land is now vacant and there are plans to build a hotel as well as an aircraft maintenance facility. So far, nothing has materialised but it seems that the developers are biding their time until they see how the airport's new plans develop.

After years of virtual stagnation, Liverpool Airport appears to be on the threshold of success and the management is determined to exploit every opportunity to generate traffic. As Bryan Trunkfield told me, 'An airport is really only a strip of concrete and our only weapon, as indeed for any airport, is to offer incentives'.

Origins

Today's achievements are built on a foundation which dates back to 1928, when Liverpool Corporation acquired land at Speke for the development of an airport. The official airport for the city was at Hooton Park, Wirral, on the opposite bank of the River Mersey. It was difficult to reach, there being no road tunnel in those days. In July 1930, Speke was granted a private civil licence and Imperial Airways began a short-lived regular service to Croydon via Manchester and Birmingham, subsidised by the three provincial corporations. It was not a success, however, and was discontinued after three months.

In 1931, Blackpool and West Coast Air Services inaugurated an unsubsidised Liverpool–Blackpool route and soon extended it to the Isle of Man. An old farmhouse became a makeshift terminal building, a hangar was erected and outbuildings were adapted as stores and workshops. Speke was made an airport in the true sense in 1933 when it was given a public use licence and customs facilities were provided. The impressive official opening ceremony on 1 July was claimed to be the greatest air pageant ever held outside Hendon.

1934 saw commercial air transport begin to establish itself at Speke, with KLM flying the first mail to Amsterdam and Aer Lingus launching a Dublin service. The Isle of Man route was now approved also for mail carrying and this and the Dublin service were to form a major portion of the airport's business right up to the present day. A Lorenz Blind Approach system was installed in 1936 to assist bad weather operations, services now being augmented by Railway Air Services and Hillman's Airways.

British Eagle Bristol Britannia G-ARKA and Cambrian Airways Viscounts (June 1964).

Liverpool Corporation pursued a policy of constant improvement, with the aim of levelling and draining the whole site. It was done so well that during the extremely wet winter of 1935/36, when almost all the other airports in England were closed, Speke remained operational. A large hangar and the distinctive control tower were built in 1937, the latter standing in solitary splendour until the terminal, based, it is said, on the one at Hamburg, West Germany, was completed in 1939.

By this time, the airport was the second busiest in the country after Croydon. An RAF auxiliary squadron had lodged at Speke since 1936 and Rootes Securities had opened a shadow factory on the perimeter in 1938 for Blenheim production. At the same time, the United States Lockheed Company set up a subsidiary to assemble Hudsons shipped to Liverpool Docks.

At the outbreak of the Second World War, Speke was requisitioned and until December 1941 housed a number of fighter units. Hurricanes of the Czechoslovak No. 312 Squadron shot down a Junkers Ju 88 over Merseyside on 8 October 1940. Speke's proximity to the port of Liverpool was the reason for its selection as the base for one of the most unusual units in the history of the RAF, the Merchant Ship Fighter Unit (MSFU).

Until long-range aircraft such as the Liberator were available to combat the menace of the prowling Focke-Wulf Condors, a stop-gap solution was required to protect the convoys. The answer was to catapult-launch Hurricane fighters from suitably equipped merchant ships and, after shooting down or driving off the attacker, the pilot would bale out and hope to be picked up. The necessary catapult training was performed at Speke.

After the MSFU was disbanded in September 1943, the airport's efforts were concentrated on Halifax production and the assembly of fighters shipped over for the US Army Air Force, as well as a variety of American military aircraft for the RAF. There was also a detachment of No. 776 Squadron, a Fleet Requirements Unit which was responsible for target towing, radar calibration and other essential duties. Apart from a short break in September 1939, the Isle of Man and Dublin services had been maintained and the suspended London service was revived in November 1944.

New Beginning

After the war, the centre of commercial aviation in the northwest gradually shifted to Manchester's Ringway Airport and, by the time Liverpool Corporation regained control of the airport from the government in 1961, the initiative had been lost. The 3-runway layout built during the war for Halifax test flying was incapable of extension, so it was decided in 1963 to build an entirely new 2,286m runway on an unrestricted site immediately to the southeast.

Opened in 1966, it was linked to the existing site by a taxiway. An ILS was installed on the eastern approach, together with the necessary high-intensity lighting to enable it eventually to operate to ICAO Category II standard, a goal which it later achieved. Liverpool was now able to accept large jet airliners on a regular basis, but the development came too late to revitalise its dwindling fortunes in the face of keen competition from Manchester.

In 1974, the operation of the airport was transferred to the newly created Merseyside County Council. The planners soon concluded that the only realistic long-term solution to the problem of what was, in effect, two airfields, was a

A D.H.86 G-ADVK of Blackpool and West Coast Air Services stands outside the big 410 by 215ft brand new hangar opened on 11 June 1937. With two sets of Esavian folding doors, it was probably the best in the country at the time.

phased programme for a transfer to single runway operation with a new terminal located alongside the new runway.

This had always been the intention, but financial constraints delayed implementation until 1981 when a new control tower and fire station were built on the southern site with considerable support from the European Regional Development Fund. The next phase was the construction of an apron, followed by a passenger terminal, fuel farm and car park which were opened in 1986.

It would be optimistic to suggest, as some have done, that Liverpool could become the 'Gatwick of the North' in its relationship to Manchester Airport, but it is certainly striving to meet the travel and leisure needs of Merseyside. □

BOOK REVIEWS

Cierva Autogiros. The Development of Rotary-Wing Flight
Peter W Brooks

Airlife Publishing, Shrewsbury, England
Hardbound, 384 pp, illustrated, £19.95.

Until the appearance of this volume, the only wide-ranging English-language book on the Cierva Autogiros has been Cierva's own account, *Wings of Tomorrow*, published in 1931. As well as being extremely hard to find, Cierva's book covers only the early stages of the story and does not give precise details.

Tragically, Señor Don Juan de la Cierva Codorniu died in the crash of a KLM DC-2 at Croydon on 9 December 1936, so it was left to others to recount the history of the remarkable aircraft he invented.

Peter Brooks' book had its origins in a two-part magazine article written by the author in 1955, but it has grown into an excellent and comprehensive history, a fitting tribute to the Spanish genius who stands as 'the greatest name in the history of rotary-wing flight before Igor Sikorsky'. Beginning with the early aeronautical influences on the young Cierva, Brooks traces the progress from his first aeroplane to the post-Second World War autogiros which are direct descendants of the primitive and unsuccessful C.1 of 1920. Many famous manufacturers in Europe and the USA come into the story – Avro, Westland, de Havilland, Weymann, SNCASE, Focke-Wulf, Kellett and Pitcairn.

Brooks identifies no fewer than four hundred and fifty-four Cierva-type autogiros of ninety-two types, built throughout the world, and all are well covered, with histories, specifications and general arrangement drawings – the latter being the work of several different draughtsmen and varying in quality. The author's research is impressive, and this is indisputably an essential and primary source of reference on the subject. Two particularly interesting chapters cover autogiro development in the Soviet Union and Japan. As well as notes, a glossary, bibliography and index, there are twelve appendices covering autogiro production, the Kay, Hafner and Weir machines and, as a bonus, a short history of helicopter development. **P J**

The Fighting Me 109
Uwe Fiest

Arms & Armour Press, London.
Hardbound, 160 pp, illustrated, £14.95.

Monographs of famous aircraft types always find a market, and this one, featuring one of the classic fighters of the Second World War, will appeal to many Luftwaffe enthusiasts.

Rather than reiterate all the technical information that has been published in previous sources, the author has attempted to present a historical survey covering the men who flew the aircraft and its operational use. Although the Bf 109 (in spite of the book's title, the aircraft is referred to as the Bf 109 throughout the text) had become obsolete in the second half of the war, it was never entirely superseded by the Focke-Wulf Fw 190, and formed the mainstay of the Luftwaffe's day fighter force. In terms of production, the tally of 33,000 Bf 109s was bettered only by the Soviet Union's *Sturmovik*.

Fiest's book is divided into three sections. The first is a development history which takes the reader from the prototype Bf 109, first flown in August 1935, to the final variants, the K, H and Z, the last being an extraordinary twin-fuselage machine combining two Bf 109 airframes to produce a 'heavy fighter'. Sadly, this variant is not illustrated.

The title of the second section, 'The Bf 109 in Action', speaks for itself. This covers the operational use of the aircraft (those who flew it, their scoring system and the corresponding rudder markings); the day when top scoring ace Hans Joachim Marseille, who claimed one hundred and fifty-eight victories before his death in 1942, downed seventeen British fighters; the defence of the Reich, and a typical Luftwaffe response to a raid.

All sections are profusely illustrated by black and white pictures, and the final section (half of the book) is a 'photo gallery' of pictures covering all aspects of Bf 109 production and operation. Three appendices list the major Bf 109 sub-types, describe the organisation of Luftwaffe units, and describe Bf 109 weapons systems.

The text appears to be secondary to the illustrations, and it will be of equal value to the modeller and the student of history. This is not the last word on Messerschmitt's classic fighter, but it is a useful addition to the literature. **P J**

Failed to Return — Mysteries of the Air 1939-1945
Roy Conyers Nesbit

Patrick Stephens, Wellingborough, England.
Hardbound, 192 pp, illustrated, £12.95.

I must confess that the title of this book deceived me. I assumed that I would read about strange and unresolved disappearances of aircrews, an assumption strengthened by familiarity with Nesbit's articles in *Aeroplane Monthly* on the disappearances of Amy Johnson and Glenn Miller.

Although the book does include these two accounts, enhanced by some later feedback, the other chapters are more of the 'failed to return' type, rather than true mysteries. Nesbit has certainly done a great deal of delving in public and private archives, however, and has uncovered some interesting stories. The information was available, but this is the first time it has been accumulated and sorted to produce well recorded accounts of the events selected.

There is a good variety. Wreckage recovery forms the theme of the chapters on the Lock Ness Wellington and the Halifax disinterred at Rozenburg Island, Netherlands, in 1967. An audacious escape attempt by two German POWs forms the basis for another story, while research into photographs of Allied aircraft crashed in enemy territory produce three more chapters. We all know that every picture tells a story, and this is a good example of the adage exploited to the full. Other chapters cover the flight of Rudolf Hess, and the sinking of the *Cap Arcona*.

There is sufficient variety here to provide absorbing reading in small doses, and the text is easy to read. There are still questions which remain unanswered, and perhaps Nesbit's work will trigger further research into them. **P J**

THE PUTNAM AERONAUTICAL REVIEW

ISSUE NUMBER TWO **JULY 1989**

Editor: John Motum

Consultant Editor: John Stroud

Assistant Editor: Linda Jones

Design: Swanston Graphics Ltd

Managing Editor: Julian Mannering

Publisher: William R Blackmore

© Conway Maritime Press Ltd

ISBN 0 85177 527 6
ISSN 0955-7822

Published quarterly by:
Conway Maritime Press Ltd
24 Bride Lane, Fleet Street
London EC4Y 8DR

Telephone: 01 583 2412

**Annual subscription for
four issues:**
£16.00 UK, £18.00 overseas,
$33.00 USA and $39.00 Canada.
All post paid.

Typeset by Swanston Graphics Ltd, Derby.
Printed and bound in Great Britain
by Page Bros (Norwich) Ltd.

CONTRIBUTORS

D G Addicott is a former Vickers test pilot, current air show and films pilot, and former chairman of the Vintage Aircraft Flying Association, during the period in which the Vickers Vimy replica was being built and flown.

Dr Norman Barfield has spent much of his working lifetime in the engineering development and promotion of commercial airliners built on the Brooklands site at Weybridge, Surrey. In this issue he completes the story of airline operations by VC10s.

Frank J Delear is a well known aviation historian, author of five books, including *Famous First Flights Across the Atlantic*, biographer of Igor Sikorsky and a former PR director of Sikorsky Aircraft Division of United Technologies.

Dr René J Francillon is an independent researcher and air historian who lives in California and is well known as a Putnam books author. His latest *Grumman Aircraft since 1929* will be published later this year.

Philip Jarrett is production editor of *Flight International* and has been researching and writing aviation history for more than twenty years. His book *Another Icarus: Percy Pilcher and the Quest for Flight* was published by the Smithsonian Institution.

Gerhard Katzsch, aviation and travel writer, lives in Hamburg. He has had long-term connections with both the British airline and aircraft manufacturing industries, and the travel industry worldwide.

John D R Rawlings is a former C P Robertson Trophy award winner for long-term writing on RAF subjects, including the development of a Service awareness of its own history. A Second World War flying instructor, he continued flying and has over sixty types in his logbook, was chaplain at the Royal Aircraft Establishment Farnborough for seventeen years, and has written eight books.

Front cover: The elegant lines of the airship R34, photographed at Inchinnan, Renfrewshire.
Back cover: Erection of Alcock and Brown's Vickers Vimy at Quidi Vidi before the first non-stop crossing of the North Atlantic by aeroplane.

Editorial

Why, three score years and ten, a life-span, after the first transatlantic air pioneers, the main subject of this issue, should we remember them?

Almost sufficient is that as stories of high endeavour, their achievements bear retelling. Many *Review* readers – most, we hope – still try to get a seat near the sides of the wide-bodied cabin in their trans-atlantic airliner to catch a first sight of pack-ice or a coastline. We recognise achievement, respect it and wonder what lessons there may be still to learn.

The first airmen to succeed in crossing the North Atlantic by air were American, their names scarcely remembered. The six-man Navy crew led by Lt-Cdr Albert C Read were at the pinnacle of an expensive and highly professional effort which foreshadowed by twenty years the first regular commercial crossings of the North Atlantic by flying-boat with pas-sengers in the summer of 1939. Both endeavours used aircraft which were massive for their day, the Curtiss NC-4 and the Boeing 314, respectively.

Next came the equally professional but quite different British effort, which was more in the nature of an obstacle race in which four teams competed seriously to win a prize of £10,000. First, aircraft had to be built in Britain and tested, then dismantled and taken by sea some 2,500 miles to Newfoundland and the procedure repeated in reverse order, except that this time suitable flying fields had to be found as well. Only after that could attempts be made and, as history records, Alcock and Brown were the first men to cross the Atlantic direct.

Passenger crossings of the North Atlantic using land-based aeroplanes did not begin until 1941 when first the British Overseas Airways Corporation (BOAC) and a little later Pan American and Trans World Airlines (TWA) crews, all on war service contracts, operated as required, subject to weather. Only in 1946 did regular commercial services with land-based aircraft begin. Mean-while, wartime services with flying-boats had continued.

Perhaps the most imaginative of all the 1919 Atlantic crossings was the two-way voyage by the airship R34. Remote as airships may seem today, it was this form of transport with which the German Zeppelin company started scheduled summer crossings of the North Atlantic in May 1936, having made an earlier beginning on South Atlantic scheduled operations in March 1932. From ancient (in air terms) history, the message for today is that challenges, though diffe-rent, continue.

To begin with there is the question of scale. Only in 1957 did the passenger crossings of the North Atlantic by air exceed those by sea; the total for each was about a million. Exactly thirty years later, some 25 million passengers crossed in about 135,000 commercial flights. Such growth creates the type of challenge on which aviation thrives.

Like many surface transport undertak-ings, the Atlantic commercial air traffic peaks, at two major periods in each twenty-four hours. The major traffic flows occur from 1100 to 1600hrs GMT westbound and 0400 to 0900hrs GMT eastbound. Managing the traffic is the function of Oceanic Air Traffic Control which, with military aircraft allotted to the higher flight levels, also serves the needs of business jets and the high-flying passage of Concordes of British Airways and Air France.

Given that air traffic above the North Atlantic continues to expand, what is likely to happen next? Twin-engined aircraft, rather more advanced than Alcock and Brown's original twin, are already in scheduled operation above the North Atlantic. As these flights must at the time of writing be operated, by regulation, within two hours' single-engine flying time of a diversion airport to cater for an emergency engine shut down, observant air passengers are notic-ing differences in their flight tracks and flight times; this is especially noticeable when bucking against a strong westerly jetstream which just happens to be in the airspace through which the more north-erly track demanded by the regulations for twin-engined operation passes.

There are now moves in the USA to extend the two hour diversion case to three hours and to express it in minutes, perhaps to make it more palatable. Many will wish to see much more statistical evidence on the safety of the 120 minute rule before proceeding further.

However, there is a strong case for the basic principle of twin-engined operation over-ocean, because it potentially opens up more destinations to more passen-gers. Boeing 747 capacity for major trunk routes between major world cities is not in dispute. In Britain, the habit of tinkering with the existing airports for London continues, and of course this is necessary, but an intelligent apprecia-tion of the contribution which could be made by British regional airports is not yet being considered seriously.

It is said with truth that the majority of visitors to Britain wish to fly to London, but this savours of defensive argument, when the majority of inhabitants of the United Kingdom live neither in London nor the vaunted southeast. In consider-ing future airport policy, the needs of those Britons neither requiring nor wishing to transit through London are a factor. The English Midlands is a major population centre and, already with good rail and motorway links, is well able to draw in and despatch air passengers from and to other parts of the country. A new or adapted airport in Midland Britain could well become United Kingdom One, rather than London's fourth. However, this is not the place to enter into a new and no doubt long-running debate about which one.

Should a policy of partial planning and 'muddle through' prevail, a large number of twin-engined airliners with the range to reach United States Gateway cities and maybe beyond, operating from Bri-tain's provincial airports, could well be needed to prevent stagnation. This type of solution, a bit of this and a bit of that, the compromise found increasingly often, may be put down to market forces. If 'Without vision the people perish', then surely all interested parties will work for better solutions. Sir John Alcock and Sir Arthur Whitten Brown, both of whom grew up in Manchester, would probably have understood. □

The first heavier-than-air commercial crossing of the North Atlantic was made by Imperial Airways on 20-21 July 1938. The aircraft used was the Short S.20 *Mercury*, the upper component of the Mayo Composite Aircraft. Capt D C T Bennett and Radio Officer A J Coster were separated from the Short S.21 *Maia* near Foynes, Ireland, flying 2,930 miles direct to Montreal, carrying mail and newspapers. *Mercury*, seen here at Montreal, continued to Port Washington, New York, returning via Botwood, Newfoundland, the Azores and Lisbon.

Curtiss NC-4: First Aircraft to Cross the Atlantic

Frank J Delear

Three Curtiss flying-boats set out to fly the Atlantic. One succeeded, the NC-4. Here is a tale of highly organised and expensive endeavour, symbolic of the effort which would be needed much later to establish commercial air traffic above the North Atlantic.

One of flying's more frightening experiences is the 'near miss', the inadvertent passing of two aircraft dangerously close to each other while in flight. On the night of 16 May 1919, one of history's most famous transatlantic flights came within a few feet and a few seconds of disaster when not two, but three aircraft experienced a near miss 4,000ft above the Atlantic Ocean. The craft were the US Navy's giant Curtiss flying-boats, the NC-1, NC-3 and NC-4.

At the time, the lumbering, four-engined biplanes were droning along at their normal cruising speed of about 63 knots and were more than 300 miles into their attempt to make the first crossing of the Atlantic by air. A little luck, a moonlit night, and some quick, evasive action combined to avert what almost certainly would have been a fatal collision. With the moment of danger past, the flight pressed on through the night, eventually to earn aviation immortality for one of the three aircraft.

In retrospect, that close call above the ocean was typical of the troubles which plagued not only the flight but the long months of preparation leading up to it. Despite difficulties, however, the imagination, desire and determination of a small group of naval aviators kept the daring project alive even when, at times, it appeared that all was lost.

The story of the NC boats had its beginning in 1917, more than a year before the end of the First World War. German U-boats had been sinking a great many Allied vessels, including some in which aeroplanes were being shipped overseas for use in seeking out and bombing the enemy's underseas craft. The Navy decided that the crisis called for aircraft large enough to fly across the Atlantic, a far-fetched idea at the time, but one that promised a means of evading the deadly U-boats.

In September 1917, the Navy commissioned Glenn Curtiss, the pioneer pilot and seaplane builder, to design a long-range flying-boat able to carry sufficient bombs, depth charges and guns to destroy enemy submarines. The order called for four such aircraft to be built at the plant of the Curtiss Engineering Corporation, Garden City, Long Island, New York. A giant hangar was erected, meanwhile, at the Naval Air Station at nearby Rockaway Beach to house two of the huge craft. Also, a marine railway was installed for launching and beaching the flying-boats.

The four aircraft, a joint project of the Navy and Curtiss, were designated NC (Navy/Curtiss) and were soon nicknamed 'Nancies'. (The name called to mind an earlier Curtiss model, the JN-4D 'Jennies', the First World War trainers in which many American pilots later won their wings.) Spearheading the NC project were Rear Admiral David W Taylor, chief of the Bureau of Construction and Repair; and Cdrs Holden C Richardson (who designed the hulls), Jerome C Hunsaker and George C Westervelt.

The idea of crossing the Atlantic by air had tempted men from the earliest days of flying. Balloons and non-rigid airships had made a few unsuccessful attempts. In 1913, Lord Northcliffe, a leading British newspaper publisher, offered a $50,000 (£10,000) award for the first transatlantic flight between the United States, Canada or Newfoundland and Ireland and England (in either direction) in seventy-two hours. Alightings on the water for repairs, and even towing, were permitted.

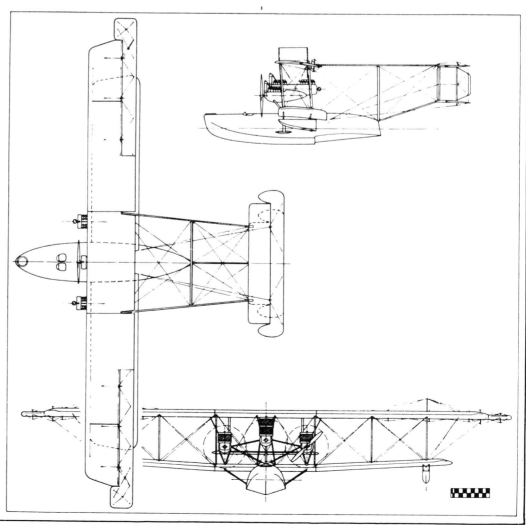

Curtiss NC-4.

The war delayed the race for over four years, but by the spring of 1919 three British aircraft were in Newfoundland being prepared for the risky flight, while a fourth was en route by sea to join the competition. The aircraft were a single-engined Sopwith biplane (Harry Hawker, pilot; Kenneth Mackenzie-Grieve, navigator); a single-engined Martinsyde biplane (F P Raynham, pilot; C W F Morgan, navigator); a Handley Page V/1500, a four-engined First World War bomber, with a crew of four; and, at sea, a twin-engined Vickers Vimy bomber (Capt John Alcock, pilot; Lieut Arthur Whitten Brown, navigator).

The NC-1 made its first flight on 4 October 1918 at Rockaway Beach with Cdr Richardson at the controls and a crew of five aboard. The craft, the world's largest flying-boat, made an impressive sight with its squat, grey hull and bright, yellow wings spanning no less than 126ft. Mounted high on wooden beams was a box-like tail assembly which was twice the size of the standard fighter biplanes of the day.

Before the flight, Richardson had expressed private doubts as to his design of the aircraft's hull. But the NC-1 performed very well, both on the water and in flight, easing the designer's fears. Further success came on 25 November 1918, when the NC-1, its three 400hp Liberty engines straining to their utmost, flew with fifty-one men aboard, a world record at the time.

With the end of the First World War, and with only the NC-1 built, the NC boats' original purpose – submarine hunting – no longer existed. The project was in danger of being abandoned when a new goal appeared because of renewed interest in an aerial crossing of the Atlantic. Why not use the NC boats to make the first crossing? Such a flight would help strengthen naval aviation and make a major contribution to air progress in general.

Cdr Westervelt pushed the idea with a lengthy report and thorough plan for the mission. Ignoring some complaints of 'publicity stunt', a Navy Transatlantic Flight Planning Committee accepted Westervelt's recommendations and assigned Cdr John H Towers, a pioneer naval aviator and already active in the NC project, to prepare for the flight.

Under the flight plan the four Nancies would fly from Trepassey Bay, Newfoundland, to the Azores, an over-water hop of 1,200nm, then on to Portugal to complete the ocean crossing, and finally, on to England. Stationed every 50 miles along the entire route would be sixty-eight destroyers to form a bridge of safety and information. Five battleships were placed every 400 miles as weather stations, along with every fourth destroyer. In daylight each vessel was to emit black smoke to show wind speed and direction. At night searchlights were to be aimed into the wind to give wind direction, and star shells were to be fired every five minutes until the Nancies had safely passed the area.

The chosen date for departure from Newfoundland was 16 May because there would be a full moon, a night flight being selected so the aircraft could alight in daylight at the Azores. The Navy was not seeking Lord Northcliffe's $50,000 prize since the competition rules had been changed and now required a non-stop flight with no alightings allowed.

Work was speeded up on the NC programme, with subcontractor companies in Massachusetts, Rhode Island and New York building the remaining hulls, wings and tail sections. Using lessons learned from the NC-1, Navy and Curtiss engineers made many design changes. For greater range, speed and payload they added a fourth engine behind the centre powerplant, so that the three original engines pulled and the fourth pushed the boats through the air. They moved the pilot's cockpit from its position high in the centre engine nacelle down and forward to near the hull front.

Each aircraft had stations for a crew of six: an aircraft commander/navigator in the bow; behind him, two pilots side-by-side in an open cockpit; and in the hull's aft section a radio officer, an engineer and a mechanic. Nine 200 gallon aluminium fuel tanks were installed amidships. Crew communication, almost impossible over the roar of the engines, was by a metal speaking tube, by headset telephone, or by crawling from compartment to compartment. There was no cabin heating, nor were there any electrically heated flying suits.

Technically, the major naval air development of the First World War was the long-range flying-boat. Among these were the twin-engined Curtiss H-12 and H-16, and the single-eingined HS. US Navy crews flew the HS and H-16 types when these American-built craft became available in 1918. Earlier, they had used French-, British- and Italian-built aeroplanes for anti-submarine patrols, convoy escort and bombing.

Thus, the larger and longer-range NC boats were logical extensions of their Curtiss predecessors, and not freak machines put together to perform a single record-breaking flight. Flying-boat is an apt term for the Nancies which were stout, seaworthy craft that ploughed through rough water at 60mph and then took to the air to reach top speeds of over 90mph.

The Nancies' double-planked V-bottomed hulls, designed to cushion the shock of alighting, offered a good combination of strength and lightness. The bare hulls, as completed by the yacht builders who served as subcontractors, weighed only 2,800lb each, yet their displacment was 28,000lb, or one-tenth of a pound of boat per pound of displacement. To provide a watertight hull and yet keep the planking thin and light, a layer of muslin was set in marine glue between the two plies of planking.

Strength and light weight were achieved elsewhere throughout the NC design. The main wing spars, for example, were hollow spruce boxes. The 12ft long wing ribs weighed only 26oz each and, on test, were required to carry a 450lb load of sand for twenty-four hours without damage. Design ingenuity was evident in this early aircraft. The wing's hinged leading edge enclosed the aileron control cables to eliminate air drag on the cables. At the same time, the cables were accessible for inspection by simply swinging up the leading edge on its hinges.

Careful balancing of all control surfaces, along with friction-free control cables running through ball-bearing pulleys, made the 14 ton Nancies easy to handle by one man of normal strength. Among the instruments were the then-new bubble sextant, a drift indicator adapted from one used in an Italian airship, and a radio direction-finding compass. The commander/navigator had a radio for talking with the other aircraft or ships within a 25 mile radius. There was also a 'telegraph' for sending and receiving messages up to 200 miles.

The principal specifications and performance figures for the Nancies included: wing span 126ft upper, 94ft lower; length overall 68ft 3½in; height 24ft 5⅛in; hull length 44ft 9in; tail span 37ft 11in; chord 12ft; gap between wings 13ft 6in to 12ft; dihedral angle zero upper, 3 degrees lower; wing area (including ailerons) 2,380sq ft; tailplane 268sq ft; elevator 240sq ft; fin 79sq ft; rudder 69sq ft; power four 12-cylinder Liberty engines of 400hp each; fuel capacity 1,890 gallons; oil 160 gallons; weight empty 15,874lb; weight loaded 28,500lb; maximum speed (at 28,500lb) 74 knots (85mph); maximum speed (at 24,000lb) 84 knots (96mph); and cruising speed 63 knots (72mph).

In early 1919, the NC project was moving along smoothly, but soon came ill omens and troubles extending eventually into the transatlantic flight itself. A late March storm tore the NC-1 from its moorings and battered the craft against the beach for three days, damaging the hull and a wing. The ship was repaired with parts from the NC-2 (which had not been flying very well) and the planned four-aircraft flight was reduced to three.

On 5 May, the date of scheduled departure of the flight from Rockaway Beach to Halifax, Nova Scotia, the NC-1 and NC-4 were damaged in a hangar fire that started at 0215hrs. Frantic work by repair crews returned the boats to service, again with parts from the NC-2.

The 6 May found a large crowd gathered to see the NC boats take off for Halifax but bad weather forced a postponement. Turning to leave, the spectators looked on in horror as a small seaplane spun out of the sky and crashed into a fuel storage tank, killing the two-man crew.

The take-off was rescheduled for 7 May, but again disaster struck: early that morning, shortly before the scheduled take-off, the NC-4's flight engineer, Harry Howard, lost a hand when he accidentally thrust it into a whirling propeller. Once more the departure was postponed.

When the three flying-boats finally took off at 1000hrs on 8 May, only a few spectators were on hand to see the departure. Cdr Towers, head of the newly formed NC Seaplane Division 1, was in charge of the flight. Towers' flagship was the NC-3, with Cdr Richardson as pilot. The NC-1 was skippered by Lt-Cdr Patrick Bellinger with Lt-Cdr Marc A Mitscher (later of Second World War fame) as pilot. Lt-Cdr Albert C Read, a quietly spoken and diminutive (5ft 4in, 120lb) New Englander, commanded the NC-4 with 1st-Lieut Elmer F Stone of the US Coast Guard as pilot. Each flying-boat carried its crew of six.

Trouble struck at 1400hrs on the flight to Nova Scotia. The NC-4, fast becoming an unlucky aeroplane, was forced to set down at sea when first one and then a

NC-4 on a test flight before its historic transatlantic flight in 1919. (*US Navy*)

second engine failed. The other aircraft flew on and reached Halifax at 1900hrs. The next day, they flew to Trepassey Bay, bouncing and shuddering through stormy weather most of the way.

The NC-4, riding an empty sea some 80 miles northeast of the Naval Air Station, Chatham, on Cape Cod, Massachusetts, taxied all night on its remaining two engines, reaching Chatham at dawn. Repairs were made and, after several weather delays, the NC-4 took off again on 13 May for Halifax, reaching that city despite the rough running of three engines.

After more engine repairs, the NC-4 took off for Newfoundland on the morning of 15 May. Within minutes, Read and his men were again on the water, as usual with a power problem. The trouble, a blocked fuel line, was quickly repaired and the NC-4 lifted from the sea at noon. Descending for their alighting at Trepassey Bay, the crew of the NC-4 met an unpleasant surprise: the NC-1 and NC-3

The epic journey of the Curtiss NC-4.

were speeding through the water on their take-off runs. Towers had decided to leave for the Azores without the tardy and trouble-plagued NC-4. At that moment, fate stepped in to change aviation history. The two flying-boats, overloaded and running in a crosswind, were unable to get airborne. The NC-4 alighted, the others returned to base, and the flight was rescheduled for the following day.

Inaccurate fuel gauges, it was found, had caused the NC-1 and NC-3 to be overloaded with petrol. It was a situation which Read, knowing the need for special interpretation of the fuel readings, had suspected as he watched the futile take-off attempts from overhead. Read ordered a new engine and three new propellers to be installed, a Herculean, overnight task which by morning had been completed. The NC-4 was now poised for the real flight, with Read determined to erase the jinx label once and for all. No more shaky flights for his particular Nancy!

Trepassey Bay and the surrounding hills resounded to the roar of twelve Liberty engines as the three flying-boats taxied out and began their take-off runs

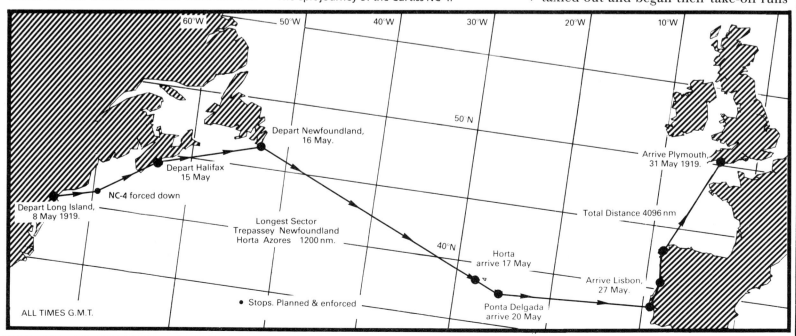

Depart Newfoundland, 16 May.

Depart Halifax 15 May

NC-4 forced down

Depart Long Island, 8 May 1919.

Longest Sector Trepassey Newfoundland Horta Azores 1200 nm.

Horta arrive 17 May

Arrive Plymouth, 31 May 1919.

Total Distance 4096 nm

Arrive Lisbon, 27 May.

Ponta Delgada arrive 20 May

• Stops. Planned & enforced

ALL TIMES G.M.T.

on the afternoon of 16 May. The NC-4, with 70lb of oil tossed overboard to lighten the load, became airborne, but the others were still on the water when they reached the open ocean. Even with the proper fuel load they were still too heavy.

Returning to base, Towers, aboard the NC-3, discarded one of his two radios, a sack of 'first cover' mail, and put ashore one of his two engineers. The NC-1 was also lightened. Read, circling overhead, finally alighted to await the others. Then, after a long run in a gusty crosswind, all three boats lifted into the air, the NC-4 climbing faster than her sister ships. Ahead lay the challenge of an overwater hop of some 1,200 miles, much of it at night, with all the risks and unknowns involved. Ahead also, as it turned out, lay an aeronautical achievement of imperishable quality and significance.

The aircraft were supposed to fly in formation, but several factors interfered with that plan. First, the NC-4 proved to have a higher cruising speed and kept pulling ahead of the others. When Read ordered a slowdown, the NC-4 became sluggish and difficult to control. Second, the NC-3's wiring had become soaked during the long take-off runs, blacking out the flying-boat's wingtip and cockpit lights. Third, the darkness and a cloud overcast made a close formation difficult and dangerous.

In addition, as the flight progressed, the question of whether the flying-boats or the station vessels on the sea were exactly on course became increasingly uncertain. As Towers would note later, even a line of destroyers 50 miles apart did not make for easy navigation. 'When we didn't find them just where we expected them', he said, 'there was always the question, are they wrong or are we?' As a result of all the foregoing problems, the Nancies soon found themselves 'on their own', which led to the near disaster mentioned earlier.

Mechanics clean and overhaul the NC-4's four Liberty engines in preparation for the transatlantic attempt. (*US Navy*)

Looking down to take a drift reading from his post in the bow cockpit of the NC-3, Towers suddenly saw the NC-4 directly below and very close. At almost the same moment, Read glanced up, sighted the NC-3 and quickly ordered his pilots to veer off and down. Only the moonlight, silhouetting the blacked-out NC-3, had enabled Read to see the other aircraft in time to avoid a collision. Seconds later, Towers caught a glimpse of the NC-1 passing above only a scant 50ft overhead. Bellinger, in the NC-1, had not even seen the darkened NC-3!

Again, the NC-4 pulled well ahead, her engines, instruments and navigation almost perfect. Rain, fog and rough air compounded the problems for all three aircraft. The stars became obscured, while the star shells and searchlights of the surface ships were no longer visible.

By dawn, the Nancies were near station vessel 14. Later, the air turbulence worsened and the crews, tossed roughly about and the pilots struggling with the controls, became unsure as to exactly where they were. They dropped lower through the mists, probing for a glimpse of the sea. At one point, the NC-4, buffeted by the turbulence, fell into a nose-down spiral, broke momentarily into clear air and recovered, barely missing a crash into the ocean.

Read, whose specialisation was navigation, sighted ship 16. Later, determining his position to be past ship 17, he continued with confidence toward the Azores. The skies cleared for a while, but then became foggy again. Read now knew that he was nearing the Azores and he began to worry about ramming one of the islands' high mountains, hidden somewhere ahead in the fog.

Suddenly, through a break in the clouds he spotted Flores, one of the western islands of the Azores. After Flores' coastal farmlands had slipped by below, Read saw surface ship 22 which he knew was about 250 miles from the flight's planned destination, Ponta Delgada on the island of Sao Miguel. The flight pressed on, but by the time the NC-4 reached the island of Fayal the weather had deteriorated again and Read knew he would have to find an alighting place as quickly as possible.

NC-4 on the seaplane ramp at the US Naval Air Station, Rockaway Beach, Long Island, New York. (*US Navy*)

After a short search, Read found Horta harbour, on Fayal's southeast tip, and the NC-4 glided to a safe alighting, 15hr 18min out of Trepassey Bay – only minutes before the weather became 'socked in' completely. Aided by the usual west-to-east tailwinds, the NC-4 had achieved an average surface speed of 82 knots (94mph) for the 1,200nm flight. Although the planned terminus of the Atlantic crossing, Lisbon, still lay more than 900 miles away, the longest and most difficult leg of the Navy's trans-atlantic enterprise had ended in success, at least for the NC-4. But what of her sister ships?

Both aircraft had strayed off course, the NC-3 to the south and the NC-1 to the north. Since passing station ship 13, the NC-3 had been flying blind through thick clouds. With only two hours of fuel left, Towers guessed he must be nearing the Azores and, like Read, also began to worry about hitting one of the islands' towering peaks. He decided to set down on the sea and await clearer weather. As the NC-3 broke through the clouds, the crew thought the sea looked fairly smooth. But ocean swells are deceptive when viewed from the air. The aircraft struck a huge swell, several engine struts buckled, and the frames of the hull were split. No one was injured and the rugged hull still floated, but the crew knew that further flying was impossible.

Bellinger, farther north, faced the same fog problem and also decided on an ocean alighting, with even worse results. As the NC-1 touched down, a 12ft wave tore off its lower tail section. The NC-1 and NC-3 now lay dead in the water and, since both were well off course, were far removed from the station ships which soon began a search. The NC-1 was some 100 miles from Flores and the NC-3 about 50 miles closer to that island. Both were more than 200 miles short of Ponta Delgada.

Though badly damaged, the NC-1 was able to limp through the rough seas, its seasick crew bailing water from the rolling hull. The ordeal ended after five hours when the battered flying-boat was spotted by the Greek ship, *Ionia*, which took the weary fliers aboard. Later, one of the searchers, the destroyer *Gridley*, reached the scene, and Bellinger and his crew were transferred to the American vessel. The *Gridley* took the NC-1 in tow, but this brought further damage to the aircraft. The destroyer then rammed and sank the crippled flying-boat to prevent her from becoming a menace to shipping. The NC-1 had become the first of a tragic list of aircraft fated to fall victim to the Atlantic in the ensuing decade and a half, although her crew, unlike many of the others, had lived to tell the tale.

Photographed from a Navy scout aeroplane, the NC-4 is seen on 8 May 1919, 60 miles out on the first leg of its long journey. This flight was interrupted by a forced alighting off Cape Cod, followed by repairs at the Naval Air Station, Chatham, Cape Cod. (*US Navy*)

Meanwhile, Towers and his men, also fighting for survival, were receiving radio calls, including the news of the NC-4 and NC-1. However, their own SOS transmissions were not getting through. From the radio receptions they knew that the destroyers had been sent many miles north of Towers' position southwest of Fayal. Their sole task now was to stay afloat and hope that the winds would carry them to one of the islands.

The wind was now of gale force and the waves washed the port wing float away. The men took turns perched far out on the starboard wing to keep the craft from capsizing. The NC-3 drifted eastward through the night of 17 May and at dawn her crew could see the peaks of mountains in the distance. Pilot Richardson held the NC-3 headed into the wind and the flying-boat continued its sternward drift to the east. This procedure, aided by Towers' skilful use of a sea anchor and one engine, finally brought the crippled NC-3 into the harbour at Ponta Delgada on the morning of 19 May. Towers refused help from a destroyer, and he and his men brought the tattered aircraft safely ashore themselves. It was a drama-tic arrival, and the gallant craft received a roaring welcome from ships' whistles and even a 21 gun salute.

The NC-3 never flew again. Yet her journey was a victory of sorts – a flight of almost 1,200 miles, followed by some sixty hours of tossing through more than 200 miles of stormy seas to become the first of the three Nancies to reach the first stop they had planned when they took off from Newfoundland. After being weath-ered in at Horta since 17 May, Read and the NC-4 reached Ponta Delgada on 20 May, a short hop of 1hr 45min.

During the stay at Ponta Delgada, the question arose as to whether Towers, as overall commander of the three-aircraft flight, should supplant Read as com-mander of the NC-4 and complete the transatlantic trip. The temporarily tense situation ended when Secretary of the Navy Josephus Daniels ruled that Read should continue as skipper of the NC-4. Thus did the course of aviation history again hang momentarily in the balance for the two naval aviators.

The NC-4 left Ponta Delgada on the morning of 27 May and, following the usual string of destroyers, alighted in the harbour at Lisbon at 1602hrs, the 800

NC-4 enters the harbour at Lisbon on 27 May 1919, to complete the first crossing of the Atlantic by air. (*US Navy*)

History made, the Curtiss NC-4 of the United States Navy, and admirers, rests on the slipway at RAF Mount Batten, Plymouth, in June 1919. (*Courtesy of Devon Library Services*)

mile flight having taken 9hr 43min. The reception at Lisbon was tumultuous and the publicity worldwide, for the big flying-boat had completed the first air crossing of the Atlantic, a giant step in aviation progress and one that had come a scant sixteen years after the Wright brothers' first faltering flights at Kitty Hawk.

The NC-4's flying time from New-foundland to Portugal was 26hr 46min, covering a span of eleven days and a total distance of 2,000nm. Or, measured from the original point of departure at Rock-away Beach, 41hr 51min in flight, cover-ing nineteen days and 3,000nm. Three days later, on 30 May, the NC-4 left Lisbon on a planned non-stop flight to Plymouth, Devon. The projected 775 mile hop was interrupted when a water leak in the port engine forced an alight-ing for repairs on the Mondego River near the town of Figueira, Portugal. The flying-boat ran aground on a sandbar but was extricated without damage. On the advice of the port's captain to wait for high tide, and to avoid arriving at Plymouth after dark, another overnight stop was necessary at Ferrol in Spain.

The NC-4 took off from Ferrol at

0227hrs the next day, 31 May. After a foggy and rainy crossing of the Bay of Biscay and the English Channel, during which the NC-4 missed four of the six destroyers stationed between Ferrol and England, the NC-4 completed its epic voyage. At the end, Read and his crew had logged 53hr 58min of flying time since leaving Rockaway Beach, travelling 4,096nm over twenty-eight days.

The NC-4 touched down on the waters off Plymouth at 0926hrs, 31 May, or, as someone put it, 'in plenty of time for lunch'. Read and his men were received by the Mayor of Plymouth during a ceremony at 'Plymouth Rock', a monu-ment marking the spot from which the Pilgrim Fathers had embarked for the New World 300 years before. The NC-4 crew, besides Read and pilot Stone, included Lieut Walter Hinton, co-pilot; Lieut James L Breese, pilot-engineer; Ensign Herbert C Rodd, radio operator; and Chief Machinist Mate Eugene S Rhoads, engineer.

The NC-4's flying days did not end

with her transatlantic achievement. Re-turned to the United States by ship, the flying-boat and her crew began a Navy recruitment tour. Between September and December 1919 they visited forty-three cities, ranging south from New York along the Atlantic seaboard and up the Mississippi River and back, with a side trip from New Orleans to Galveston, Texas. Thousands cheered the NC-4 crew who, with other members of Sea-plane Division 1, enjoyed a few passing moments of fame.

Which of the pioneering transatlantic flights was the most significant from the historical perspective? For years the answers have wavered between Lind-bergh's daring solo flight from New York to Paris in 1927 which gave aviation an unmatched emotional boost, and the historic and equally dramatic achieve-ment of Alcock and Brown eight years earlier. But for overall impact and signifi-cance, the stronger contender may well be the Nancies of 1919.

First of all, the Navy's transocean endeavour was thoroughly prepared from the viewpoints of aircraft design and construction, navigation, com-munication and safety. The NC-4 alone

Cdr Holden C Richardson, the Navy's first engineering test pilot, designed the huge hulls of the NC flying-boats and was pilot of the NC-3. (*US Navy*)

Lt-Cdr Albert C Read commanded the NC-4 on the epic transatlantic flight. (*US Navy*)

succeeded, but only because in the final 200 miles fog prevented the other two from also completing the vital first leg of the flight. The successful flight of the NC-4 was significant not only because it was a 'first', but because it showed how the Atlantic had to be flown in 1919, and perhaps for all time, if it was to be flown on regular schedules. Richard Smith, author of the book *First Across*, expressed it well. He wrote in 1973:

A half century after the NC-4, every Douglas DC-8, Vickers VC10, Ilyushin Il-62 and Boeing 747, not to mention all the military aircraft operating through

the North Atlantic's busy air corridors, were flying under what may well be called an 'NC-4 system' – good radio communications, adequate weather intelligence from the North Atlantic rimland supported by ocean station vessels, and an air-sea rescue network to provide assistance in the rare event of a ditching. This is the legacy of the NC-4.

The NC crews, probably because their flight lacked the individual heroics of Alcock and Brown, and Lindbergh, and many of the other early flights, did not receive due recognition in the long term, especially in the United States. Eleven years passed before a small handful of the NC fliers received decorations from President Hoover. The flight of the NC-4 has long been commemorated by plaques at Plymouth and Lisbon, but in the USA it was not until fifty years after the flight that a small marker was placed at Rockaway Beach, a modest tablet donated by the citizens of that community.

In contrast, the flight of Alcock and Brown is commemorated by plaques at St John's, Newfoundland, and Clifden, Ireland, and by a magnificent statue at London Heathrow Airport. Their Vickers Vimy has been on display for seventy years at the Science Museum, London.

In the USA, the NC-4 gathered dust in Smithsonian Institution warehouses for half a century, surviving largely through the efforts of the Smithsonian's Paul Garber. In 1968-69, in preparation for the fiftieth anniversary of her flight, the NC-4 underwent restoration in a Smithsonian shop at Silver Hill, Maryland.

In 1969, the restored NC-4 was displayed for a while on The Mall, Washington, DC. She was then loaned to the Naval Air Museum, Pensacola, Florida, where she stands to this day, on view for thousands of visitors each year, most of whom express awe at her size and a new appreciation of her special place in the history of aviation.

In 1937, commenting on the courage of those who pioneered transoceanic flight, the late columnist Walter Lippmann wrote:

The world is a better place to live in because it contains human beings who will give up ease and security and stake their own lives to do what they themselves think worth doing ... And what they prove to themselves and to others is that man is no mere creature of his habits ... but that in the dust of which he is made there is also fire, lighted now and then by great winds from the sky.

Words that surely apply to the gallant men of the Nancies and all who followed them. □

With her 24½ft height and 126ft wing span, the massive NC-4 dominates the Naval Air Museum at Pensacola, Florida, final home of the history-making flying-boat. (*Pratt & Whitney Aircraft*)

Voyage of His Majesty's Airship R34

Philip Jarrett

The First World War made it clear that the military potential of the large rigid airship, especially as an offensive weapon, was very limited owing to its vulnerability. As a commercial aircraft, however, it seemed to offer great potential when peace dawned late in 1918.

R34 takes shape at Inchinnan, Renfrewshire, in 1918. The new airship shared her shed with R27, and the final assembly of Handley Page V/1500s (serial E8287 is visible in right foreground), which William Beardmore built under licence, also took place at one end of the massive building.

With a few notable exceptions, heavier-than-air aircraft were still only able to carry very small payloads over limited ranges, and they were not a great deal faster than the latest Zeppelins. While the airship was ponderous and cumbersome by comparison, it had great weight-lifting potential. Because of the atmosphere's low density, usable amounts of lift demand large volumes of gas, and because lift increases with size more rapidly than does structural weight, the airship's efficiency improves with size. Unfortunately, it also becomes more unwieldy and presents great problems of handling – both in the air and on the ground – and hangarage.

Nevertheless, the Zeppelins of Germany's Deutsche Luftschiffahrts AG (Delag), formed in 1909 as the world's first airline, were the first transport aircraft. Although no scheduled services were operated before the 1914-18 war, seven Zeppelins had made local and inter-city flights totalling 172,535km (107,205 miles) in 1,588 flights. A total of 33,722 passengers and crew had been carried, without injury to passengers.

Britain lagged well behind in rigid airship development, acquiring most of its expertise from the study of German vessels which were captured or forced down. The 33 Class of British rigids, which were completed at the end of the war, were based on the design of the Zeppelin L33, which had been brought down in Essex in September 1916, but incorporated modifications resulting from the capture of L48 in June 1918. They were the largest airships built in Britain to that date.

The second airship in this class, R34, was destined to become famous as the first aircraft to make a return flight across the North Atlantic, and the first to make the East-West crossing. Laid down by William Beardmore and Co Ltd at Inchinnan, Renfrewshire, on 9 December 1917, R34 was finished on 20 December 1918. She was 643ft long, had a maximum diameter of 78ft 9in, and a capacity of 1,950,000cuft, giving a gross lift of 59.2 tons. The disposable lift (discounting the weight of the structure) was 30 tons.

The powerplant comprised five Sunbeam-Coatalen Maori IV engines each rated at 250hp. One of these engines was mounted in the rear of the control car, two with reversing gear were in small amidships gondolas, and a pair were geared to one propeller in the after gondola. Clutches allowed the engines to run without turning the propellers.

The R34's hull was built up of duralumin girders, being a 13-sided polygon in cross-section with twenty main transverse frames, one every 10m, with intermediate transverse frames. All frames had radial wire bracing. Connecting these frames were thirteen main longitudinal girders, separated by intermediate longitudinals which gave the finished hull the appearance of a 25-sided polygon. Extensive wire cross-bracing stiffened the structure. Incorporated inside the hull was a triangular keel which acted as the load carrier between the frames. Within the keel was a 1ft wide walkway extending the length of the ship and widened at the crew space, towards the front. This walkway allowed access to the engine cars, as well as accommodating the eighty-one petrol slip-tanks which fed the main tanks above each car, the water ballast, and provision for bombs.

The hull structure formed nineteen chambers for gasbags made from a single thickness of rubber-proofed cotton fabric lined with goldbeater's skin (the outer layer of the gut of an ox). Each bag had an automatic release valve, and some had hand-controlled manoeuvring valves. The outer cover was of doped linen, tightly stretched between each of the main frames. The cruciform tail fins and balanced control surfaces were built of duralumin girders and covered with doped linen.

Although the forward gondola appeared to be one unit, it was built in two parts, the front section housing the control room and 'wireless cabinet'. The control room contained the steering controls – a rudder wheel and elevator wheel, plus a magnetic compass, a clock and aneroid barometer, an inclinometer to measure the tilt of the ship, a height recorder, a rise and fall indicator, gas valve controls, engine telegraphs and water ballast controls. A chart table was also provided.

In case the main car controls failed, auxiliary rudder and elevator controls were fitted to the rear car. Both front and rear cars had bumping bags of compressed air beneath them. Two canvas tubes containing rope ladders led off from the keel to a machine gun platform on top of the hull and a gun pit behind the upper fin. When they were not on duty in the cars, the crew lived within the keel and slept in hammocks slung from the frame directly over the outer cover. A crew of twenty-two was the normal complement.

Bad weather kept the completed R34 in her shed at Inchinnan until 14 March 1919, when she made a maiden flight lasting almost five hours. This was followed by an extended trial flight on 24/25 March, during which some damage was caused by a jammed elevator.

R34's third flight, at the end of May, was a twenty-one hour cruise over the

Partially inflated gasbags in position in R34's hull.

The outer cover was laced on, and the gaps covered by sealing strips. The front car is being fitted.

North Sea, forcibly extended to Yorkshire by fog, with a final landing at East Fortune to the east of Edinburgh, which was to be her permanent service base.

After repairs, a six hour training flight over the Firth of Forth was made on 15/16 June, and this was followed on 17-20 June by a fifty-six hour, 2,000 mile trip over the southern Baltic, partly as a flag waving exercise to persuade the Germans to sign the peace treaty.

On 30 December 1918, the Admiralty had offered to lend the Air Ministry airships and stations to allow an assessment of the airship's potential as a civil transport aircraft. R34 had specifically been offered for a transatlantic flight. Early in 1919 Cunard, Armstrong Whitworth and Beardmore proposed the setting up of a company to undertake transatlantic services. After further negotiations it was decided to transfer all airships to the Air Ministry, and to send

Journey of the airship R34.

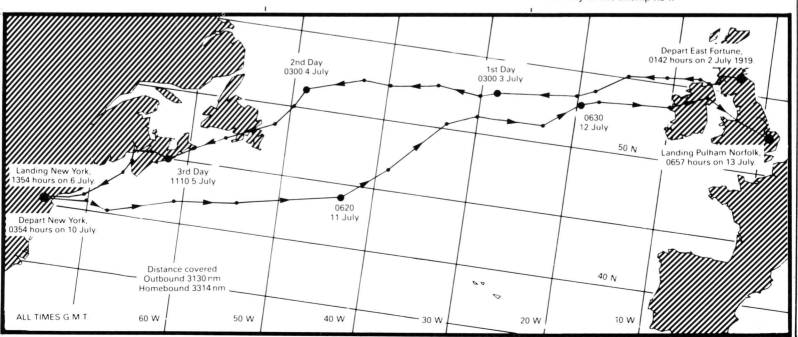

R34 is walked out of her Inchinnan shed for the first time, 14 March 1919.

R34 to America if the cost could be kept to about £2,000.

By mid-May, the estimated cost had risen to £3,000, excluding petrol and the cost of sending a small handling party to the United States, but Maj-Gen Sir Hugh Trenchard, chief of the air staff, believed that it was now too late to withdraw, as stores for the return voyage were waiting at White City, London, to be taken to the USA by the handling party.

Basing problems then arose which further delayed the flight, and the US Navy and Army argued over which of the Services should service the airship in the USA. In mid-June, the basing problem was resolved by basing R33 at Pulham, Norfolk, and R34, R29 and two non-rigid airships at East Fortune, to maintain the blockade if Germany refused to accept the peace treaty. An engine replacement and various modifications after this episode further delayed the transatlantic flight until July at the earliest.

Eight extra crew members were added for the flight, bringing R34's complement to thirty. She was captained by Maj G H Scott, an experienced airship com-

mander, and the second officer was Capt G S Greenland. The engineering officer was 2nd-Lieut J D Shotter, and the admiralty observer was Maj N E M Pritchard. The aviation pioneer Brig-Gen E M Maitland (later Air Commodore), Britain's senior airship officer, was the senior officer aboard. In true naval style, the crew was divided into port and starboard watches. Parachutes were provided for all of them.

A representative of the US Naval Airship Service, Lt-Cdr Zachary Lansdowne, came over to take part in the flight at the Air Ministry's invitation, and the cruisers *Renown* and *Tiger* were lent by the Admiralty to provide R34 with weather forecasts by radio from positions southwest of Ireland. In the event of the airship having to make an emergency descent, a destroyer was positioned southwest of Ireland to take her in tow, and a reserve landing party was prepared at Fermoy, County Cork.

The airship's elegant lines are revealed on her first airing at Inchinnan.

On the evening of 1 July, the final preparations were under way. R34 had aboard 4,900 gallons of petrol, 230 gallons of oil, 3 tons of water ballast and almost a ton of food, and her gasbags were 99 per cent full.

At 0100hrs (GMT) on 2 July, the crew, wearing heavy-duty flying suits with lifesaving collars and incorporating parachute harness, boarded R34, or 'Tiny', as they had nicknamed her. The shed doors slid aside at 0123hrs and the ground crew of 400 airmen, 80 women, and 150 men of the Black Watch walked the airship out stern first and held her down. Once she was clear of the 70ft high windscreens she was turned round into the wind, and then the engines were started and warmed up. At 0142hrs, 18 minutes early, Scott gave the command and a bugler signalled 'let go'. In slight rain and mist, the airship rose into the night sky, the clutches were engaged, and R34 quickly disappeared from view in the low cloud, cheered on by the ground crew. Releasing ¼ ton of ballast, she rose rapidly to 1,500ft, emerged from the clouds and headed up the Firth of Forth toward Edinburgh.

Because of the likelihood of losing gas by expansion, Scott could not let the heavily loaded airship rise very high in the early stages of the flight. However, it was necessary to ascend above 3,000ft to clear the Scottish hills, and 2 tons of lift was lost in the process.

The air at 1,500ft was 'most disturbed and bumpy, due to the wind being broken up by the mountains ... and causing violent wind currents and air pockets'. By 0420hrs, however, conditions were calm, and the forward engine was rested. The last land to be sighted before R34 headed out into the bleak Atlantic was Inishtrahull Island, off the north coast of Ireland.

Breakfast for the next watch came at 0730hrs to the accompaniment of jazz records. The watch then changed for the relieved crew to breakfast at 0800hrs. Cooking was done on a plate welded to the exhaust pipe of the starboard engine.

At 1100hrs, the airship was still between clouds, and Maitland noted that there was 'no visibility whatsoever'. The R34 was now running on only the two wing engines at 1,600rpm, giving her a speed of 29 knots – her most efficient speed because she consumed only 25 gallons per hour on the two engines. In her loaded state, the airship's static lift was supplemented by a large amount of dynamic lift provided by the vessel's streamlined shape – as much as 2½ tons when flown at an angle of 6 degrees at about 39 knots. As petrol was burned, so the static lift improved.

Although the airship was in radio contact with East Fortune, Clifden, and Ponta Delgada, it was clear that directional wireless was not yet reliable enough for determining position, so Maj G G H Cooke, the navigator, had to take readings whenever possible, sometimes using a cloud horizon. After an 'excellent' lunch of beef stew and potatoes, and chocolate and cold water at 1145hrs, the watch changed again to give the old watch four hours' sleep.

A little before 1400hrs Sgt H M Watson, one of the riggers, presented Maj Scott with a stowaway, AC2 William Ballantyne, a crew member who had been excluded from the final transatlantic crew but had determined that he was not going to miss a unique opportunity. Ballantyne at first hid behind the cur-

R34 soon after arrival at East Fortune, her permanent service base. (*Trustees of the National Museums of Scotland*)

The airship's control car. Note the bumper bags.

tains in the crew's quarters, then climbed up to lie on top of a girder between the sixth and seventh gasbags. Although his plan had been to remain hidden until the USA was reached, he was positioned above a gas vent and inhaled hydrogen every time the gas was vented by expansion. This made him extremely sick, and he had to beat a retreat to the keel, where he was discovered. It was too late for R34 to turn back, so after a reprimand from Maj Scott and a spell of recuperation, he was put to work on his routine duties.*

A glimpse of the ship's shadow on the water at 1515hrs enabled the drift angle to be determined – 21 degrees. The two wing engines were now being rested, and the wireless officer managed to receive St John's, Newfoundland, faintly but clearly, and speak to them, though East Fortune and Clifden were still in touch. A remarkable vivid rainbow effect was experienced, one complete rainbow en-

* When Peter Masefield, then chairman of the British Airports Authority, gave a fiftieth anniversary dinner at Heathrow to commemorate the events of 1919, William Ballantyne was a guest and given a great welcome.

circling the ship and another encircling its shadow.

Tea was served at 1545hrs, and the second officer, Capt G S Greenland, was trying in vain to discover who used his toothbrush to mix the mustard for lunch. After tea, Gen Maitland discoverd yet another stowaway: 'Wopsie', the ship's cat, had been adopted by LAC G Graham and became an unofficial crew member. At 1630hrs, Maitland noted that they had seen barely anything of the Atlantic so far, and were still in low cloud and fog.

At 1800hrs, Scott took the R34 to 2,000ft, where blue skies were found above the clouds, providing an 'enchanting' view, and also signs of a depression coming from the south. Scott exploited

R34's rear gondola housed two Sunbeam Maoris geared to a single propeller – hence the twin radiators. Auxiliary rudder and elevator controls were fitted in this car.

this by steering across its path so the winds enhanced the airship's speed.

By 1900hrs, the clouds were up at 3,000ft and R34 was travelling on all five engines, nose-down because the airship was light at night. At 2100hrs, bumpy flight between cloud strata was being maintained and it was very cold. Scott eventually descended to 1,500ft in an effort to improve the vessel's performance. By the early hours of 3 July, the halfway point between Ireland and Newfoundland was reached. 2nd-Lieut J D Shotter, the engineering officer, was finding his task demanding and suffering from fatigue and a headache.

The morning of 3 July passed fairly uneventfully, but at 1400hrs the starboard amidship engine developed a

The port amidships gondola. The engines in these gondolas had reversing gear. (*Via A C Jack*)

Maj George Herbert Scott, commander of R34 on the transatlantic flight. (*Trustees of the the National al Museums of Scotland*)

cracked cylinder waterjacket. Shotter effected a quick and good repair by plugging the hole with chewing gum (issued to the crew as a tobacco substitute) and covering it with a piece of copper sheeting.

Between 1800 and 1900hrs, R34 moved steadily further into the shallow depression, encountering 39 knot winds whipping up rough sea below, and 'torrents of rain' which restricted visibility to half a mile. Maitland recorded that the airship was 'remarkably steady' in these conditions, but at 2000hrs Scott took her over the top of the weather at 3,400ft.

Maj-Gen Edward Maitland, right, talks with a crew member during the outward voyage. (*Trustees of the National Museums of Scotland*)

Although no photograph has been found depicting the complete transatlantic crew of R34, most are shown here. Maitland, Pritchard, Durrant and Ballantyne are absent, and Heath was not selected. The identities are as follows: Front row: L to R Cpl F Smith, AC2 R Parker, Sgt A G Evenden, Cpl R J Burgess. 2nd row: 2nd-Lieut J D Shotter, Capt G S Greenland, Lt-Cdr Z Lansdowne (US Navy), Maj G H Scott, Maj G G H Cooke, Lieut G Harris, 2nd-Lieut H F Luck. 3rd row: AC1 W J Edwards, Cpl E P Cross, LAC G Graham (& Wopsie), Flt Sgt N A Scull, Flt Sgt W R Gent, WO2 W R Mayes, Flt Sgt R W Ripley, Flt Sgt W J Robinson, LAC J N Forteath, LAC F P Browdie. 4th row: Cpl H R Powell, Cpl J H Gray, Sgt H M Watson, Sgt J Thirlwall, LAC J S Mort, AC2 J Northeast, Heath. (*Trustees of the National Museums of Scotland*)

Half an hour later the centre of the depression was behind, the rain had ceased, and flight was once again smooth, amid an impressive and colourful sunset.

'A wonderful sunrise and glorious soft colourings' greeted the watch at 0430hrs on 4 July. At 1,000ft there was a bright blue sky with fog beneath and a fog bank ending 10 miles away to the south. By 0900hrs they were over a large icefield, and there was some concern about the fuel supply. Only 2,200 gallons remained, and strong headwinds were expected on the east coast of the USA. Messages of congratulation were already being received, but Maitland deemed them premature. Scott took R34 up to 4,000ft to reduce the loss of gas through excessive superheating.

At 1230hrs Scott sighted land on the starboard beam – a few small rocky islands glimpsed briefly, then, after a closer look, identified as the northwest coastline of Trinity Bay. The crossing from Rathlin Island to Newfoundland had taken exactly fifty-nine hours.

By 1430hrs the airship was over Newfoundland at 1,500ft, but half an hour later dense fog had again enveloped her. She was making 38-40 knots and heading for Fortune Harbour. Near Fortune, letters were dropped by parachute and, although the parachute burst, they were eventually forwarded and delivered.

The night was very dark and clear, and at 0230hrs on 5 July the lights of Whitehaven showed up brightly to starboard. A strong headwind prevented R34 from making any appreciable headway, so at 0700hrs Scott turned inland to avoid a southwesterly 'wind barrage' blowing off the coast. Crossing the coast

at Goose Island and Country Harbour, Nova Scotia, the ship passed over miles and miles of sparsely populated forest. Three and a half hours later there were still 'huge forests' 800ft below.

At lunchtime, 1230hrs, Maitland was regarding the petrol question as 'a distinctly serious matter'. With 500 miles to go to New York, he reckoned that, with no adverse wind or weather, it could be done on two engines, with occasional help from a third. All five could not be used because consumption would be too high. As a precaution, Lt-Cdr Lansdowne signalled the US Naval authorities in Washington and Boston, requesting that a destroyer be on hand to take the R34 in tow if she ran out of petrol in the night. At dawn, after towing the airship during darkness, the ship would cast off the airship to fly to Long Island under her own power.

Three hours later, evidence of electrical disturbance was around, and atmospherics became very bad. Seeing a severe storm moving southwards down the Canadian coast, Scott ran all five engines and turned east to avoid it. Although he succeeded and the airship passed through its outer edge, Maitland recorded that 'we had a very bad time indeed; it is quite the worst experience from a weather point of view that any of us have yet experienced in the air'.

Fortunately, R34 was back in clear weather by 1930hrs, well away from Nova Scotia and heading for New York. The crew witnessed a 'particularly fine electrical disturbance type of sunset'. At 2100hrs another thunderstorm necessitated another change of course. 'As every gallon of petrol is worth its weight in gold', wrote Maitland, 'it almost breaks our hearts to have to lengthen the distance …'.

At 0400hrs on Sunday 6 July, the weary crew sighted the US coast at Chatham. Scott began to wonder whether there was sufficient fuel to reach New York, or whether he would be forced to land at Montauk, and he even confirmed his intention to do so by radio. However, Lieut Shotter took a party on a scavenging tour of the R34's eighty-one fuel tanks, and succeeded in draining off enough fuel to enable Hazelhurst Field, Mineola, New York, to be reached, though a flight over the city was not possible.

When Hazelhurst Field was reached, the airship circled while Maj Pritchard parachuted to take charge of the mooring and landing. This task was to have been performed by Maj H C Fuller of the advance party, but he had rushed off to Boston to be on hand in the event of an early descent. Apart from eight English airmen, the landing party comprised American soldiers and sailors who had no experience of handling large rigid airships, so Pritchard's guidance was essential.

Finally, at 1354hrs GMT – 0945hrs local time – R34 nosed down to land on Long Island after a world endurance record flight lasting 108hr 12min. There remained only 140 gallons of petrol, enough for some two hours' flying on reduced power.

After formal greetings, a banquet was held at the Ritz-Carlton Hotel. The airship was left in the care of a thousand men of the US Army Air Service, and grandstands had been set up to allow the public to view the vessel. She had to be moored in the open using the 'three-wire system' at night, but was hauled down and held by relays of five or six hundred men during the day, when the airship became lighter.

The engineers were soon back at work, checking the engines, changing plugs and soldering up vents in the slip-tanks which had allowed petrol to leak out when the airship was flown at an angle.

Although the weather was good when R34 arrived at Hazelhurst Field, conditions soon worsened and a 'terrific thunderstorm' broke overhead at 1300hrs. Regassing had begun, but there was insufficient gas aboard and the airship was dashed to the ground by the stern, which took the handling rails from the rear car. When the rain stopped, just before 2100hrs, R34 was let up on her mooring, where she rode out the night safely. On Monday morning, the ship was almost carried away by a strong breeze after she had been regassed. Some damage was sustained when the casting holding the main mooring point in the

bows gave way, but a temporary repair was effected.

The crew worked in the mornings only, the rest of the day being spent on leave to allow sightseeing in New York, using cars placed at their disposal. Everywhere they went the crew members were feted, and the stowaways Ballantyne and Wopsie attracted particular attention. Although the cat was allowed to make the return flight, Ballantyne was ordered to return by sea with the advance party. Flt-Sgts E E Turner and W Anders replaced both him and Wireless Operator Edwards, thus spreading the workload for the hard-pressed engineers. Lt-Cdr Lansdowne of the US Navy was to be replaced on the return flight by Lt-Col W N Hensley, representing the US Army Aviation Department.

The final maintenance work was done on Tuesday 8 July, and mail for delivery in Britain was collected. A new gramophone with a fresh supply of records was also taken aboard. Departure was set for dawn on Thursday the 10th, so petrol, oil and food were loaded on Wednesday afternoon. However, at 2100hrs Scott received warning of approaching high winds and decided, in view of the unfavourable mooring conditions, to leave that night, and alerted the crew to be at their stations at 2230hrs. After hurried leave-taking and some last minute provisioning (including a case of rum – despite the prohibition laws), gassing was continued in a strong wind until the last moment. At 2354hrs the order to ballast-up came, and R34, with 4,600 gallons of petrol aboard, ascended with three engines running in a wind gusting to 26 knots.

R34's gasbags are topped up for the return flight while the airship is held down by men of the US Air Service. (*Trustees of the National Museums of Scotland*)

Rising to 2,000ft, the airship proceeded to overfly New York, reaching Fifth Avenue, Times Square and Broadway at 0100hrs on the 10th. Ten minutes later Scott turned his charge about and steered for East Fortune, over 3,000 miles away, a following wind giving the airship a speed of 64 knots. Weather conditions for the return flight were distinctly favourable, with a depression west of Newfoundland and another large one centred to the north of Ireland, and an anticyclone over the eastern Atlantic and the United Kingdom.

The US coast was crossed at 0217hrs, R34 running on four of her five engines and 'resting' the other. By 0930hrs, 430 miles had been covered, and the crew were sorting the mail. At 1045hrs a speed of 72 knots on four engines was recorded. Scott had decided to head for London, intending to time the flight from New York's Broadway to Piccadilly Circus.

At noon, a lunch of cold Bologna sausage and pickles and stewed pineapple was greatly enhanced by the rum ration. An hour later, the airship, under Lt-Col Hemsley's steerage, was proceeding 'steady as a rock' over very rough seas. 'Unless one looks out of the windows', wrote Maitland, 'we would hardly realise we were travelling at all.' Many messages wishing success were being received from Canada and the USA.

By 1650hrs, R34 was a third of the way across, 900 miles from New York and 1,850 miles from Ireland's south coast. The weather was clear with good visibility, and the sea was a deep blue. By 1815hrs, however, it was getting much colder. A five-masted schooner under full sail was seen on the main eastbound steamer route – one of the few ships the crew had sighted on the open sea.

At 0420hrs on the morning of Friday 11 July, an engineer in the aft car, thrown off

R34 returns to her East Fortune base. (*Trustees of the National Museums of Scotland*)

balance by the airship rolling, fell against the clutch lever, disconnecting the engine from the propeller. As the engine was on full load it immediately raced, and the resulting stresses caused bolts to shear and the connecting rods to fracture, punching holes through the water-jacket and sump. This rendered the engine beyond repair, and consequently Scott abandoned his plan of flying directly to London and altered course to head for Scotland.

A descent to 600ft was made at 0640hrs to get below clouds, but this put the airship in a northerly wind, while at 3,000ft the wind was southwesterly, because of the effects of the Gulf Stream. Scott therefore returned R34 to 3,000ft. In fact, an even more favourable wind was to be had at greater altitude, but at this stage of the flight gas was too precious to lose by expansion. Greater heights could safely be reached the next day, when another twenty-four hours' petrol had been burned off, making the vessel much lighter.

At noon, the Air Ministry notified the airship of an anticyclone off the south-west of Ireland, so Scott changed to a more northerly course to get into the westerly wind blowing on its northerly side.

By 1530hrs, the R34 was cruising at 32 knots on three engines, still at 3,000ft, and passing in and out of clouds at intervals. The crew had not seen the sea since 0830hrs that morning. An hour later, Scott brought the ship down from 3,800ft in an attempt to see the water and gain some impression of their speed. In the process the ship passed through five distinct, separate cloud layers, but at 900ft the cloud was still thick, and he had to abandon the attempt.

It began to rain heavily at 1700hrs, and was still doing so two hours later, when Scott took R34 up to 5,000ft in an unsuccessful attempt to rise above it, but

returned to 3,000ft. It was very cold and dark, and all the windows and doors were shut. Not until 0045hrs on Saturday 12 July did the weather begin to clear, with the sea becoming visible from 2,500ft. 'As we lay in our hammocks we listened to the rain beating pitilessly down on the outer cover of our trusty ship of the air', Maitland recorded, 'and our feelings, despite the weather, were those of complete confidence and security.'

A magnificent sunrise greeted them at 0300hrs, and three hours later they were 760 miles from East Fortune, flying on three engines. The remaining aft engine had suffered broken valve springs, and they had to be replaced.

Because the crew were confident that the breakfast that morning would be their last aloft, the rations were rather more generous than usual at 0800hrs.

At 1055hrs, R34 was making 35 knots, from which it was estimated that East Fortune would be reached at daybreak on the 13th, but by noon a northeasterly wind had reduced speed to 28 knots, which meant that there would be another in-flight breakfast after all. Half an hour later the clouds had cleared, and from 5,000ft the crew could see over an area of more than 19,000 sq miles. Despite this, there was not a ship in sight in the vast Atlantic waters.

The first indication that the crossing was almost completed came at 1730hrs, when a pair of trawlers were sighted off the airship's starboard beam. By now, the R34 was at 3,000ft, the difference in temperature between this height and 5,000ft being 'most marked'. Her speed had risen to 32 knots. The navigator, Maj Cooke, believed that, even in clear weather, an accuracy of better than 20 miles could not be guaranteed when estimating an airship's position in mid-Atlantic. It was thought that the perfection of directional wireless would eventually make much greater accuracy possible.

Just before 1900hrs, the airship passed

through a squall but remained steady, making 1,600rpm on four engines. Then Clifden, County Mayo, was obtained on directional wireless, revealing that the Irish coast was not far away. Twenty-five minutes later, Lt-Col Hemsley was the first to spot land, on the starboard bow. Two small islands were identified as those which had greeted Alcock and Brown as their Vickers Vimy approached the end of its historic flight, and the Clifden wireless mast was sighted.

At exactly 2000hrs, R34 crossed the coastline a little to the north of Clifden, the coast-to-coast elapsed time from Long Island being 61hr 33min. Climbing to 6,000ft to cross the Irish mountains, the crew were presented with a magnificent cloud panorama and glimpses of lakes, harbours and green fields. By 2110hrs, R34 decended to 2,000ft and was making 38 knots over flat country. A small two-seat aeroplane from the Castlebar area flew past below, its occupants waving a welcome.

Capt Scott received a message from the Air Ministry at 2320hrs, instructing him to change course for a landing at Pulham Airship Station, Norfolk. As the crew had hoped to return to their original point of departure, and a reception was in preparation at East Fortune, this caused great disappointment, and Scott queried the order because the weather was good at the Scottish base. Unfortunately, the powers that be insisted that the airship went to Pulham, so Scott requested that the officers' and crew's kit be sent on quickly. At this time, R34 was at 5,000ft over the Isle of Man.

By 0415hrs, the airship was over Nottingham and heading directly for Pulham, and at 0500hrs a congratulatory message from His Majesty King George V was received, along with others.

Finally, at 0620hrs on Sunday 13 July, R34 arrived over Pulham with only two of her Sunbeam engines still running, although the propellers of the other two units were allowed to windmill to make a more impressive spectacle. As the airship circled twice, Scott signalled the 400 man ground crew by flashlight and they took up their positions. Apart from officials and newspapermen there were few spectators to welcome the triumphant team.

As R34 nosed down to land, a small RAF band struck up *See the Conquering Hero Comes*, only to be doused by several gallons of water ballast released from the nose by Scott. Gallantly, the band played on. The touchdown was made at 0657hrs, 75hr 3min after leaving Long Island. In her transatlantic round trip, R34 had covered more than 7,000 miles. After the return flight 1,000 gallons of petrol remained in the tanks – a rather more comfortable margin than at the end of the outward journey.

If the arrival had been low key, the succeeding events were little better. Maitland, Scott and Shotter compiled reports on the voyage. While Shotter made no criticism of the engines which had been in his charge, Maitland and Scott considered them unsuitable and said that R34 was underpowered.

The crew received little reward for their efforts. Air Force Crosses were awarded to Maitland, Cooke, Guy Harris (the meteorological officer) and Shotter, and Scott was given a CBE. The rest of the crew received little from officialdom. The other officers received platinum cigarette cases from the US Navy, and six warrant officers and NCOs were awarded Air Force Medals. The lower ranks were given silver mounted propelling pencils by the New York Fire Brigade.

Ironically, included in the 40lb of mail on the return flight were two gold medals awarded by the Aero Club of America to Capt John Alcock and Lieut Arthur Whitten Brown, both knighted for their recent one-way non-stop transatlantic flight in a Vickers Vimy.

Stowaway Ballantyne was not court-martialled, and a year later returned to flying duties on the Vickers R80 airship.

R34 survived the epic flight in good shape, though her envelope was rather slack and worn. On 31 July/1 August, she returned to her East Fortune base, making a detour to fly over London on the way, and once home underwent a long refit. She next flew on 4/5 February 1920, making a seven hour flight to her new base at Pulham. After another six weeks grounded she was permanently assigned to Howden, and flew there at the end of March.

On 27 January 1921, with Flt Lt Hedley V Drew in command, R34 made her first flight for several months, to check out repairs and train eight navigating officers. Owing to a combination of errors by her crew and officers on the ground, she became lost over the North Yorkshire hills and brushed the ground in darkness. Luckily no one was hurt and R34 suffered only minor damage, but her fore and aft propellers were damaged. After struggling back to Howden she was landed and hauled down, but the ground crew could not get her into the shed. More damage was done to the fore and aft cars in the ensuing battle with the wind, and then her rudder jammed hard over, making her uncontrollable. Eventually, she was moored in the open using the three-wire system, but the storm continued to batter her overnight, puncturing her forward gasbags and smashing in her bow. By the morning, R34 was beyond repair, and after the valuable equipment was removed she was reduced to scrap. It was an ignominious end for a famous aircraft.

The R34's great flight fuelled a fire of speculation on the airship's great potential as a commercial transport. C G Grey, Editor of *The Aeroplane*, wrote that the airship would become 'the recognised vehicle for high-speed, long-range passenger transport', and in a special airship issue published shortly after the flight, the journal stated, 'In the present state of development of aircraft it can safely be said that for carrying large cargoes for long distances without stops at speeds up to about 70mph, the airship cannot be rivalled by the aeroplane either in cost or in reliability'.

'On the other hand', the writer continued, 'for relatively small cargoes, to be carried at considerably greater speeds … the aeroplane need not fear the competition of the airship.' However, a distinct 'probability' that the aeroplane's efficiency as a weight carrier would improve 'very considerably' was noted, and it was expected that the aeroplane would 'gradually encroach upon what is now regarded as the airship's special sphere of usefulness'. How right he was.

There was also an enormous difference in cost. The R34 had cost £350,000 to construct, and its shed at East Fortune added another £166,000. Although it is not a fair comparison, an unarmed Eagle-engined Vickers Vimy cost £8,390, and a Handley Page V/1500, with four Eagles, cost £18,990.

However, it was the airship that inaugurated the first regular commercial transatlantic service, between Friedrichshafen, Germany, and Recife, Pernambuco, in South America. Flown by the Zeppelin LZ127 *Graf Zeppelin*, these operations began on 20 March 1932, after several trial flights in the latter half of 1931. They ended following the *Hindenburg* disaster in May 1937. The *Hindenburg*, which was the biggest aircraft to fly, had three and a half times the capacity of the R34, was 803ft 10in long, had 7,062,100cuft capacity fully inflated, and could carry fifty passengers in luxurious accommodation.

The first scheduled transatlantic passenger flight by flying-boats did not begin until June 1939, when Pan American Boeing 314 Clippers designed to carry seventy-four passengers on short journeys began operations. Apart from a trial return flight from Berlin to New York by a Focke-Wulf Fw 200 Condor in August 1938, commercial transatlantic flights by landplanes did not become a reality until after the Second World War.

Seen in this light, R34's feat was a great achievement. Though it did not represent the latest in airship design, R34 proved that transatlantic air travel was feasible, and blazed the trail for later successes. It now seems doubtful that a commercial airship service could ever be a practical proposition – it could not compete with modern airliners on manufacturing costs, operating cost, regularity or speed, and ground handling could still pose great problems. If large airships ever take to the skies again, they will probably serve as sedate aerial cruisers for the rich or as bulk-freight carriers. But who could now afford – or risk – the enormous initial outlay? □

The end. R34 lies on the ground at Howden on 30 January 1921, after being battered by a storm on the night of 27/28 January. She was beyond repair and reduced to scrap. (*R F L Gosling*)

Alcock and Brown: First Across the Atlantic Direct

This flight made world news, pilot and navigator being knighted by King George V shortly afterwards. Subsequently, all the 1919 air crossings of the Atlantic were put into the shadows by a 3,600-mile direct solo flight, made in 1927 by Charles Lindbergh, between New York and Paris. Only in 1969 was Lindbergh's personal achievement in turn eclipsed by the Americans landing on the moon. That same year, Alcock and Brown were to be particularly remembered. Twenty years on, on the seventieth anniversary, the Editor retells the story as an example of high professional skills in perhaps the first truly reliable, as proven, long-distance aeroplane.

At 0940hrs one Sunday in June 1919, a large aeroplane came in to land on what appeared to be a field in the west of Ireland some 40 miles west-northwest of Galway. The landing run was exceptionally short, a controlled crash almost, the aircraft nosing over into a bog. Staff from the Marconi wireless station at Clifden ran over to help, the aircraft having circled their masts, firing off Very lights. The two men in the aeroplane said they had flown the Atlantic.

Soon the news was going out worldwide while Capt John Alcock and Lieut

Vickers Vimy Mk II (F.B.27A)
Basis for the Vimy Transatlantic

Accommodation	Pilot and 2 gunners
Engines	Two 360hp Rolls-Royce Eagle VIII
Wing span	68ft
Length	43ft 6½in
Height	15ft 7½in
Wing area	1,330sq ft
Empty weight	7,101lb
Gross weight	12,500lb
Max speed	103mph at sea level
Climb to 5,000ft	22min
Service ceiling	7,000ft
Absolute ceiling	10,500ft
Endurance	11hr
Armament	Two Lewis guns and 2,476lb of bombs max

The variations in the transatlantic aeroplane from the standard are noted in the text. Operationally, the high gross take-off weight, maybe as high as 14,000lb, which included 865 gallons of petrol and 50 gallons of oil, reduced both the initial service ceiling and the cruising speed significantly.

Construction: the Vimy had a wire-braced wooden wing structure, as did the tail unit, also of biplane form with twin fins and rudders. Steel tube was used in the front fuselage and wood in the rear. The fuselage longerons were hollow spars. The whole structure was fabric covered.

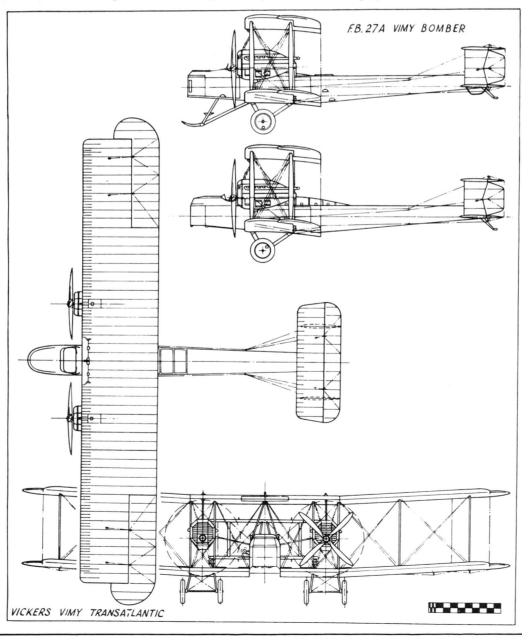

F.B.27A VIMY BOMBER

VICKERS VIMY TRANSATLANTIC

Arthur Whitten Brown were taking breakfast. By their watches set to GMT an hour behind BST, the landing had come at 0840hrs on 15 June, just 16hr 28min after take-off from Newfoundland. The coast-to-coast time quoted by the pilot had been 15hr 57min, while the navigation log gave take-off to touchdown. The groundspeed, point to point for the 1,890 miles, had been 118½mph.

The landfall at Clifden had been remarkable, being only some 20 miles north of the direct track to Galway, their aiming point. Sufficient fuel remained in the tanks to carry on to London but the low cloud spilling untidily down the side of the local mountains hiding the peaks was, and still is, a perpetual warning to the unwary or, in this case, a tired crew. To collect their reward, of which more later, they had only to land safely anywhere in Europe and that condition the west of Ireland fulfilled.

Vimy Transatlantic.

How the Vimy Came About

With the third year of the First World War nearing its close, it was all too evident that although the pace of technical development and, indeed, the volume of supplies had increased enormously, there was little co-ordination between the Navy and the Army. Each service ran its own air arm, the Royal Naval Air Service (RNAS) and the Royal Flying Corps, respectively. One effect of this was apparent when the Germans started raiding London in daylight towards the end of May 1917, using formations of heavy bombers in groups of about twenty aircraft. At first, the defences were completely ineffective. The repercussions arising led to the appointment by the prime minister, Lloyd George, of Gen Smuts, a member of his war cabinet, to advise and report on the whole future of the air Services. The result, on 1 April 1918, was the formation of the Royal Air Force.

At only slightly lower levels, in that summer of 1917, detail decisions were being taken to improve the use of resources and better to regulate their flow. No part of that task was more important or seemed more difficult than aero engines. Some engines in the 200hp class were beginning to come into surplus. Yet of the more powerful and more reliable Rolls-Royce Eagle, and its smaller derivative, the Falcon, there were never enough.

There was now plenty of evidence to support the view that the big bomber could have a bright future – and not only as a result of raids by the enemy. RNAS initiatives had been behind the creation of the twin-engined Handley Page O/100 bomber and then the creation of the successful O/400. Now the Air Board, predecessor of the Air Ministry, decided to increase the supply of this class of aeroplane by calling in another contract-

or, Vickers. Initially, on the basis that they were available, two 200hp geared Hispano Suiza water-cooled vee-eight engines would be specified.

Vickers' new chief designer, 26-year-old R K Pierson, later told how he sat down with Maj J C Buchanan of the Air Board while together they schemed the new bomber on a sheet of foolscap paper. Pierson added, 'I can only vouch for the fact that the aircraft, as first flown, was reasonably like that preliminary free-hand sketch'.

A contract was signed on 16 August, the prototype Vickers F.B.27 making its first flight on 30 November 1917, four months after the start of detail design. On official tests at Martlesham Heath in the following January, the prototype did well, outperforming the Handley Page O/400 for load uplifted, on little more than half the power.

Two more F.B.27 prototypes followed, with power provided by two 260hp Sunbeam Maori engines or two 300hp Fiat A-12bis. However, it was to be the availability of the twelve-cylinder water-cooled 60 degree vee, 20.32 litre capacity, Rolls-Royce Eagle for the fourth prototype which would provide the power to shape history. The Eagle was the first aero engine Rolls-Royce had designed. It ran on the bench in October 1915 at Derby, less than six months after work on the drawings had been started. Initial power was 225bhp; for production Vimys, the Eagle VIII gave 360hp. The Eagle-powered Vimy went to Martlesham on 11 October 1918, a month before the Armistice, and immediately set new standards of performance, as indicated in the table.

Production orders for the Vimy, using other engines, had already begun to come through in March 1918. Soon, orders placed on a number of contractors, based on the performance of the first

Vimy's designer, R K Pierson. (*M Pierson*)

prototype, had reached 1,130. At that point, the Vimy was the undisputed leader for the future. Following the Armistice, orders were reduced to less than 300. However, Vimys would now go on to win export orders, help to start commercial aviation, be first to fly to Australia, and deep into Africa, as well as helping to open the RAF's Cairo–Baghdad Air Mail. As trainers for crews and parachute jumping, Vimys would continue in service until the 1930s, following the end of their Service life as front-line bombers in 1924.

However, perhaps the Vimy's greatest legacy was to begin the undramatic, unglamorous line of Vickers heavy bombers and bomber transports, under Pierson's design leadership, which served an RAF tasked with difficult jobs in hot countries, on a pittance of a budget.

The direct flight by the Vickers Vimy.

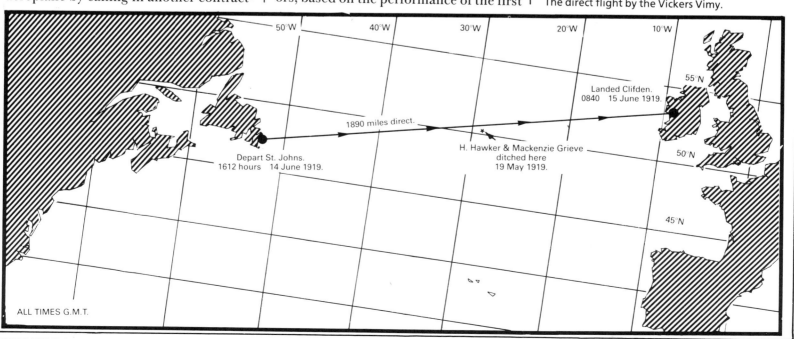

The Race To Be First

When Lord Northcliffe, proprietor of the British newspaper the *Daily Mail*, offered a prize of £1,000 to the first person to cross the English Channel by aeroplane, it was won within months, on 25 July 1909, by Louis Blériot. What now seems remarkable is that within less than four years, Northcliffe was offering £10,000, almost £400,000 in 1989 purchasing power, in an age when much less was available to be bought, for the first direct flight across the Atlantic. Although by 1913 the rate of reward per mile had gone down from £47.62 per mile to something like £5.26 per mile, Lord Northcliffe's money looked pretty safe for many years to come.

When offering his prize on 1 April 1913 (was there a significance in the chosen date?) Northcliffe stipulated that the flight had to be completed within seventy-two hours. Almost automatically this set the minimum standard to be achieved as a sector of some 1,900 miles between Newfoundland and the west coast of Ireland, although in the original rules intermediate alighting on the water was permitted.

Before the First World War, the combination of putting sufficiently reliable engines into an adequate airframe looked too difficult for most designers even to contemplate, although the American pioneer Glenn Curtiss appeared to be a possible runner. From the outbreak of war in 1914, the contest lapsed but was revived in July 1918 when Northcliffe re-offered his prize. With the conclusion of hostilities in November, a new target date had in effect been set for attempts on the first direct crossing. And there would be all the winter to prepare for it. The Atlantic racing season could really only be contemplated in the period May to September. With no blind flying instruments, the choice narrowed to perhaps six nights per month when there would be a moon to provide a visible horizon; that is, if the North Atlantic weather allowed. Apart from weather 'actuals' in Europe and North America, the weather in between was a largely unknown quantity, the significance of the value of an organised system of weather reporting by ships having not then been realised. After all, there had been no requirement for it and such reporting would begin only six or seven years later.

One potential aid for Atlantic fliers in 1919 were simple airborne wireless transmitters and receivers for passing messages, by morse code. There was little effective wireless directional finding.

The biggest assets of the day were professional people who had come through the war – a reservoir of talent. Now behind the pilots and navigators were skilled designers and engineers on the ground, a whole industry, with at its heart skilled fitters and riggers. Beyond that had to lie the core of every successful aircraft endeavour, a reliable aero engine, envisaged, designed and ministered to by practitioners highly skilled in that mechanical art.

Here we need to mention all the British competitors for it was, after all, a race to be first, for likely commercial advantage to the winning firm and personal reward for the winning crew. That all would share a niche in history was probably not envisaged by any of the participants. Despite the stakes and the sharp competitiveness, a great sporting spirit prevailed at that time.

There were to be four British teams apart from the Vickers entry. The first away on 18 April was Maj J C P Wood piloting a Short Shamrock with Capt C C Wylie as his navigator, out of Eastchurch, Kent, on a positioning flight to The Curragh in Kildare. Twelve miles out over the Irish Sea the engine failed, the aircraft being ditched about a mile off Anglesey, and the crew picked up. Overall, perhaps this was almost the best ending for the flight. At that time, and indeed for the next twenty years, the prevailing westerly wind component on the North Atlantic was consistently underestimated. Not until 1928 would an aircraft cross the Atlantic from east to west.

The other contenders backed by their companies, but not by government, and all Newfoundland-based, were F P Raynham and Maj C W F Morgan with the Martinsyde Raymor; Maj H G Brackley and his crew of three with the four-engined Handley Page V/1500; H G Hawker and Lt-Cdr K Mackenzie Grieve with the single-engined Sopwith Atlantic; and finally Alcock and Brown with their Vickers Vimy.

Hawker and Grieve took off on Sunday 18 May, over four weeks ahead of the Vimy. Their single-engined Sopwith Atlantic (one Rolls-Royce Eagle VIII) featured side-by-side seating, and had boat-shaped decking aft of the cockpit for use as a lifeboat in case of ditching. Unhappily, probably due to a trifling cause, a persistent overheating radiator problem with their single water-cooled engine, out in front and inaccessible, finally forced them to ditch, well past the halfway point at 29° W. With plenty of time to assess their predicament, they

Vimy Transatlantic. New and yellow doped for visibility. Brooklands, April 1919. (*Vickers*)

had been able to come down near and were rescued by a small merchant ship, which did not carry wireless. For over a week they were missing and could only be presumed dead.

Bad weather had kept both Hawker and Raynham grounded and it had been touch and go who would get away first. The Martinsyde team had their aircraft assembled and test flown by 18 April, just eight days after being unloaded as crates on the dockside at St John's. The Raymor was a scale-up on the Martinsyde F.4 single-seat fighter. With an enlarged fuselage seating pilot and navigator in tandem, two-bay wings and extra fuel, the 275hp Rolls-Royce Falcon powered aircraft had flown first at Brooklands on 19 February. Martinsyde, the smallest company involved, had in fact an Atlantic aeroplane under construction at Brooklands in 1914 but, inevitably, the war had put a stop to that work.

With Hawker on his way and the American NC-4 flying-boat already at Horta in the Azores, Raynham too decided to leave the rough and ready 'airfield' at Quidi Vidi where the Raymor had been assembled. Just at lift-off the wind changed direction, bouncing the aircraft back onto the ground, smashing the undercarriage and the propeller and injuring Maj Morgan. Raynham came off better. The extensively damaged aircraft was repaired and on 17 July, a month after the race had been won, the aircraft, renamed and with a new navigator, was ready to try again. Again, the undercarriage collapsed and with it Martinsyde's hopes of making the fastest crossing, if not the first.

To mention briefly all the competitors in this race is an attempt to recreate something of the atmosphere of the year. The Sopwith, Martinsyde and Vickers crews were all known to each other, for their mounts had all made their first

Following reassembly at St John's, Newfoundland, the Vimy forms a focal point for a Sunday afternoon out on 8 June. (*Vickers*)

flights and United Kingdom testing from the hallowed grounds of Brooklands where the aircraft companies were tenants and tolerated within the perimeter of the Motor Racing Circuit, Brooklands' primary purpose.

The outsider in this event, on the basis that it came from Cricklewood, was the big and impressive four-engined, 126ft wing span, Handley Page V/1500 which weighed in at 30,000lb laden, versus the normal 12,500lb gross weight for a Vimy.

Power again came from Rolls-Royce, again Eagle VIIIs to provide something like 1,400hp for take-off – in those days a lot. However, size made it less easy for the H P team to unpack their aeroplane, reassemble, rig and test it quickly, in the open in conditions more like an English midwinter, though in fact a Newfoundland spring. When the flight testing came there were four sets of engine snags to clear instead of two with the Vimy.

On that same Sunday, six days before the Atlantic flight, John Alcock (left) and Arthur Whitten Brown are as smartly turned out as the local population. (*Vickers*)

Reassembly and testing had taken a month, versus twelve days for the Vimy and eight days for the Raymor, small enough to be put together inside a temporary hangar. It is a matter of history that the Vimy made it while the others did not, the big Handley Page going on instead for an adventurous series of demonstrations in Canada and the USA.

The Men Involved

The British attempts to cross the Atlantic by air in 1919 were a remarkably close knit business, none more so than the men involved with the winning Vimy.

It is logical to start with the Vimy's designer, the aforementioned Reginald Kirshaw Pierson. In 1908, at the age of seventeen, Rex Pierson entered the Erith works of Vickers Ltd as an engineering pupil apprentice working on naval gun breech mechanisms, transferring to the aviation section when it was formed in 1910. Three years later he learned to fly on a Bristol Boxkite, his instructor being Archie Knight who, together with his colleagues Harold Barnwell and Maj H F Wood, was to be bound up with the future of Vickers Aviation. Pierson became involved in the evolutionary development of the Vickers Experimental Fighting Biplane series, the Gunbus variants, becoming chief technician as early as 1914. As recounted, by 1917 he was chief designer, responsible for the Vickers F.B.27 (to be named officially in 1918 the Vimy) – nine years from raw apprentice to the top, a position he would hold until 1945 when he was appointed chief engineer.

Another who was making progress at Brooklands before the First World War was Manchester-born John Alcock, always known to his contemporaries as Jack, who at the age of eighteen had gone to work as a mechanic at Brooklands in 1910, learning to fly there two years later. Shortly after the outbreak of war he joined the RNAS as an instructor, where he excelled, not only in teaching his pupils to fly but in his ability to motivate them as well. John Warneford, the Royal Navy's first holder of the Victoria Cross, was Alcock trained.

Finally, after two years, Alcock was released for active service and posted to No.2 Wing at Mudros on the Greek Island of Lemnos in the Eastern Aegean, about 50 miles southwest of the Dardanelles. Although by now a fighter pilot, or a scout in the naval terminology, Alcock achieved a number of 450 mile round trip bombing missions to Istanbul (then Constantinople), by Handley Page O/100. On 30 September 1917, he shot down two out of three enemy seaplanes in the morning, before setting off that evening on another long-range bombing sortie. Unfortunately, one engine failed, outbound over the Gallipoli peninsula. Forced to turn back, and unable to prolong the glide beyond 60 miles, the aircraft ditched in Suvla Bay. The crew of three managed to get out and fire off Very lights to attract any British destroyer that might be near. Out of luck, they had to swim ashore where the Turks soon captured them. Repatriated at the end of the war and demobilised, Jack Alcock, the complete aviation man of his generation, was knocking on Vickers' door at Brooklands the very next day.

What, then, of his navigator, Arthur Whitten Brown, who was as shy and retiring as Alcock was extrovert? It has been suggested that this made them an ill matched pair, but many, including the writer, would dispute this. Such combinations of personality have been wont to operate long-distance aeroplanes with minimum aids, or no aids at all, extremely efficiently. So it was to be with Alcock and Brown.

Arthur Whitten Brown's parents were American of British stock and lived in

Take-off to a place in history. Lester's Field, Newfoundland, 1612hrs GMT, Saturday 14 June. (*Vickers*)

Glasgow where his father was an electro-mechanical engineer. In turn Arthur, born in 1886, took an electrical engineering apprenticeship at Trafford Park, Manchester, with the company which later became Metropolitan Vickers Electrical. Shortly after the start of the war he was commissioned in the Manchester Regiment, and transferred to the Royal Flying Corps where he was trained as an observer.

Shot down in November 1915, Brown sustained a permanent leg injury and was taken prisoner but was repatriated two years later while the war was still in progress. During this time in captivity Brown applied his mind to solving the then quite new problems of air navigation over long distances, other than by map reading or something even more elementary such as following a railway line. Observations of the sun, the Pole star and the other navigational stars were clearly a possibility and the moon appeared suitable, though innacurate, for a good position line because it lay so close to earth. With a marine sextant requiring a visible horizon, none of it was very practical at night. With a simple drift sight it should be possible to measure an aircraft's drift angle quite easily. Two or three drifts taken on different and known compass headings, reduced to a True heading, would give an accurate wind velocity. By some such reasoning, and by developing his own methods of plotting courses and winds in relation to a track to be flown, Brown's analytical engineering mind evidently became attuned to a job which was a blend of art and science.

Brown cannot have had any practical experience of long-distance air navigation to put his theories into use before embarking on the Atlantic trip. He had, however, continued his theoretical studies into long-range air navigation problems after his return to England, as had a number of others of his generation. The *Daily Mail* prize was there to be won and pilots would need navigators. Visiting Vickers Weybridge one day in the spring of 1919 Brown, consciously thinking in terms of engineering rather than navigation, was talking to the works superintendent in his office when John Alcock, newly on the payroll, walked in. The great, historic and ultimately short-lived partnership had begun.

Both Sir John Alcock and Sir Arthur Whitten Brown went back to work at Vickers Weybridge after their triumph. A few months later, on 18 December, Vickers' chief pilot Sir John Alcock was flying the company's new Viking amphibian over to Paris for display at the Salon. The weather was poor, deteriorating into wintry misty murk near Rouen. Alcock's enforced arrival in an orchard became a crash as the aircraft nosed over, fracturing his skull. He died without regaining consciousness – a tragedy. Sir Arthur Whitten Brown returned to Metropolitan Vickers, served again in the RAF as an instructor in the Second World War, and died in Swansea at the age of sixty-two in 1948. Sir Arthur's only son, also Arthur, a flight lieutenant in the RAF, was killed in action on D Day, 1944, at the age of twenty-two.

Sir John was not married but had a younger brother, E S J Alcock, also, confusingly, John. As Capt John Alcock, of Imperial Airways and later BOAC, by the age of forty-two in 1943, he had already flown 2 million miles, mainly on the company's flying-boats. By the time he retired, his total had risen to 5 million. In retirement and within a year in 1969-70, he had been able to appreciate at first hand both the Vimy Replica and the Concorde. His brother, the other John, would by then have been seventy-eight, not an outrageous age by any means these days. It shows how short the span between Vimy and the early Concorde was. By then, Alcock's son was flying English Electric Lightnings.

These three families, Alcocks, Whitten Browns and Piersons, part of our air heritage, could hold their own in any company, in any decade. All carried their families down to the next generation of aviation, as did many of the Vimy ground staff who prepared that aircraft in the open seventy years ago for a flight that made history.

Vimy Prepares

In March 1919, immediately following his demobilisation, Alcock had been able to persuade Vickers that to fly an aeroplane for the very long period, perhaps twenty hours, an Atlantic flight might take was entirely reasonable. He would be able to do it. There and then the thirteenth Vimy on the production line was selected for the modifications required. Instead of the standard Vimy's

Brown's marine sextant. (*Vickers*)

fore and aft cockpit a side-by-side arrangement was agreed. Behind this a turtle back fairing was arranged which would both double as an emergency lifeboat, having 600lb buoyancy, and also fair over the additional fuel tankage. Provision was also made for emergency rations and a first aid kit to be stowed in the tail. Instead of the standard Vimy's nose skid, a nose wheel was chosen, but was not fitted on grounds of weight and drag reduction.

On 18 April, the Vimy was airborne at Brooklands for the first time, as the Martinsyde team in Newfoundland was just completing the reassembly of their contender before flying it. Like the children's game of snakes and ladders there were to be no certainties about this race. While the Martinsyde and Hawker teams had their aeroplanes complete, but not the weather to fly them, the Vickers team prepared, the ground party sailing from Southampton on 4 May. The Vimy itself left separately from the London Docks, where it had been held up by a dock strike, on the 13th. Meanwhile, in the barren rocky Newfoundland landscape, swept by rain and icy winds and where the last snows of winter still lingered, there were just two fields near to St John's where the Vimy might be put together again. But the competition

Brown's Mercator chart, arranged for convenience on rollers. On this print the track, with its markings at 10nm intervals, is clear as well as the arcs denoting changes in magnetic variation. On the original, plotted position lines from observations of Vega and the sun are visible. (*Vickers*)

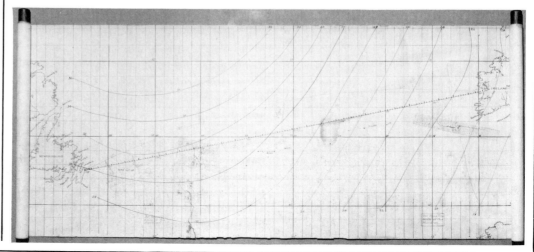

had got there first. On 26 May, the Vimy crates arrived at the docks. At this point Raynham, with great sporting spirit, his own first transatlantic take-off having failed disastrously on the 18th, offered his leased cricket field at Quidi Vidi for the reassembly and first flight.

Now it was the turn of the Vickers team to face gales and snowstorms. Despite all, the Vimy was ready for flight on 9 June. In the meantime, a frantic search involving hundreds of miles of driving had been going on for a suitable field with a clear run of 400 metres for the heavily laden Atlantic take-off. Mr Lester, whose horses had towed the Vimy crates to the first field, now provided the answer to the second. He had land at Monday's Pool which might be suitable. Dynamite, thirty men, and sheer hard graft removed the rocks, walls and fences along a narrow 1,500ft long uphill strip.

On Friday 13 June, the Vimy was prepared for the big flight, the morning being taken to hand pump the 865 gallons of fuel into the tanks in the fuselage fore and aft of the cockpit, every drop being passed through chamois filters. After lunch it was noticed that one undercarriage shock absorber had collapsed which meant defuelling the aircraft to repair it. By the time the aircraft was refuelled again it was far into the night, that second operation having brought its own problems when a rigid fuel pipe was buckled by a tow rope. An improvised replacement section was incorporated without emptying the tanks. History does not relate exactly how this was done, though one man to hold back the fuel over the end of the cut pipe has been mentioned.

On the following morning all was ready but by now a strong wind was blowing across the strip, making take-off impossible. After an early lunch taken on the grass under the wings of the aircraft, the decision was taken to go. As Alcock and Brown got into their flying suits, to be heated electrically in the cockpit by a battery positioned between them, the Royal Aero Club official, the official observer, was affixing seals to the Vimy to provide its authentic status as the aircraft which would arrive in Europe – if it did. Just before the engines were started at 1324hrs local time, a small mail bag containing 197 letters was handed aboard. Twenty minutes later, still with a crosswind component on the uphill take-off run, the Vimy became airborne with 100 metres to spare, the first danger point of the journey passed.

En Route

At 1350hrs local time, 1620hrs GMT, Alcock turned the Vimy onto a heading of 124 degrees on the compass, allowing

Sheet 1 of a fair copy version of Brown's navigation log. (*Vickers*)

for compass deviation, magnetic variation and a wind assessed as from 220 degrees T at 30mph to maintain a track of 078 degrees T for Galway. Eight minutes later the aircraft crossed the coast at 47°30′N 52°30′W, the last positive fix and the last land the crew would see for nearly sixteen hours.

Many pioneering long-distance fliers have recounted how, when the problems of the preparation have been left behind, in this case including rebuilding a large aeroplane and making an airfield, the frustrations disappear, and the need for increased and constant vigilance begins.

At this point the crew were fresh, snug and able fully to communicate. The two-way wireless worked, as did their electrically heated flying suits and their intercom, via throat microphones and earphones. Within a couple of hours these conveniences had largely disappeared. The propeller for the wind-driven generator that provided power for the wireless had sheared; transmission was no longer possible. Alcock had to remove his earphones because they hurt his ears too much. Gradually, the battery between them for the flying suits ran down. Before then, signalling the end of what might be called the first act of a three act play, came a dramatic noise

NAVIGATION LOG — R.A.F. Form 441.D — ROYAL AIR FORCE
AIRCRAFT VICKERS VIMY DATE 14-15 June 1919 NAVIGATOR Lt. A.W BROWN SHEET No. 1

G.M.T.	TRUE TR. / INS USED	TRUE HEAD	VARN W / DEV	COMP HEAD	OBSERVATIONS	SAFE ALT / R.A.S mph	ALT TEMP°C / T.A.S mph	G.S mph / DIST RUN	TIME E.T.A.
612	240 35				A/B St. JOHNS Message to BZM Windy strong and gusty	Wings not shaking			
620	078 210 30	078	30	108	S/C GALWAY Weather Cloudy	70 1000		120	
628					Posn. 47.30 N 52.30 W. Over coast	1200			
654						1600		120	
700					Pre-compued Could not get sun Sun alt 56.14 as over starbo. wing				
720	078		30	124	D.R. posn 47.52 N. Impossible to get 50.00 W. obs. of sun between fog and high clouds	70 1500		104	
820	078	094	31	124	D.R. posn 48.16 N. Cloud above and 47.00 W. below readings impossible.	65		120 224	
820	078 11 R	089	31	120	D.R. posn 48.37 N. Climbing but still 41.32 W no chance of sights	65 3000		100 324	
940					Wireless busy no message for us yet. We have 4 hours yet to Jime (to darkness)				
1020	078		31	120	D.R. posn 48.54 N. Too cloudy for 42.32 W. drift observations	65 4500		80 404	
1031					FIX. 49.30 N by Sun obs alt 17.50 38.35 W and D.R.			143 580	
1120					Dense cloud above and below. Sun obs shows that D.R. is badly out.				
1220	078		29	120	D.R. posn 50.23 N. This seems to be 32.00 W. too far annuts stars No obs and D.R. apparently out	3600		140 834	
1260					Someone is trying to talk but is being jammed				
1320	078			116	Climbing to get above clouds for star fixes. No sights up to present	65 5200			
June 15 2017					Vega obs. alt 57.30 Sky cleared. Polaris / 49.40: Q=+27 Practically no dip as cloud horizon not far below				

OPS REC / E 351.1952

Signed A.W.Brown ....Navigator

rather like machine gun fire signalling the breaking away of part of the starboard engine exhaust system.

For the next fourteen hours, inter-crew communications were to be my means of hand signals and pieces of paper. The former can only be imagined but the latter have survived. One reads, 'This is a great trip – no ships or stars or anything. Have a sandwich?'

During the early hours of a transocean flight made low down, nothing so increases the confidence of the navigator as the ability to determine the aircraft's drift by observing the passage of white caps on the waves, through his drift measuring instrument. Ground speed, too, can then be determined using a stop watch. By flying multiple courses, turning first, for example, 60 degrees to port for a minute then 120 degrees to starboard for a similar period before resuming the original course, it is possible to obtain a very accurate wind speed and direction. Whether that simple three course wind method had been devised in 1919 is doubtful, but whatever method Brown planned to use he could not, for there was fog below and cloud above.

Six hours into the flight Brown was able to snatch one sun sight using his marine sextant, the only observation made outside the aircraft in all that period. Together with his position by dead reckoning, which relied on the pilot maintaining an accurate heading on the magnetic compass, and the accuracy of the last known wind, the log entry at 2031hrs records, 'Fix 49°30'N 38°35'W by sun obs alt 17.50 and DR' with some confidence.

After nearly another four hours, at 0017hrs, Brown managed to get a sight on Polaris, the Pole star, which immediately gave his latitude. A second observation, on Vega, gave him a two-star fix recorded in the log at 0025hrs as 50°07'N 31°00'W. The log entry records an almost triumphant 'Halfway', then notes, 'South of track; wind stronger than forecast. Progress better than plan'.

To secure accurate position lines derived from sights taken with a bubble sextant in an aeroplane is not easy. Brown did not have such a simple luxury, although Cdr Byrd (later to become Rear Admiral) of the US Navy had by then invented one. Brown's marine sextant required a visible horizon, a good subsidiary reason why the pioneers flew on moonlit nights. In emergency, a special spirit level would be a poor substitute for a real horizon. The North Atlantic, with its habit of providing cloud cover both smooth and turbulent up to levels at least three times as high as the Vimy's maximum ceiling of 11,000ft, was in fact no place to practise astro navigation. The

After the Sunday morning arrival in an Irish bog, the purchasers of *The New York Herald* could for 2 US cents read all about it on the following morning. *(Vickers)*

Vimy ploughed along at its steady 70mph at around 4,000ft. There was still nearly an hour to go before sunrise. Flying conditions were smooth, although of no use to Brown for star shots due to the lack of a horizon in the hazy misty light, from the moon, in which they flew. Occasionally, the Vimy would run out into clear air, the moonlight strong enough to cast the aircraft's shadow onto the cloud below.

The next bank of cloud the Vimy entered spelled trouble. It was thick, there was no horizon and certainly no artificial horizon instrument in the cockpit, only a lateral clinometer, engine revolution counters on the nacelles rather than in the cockpit, an airspeed indicator and an altimeter. As the engine revolutions increased, the air speed rose to 90mph and then stuck. Alcock disorientated, felt the aircraft stall and go into a spiral dive, as Brown watched the compass spinning. Recovery was not possible until the Vimy emerged from the low cloud base at what Alcock later described as 'an extremely dangerous angle'. With a visible horizon there was just sufficient height to level the wings and climb slowly back to 7,000ft. It was now just before sunrise.

After the experience with the frozen pitot head, which had cleared in the warm air at sea level, the Atlantic weather still had plenty in store. By 0420hrs the Vimy was estimated to be at 21°45'W, having risen to 6,500ft and still climbing.

For the next several hours, as the Vimy ran in towards the Irish coast, the weather worsened with more thick cloud, sleet and snow. Evidently not burdened with airframe icing to any significant degree, the aircraft continued to climb, Brown logging '9400' cold' at 0620hrs. At 0703hrs he had been able to snatch his second sun shot through a break in the cloud. By 0720hrs the Vimy had reached 11,000ft and Alcock decided to bring the aircraft down to 1,000ft to enable Brown to obtain an accurate wind which was 35 knots from a direction of 215 degrees T, by drift measurements.

Within 25 minutes the Vimy had passed over two small islands, Eeshal and Turbot. The final 20 minutes were spent in selecting a suitable landing place marked today by a cairn of white stones. A fine monument in the shape of an aircraft fin was erected on firmer ground about 1½ miles away by Aer Lingus-Irish International. On 15 June 1959, Mr Sean Lemass, then Ireland's deputy prime minister, performed the unveiling ceremony, in the presence of some 300 visitors from far and wide, including Sir George Edwards who followed R K Pierson as chief designer at Weybridge, Mr Harry Couch, the Vimy chief rigger in Newfoundland, and even the writer. Appropriately in the tradition of Vickers' chief designers who were also pilots, Sir George had flown the Vickers communications flight D.H. Dove for most of the flight over to Dublin from Surrey on the previous evening.

Mr Winston Churchill, Minister for War, presents the *Daily Mail* cheque for £10,000 to Capt John Alcock on the following Friday, while Lieut Arthur Whitten Brown looks on. (*Vickers*)

Today this fine statue of Alcock and Brown, the first two men to cross the Atlantic direct by air, stands at London Heathrow opposite Number 2 car park, near the control tower. (*Vickers*)

Footnote

It is undeniable that pioneer transatlantic flight for forty years after the 1919 flights conjured up in the public imagination just one name, Charles Lindbergh, who on 21 May 1927 arrived at le Bourget Airport, Paris, after a 33½ hour direct solo flight from New York. This epic achievement, linking cities epitomising the best in the new world and the old, not only captured the public imagination, as had Alcock and Brown's crossing eight years before, but held it until Neil Armstrong and Edwin Aldrin walked on the moon in July 1969 while their crewman Michael Collins on board the moon orbiting command module awaited their return. At Cape Kennedy, Charles Lindbergh had watched their departure from earth aboard Apollo 11 for their 230,000 mile voyage out to the moon as others had stood squelching in the mud for his departure to Paris, 3,600 miles away and forty-two years before.

It is worth pointing out that Alcock and Brown's flight, eight years before

that, was made in a large aeroplane, in production in quantity, standard except for increased fuel tankage, revised crew arrangements and the removal of the nosewheel fitted originally on the transatlantic Vimy in place of a nose skid. Other Vimys, with bulbous bodies for passengers rather than the slimmer lines of the bomber version, were used in the early years of commercial air transport following the end of the First World War.

In their adapted Vimy bomber of 1919, Alcock and Brown had flown 1,900 miles, almost a typical Atlantic commercial airline sector of the early post-Second World War period. True, their payload had been limited to one old horseshoe nailed under Alcock's seat, 197 letters and two good luck mascots. Lindbergh, planning later than the 1919 pioneers, felt able and had the courage to choose the single-engined formula for his special Ryan NYP high wing monoplane *Spirit of St Louis* (220hp nine-cylinder Wright Whirlwind air-cooled radial engine) enabling him to fly virtually twice the distance, between the major cities direct.

Both epics foreshadowed the commercial future with remarkable accuracy. Both teams chose landplanes and flew broadly similar tracks to those which would be operated commercially. Before then transatlantic crossings for wartime purposes could be numbered in tens of

thousands. BOAC, for example, completing 2,000 on behalf of the British Government, and Trans World Airlines' (TWA) Intercontinental division making nearly 10,000 crossings of the North and South Atlantic. En route navigation still relied largely on celestial observations with the altitude of the navigational stars, sun or moon by now observed through a clockwork driven bubble sextant able to average many sights to reduce observer errors. Radio was used only for homing onto a radio range station or medium frequency beacon or for strictly necessary communications.

Since the earliest pioneering days the aircraft had changed rather more. When TWA inaugurated the first commercial air service between New York and Paris via Gander, Newfoundland, and Shannon on 5 February 1946, the aircraft was the Lockheed Constellation which the airline had itself done so much to initiate eight years before.

When in turn BOAC placed Constellations in service on the London to New York route in July of the same year, among the passengers aboard the inaugural flight was Sir Arthur Whitten Brown. We may be certain that the need to make one or even two technical stops between London and New York if a reasonable commercial payload was to be carried would not have been lost on Sir Arthur or on the Constellation's captain, O P Jones, better known as a pilot of the four-engined Handley Page 45s on the 200 mile London to Paris run of Imperial Airways in the 1930s.

In analysing future prospects for transatlantic air travel by fare paying passengers, shortly after his 1919 flight, Sir Arthur had come down firmly in favour of airships as being the only means of doing the job economically. And for many years between 1928 and 1937, the Zeppelin Company appeared to prove his point. □

July 1946. When BOAC's Lockheed Constellation Model 049 landed at Gander Airport, Newfoundland, for refuelling – on the airline's inaugural commercial passenger flight with landplanes from London to New York – among the passengers was Sir Arthur Whitten Brown. (*Popperfoto*)

Another Vimy, Another Time

D G Addicott

D G 'Dizzy' Addicott, former Vickers and British Aircraft Corporation (BAC) test pilot, undertook all the flying on the Vimy replica before its time in the air was ended by a disastrous fire on the ground at Manchester Airport. Dizzy, who thinks he must be the only pilot still flying who has flown a Vimy, recalls the summer of 1969, and the events leading up to it.

July 1966. In a ceremony on the BAC test airfield at Wisley, Surrey, the first Vickers VC10 strategic transport was handed over to the Royal Air Force. On display that day was the recently completed replica of the Vickers F.B.5 Gunbus built by members of the Vintage Aircraft and Flying Association (VAFA). A two-seat 100hp Gnome rotary-engined pusher biplane with the gunner first at the scene of the accident, it had no throttle, just an ignition blip switch atop the joystick – full bore or nothing.

At the end of its flying life or when the old rotary engine finally gave out, the Gunbus was to be presented to the RAF Museum, Hendon, and it was suggested that the constructors (of the Gunbus) tackle another project, the Vickers F.B.27 Vimy, eventually to take its place alongside the Gunbus. Had it been possible to look into the future and foresee the sweat and tears that were to follow, the Vimy replica might not have been started, which would have a been a great pity in view of the significance of this remarkable aeroplane.

At 12,500lb gross weight, the Vimy was six times the weight of the Gunbus, twice its size and much more complicated. It would be a major task for a band of volunteers, but under the very able direction of Geoff Gregg, VAFA's project design chief, work began almost immediately, aiming for a special category Certificate of Airworthiness (C of A) rather than a Permit to Fly. A few original general arrangement drawings were found at Weybridge, but to build a new flying Vimy involved a great deal of analysis and much detail design work for which many hundreds of new drawings had to be prepared. Original construction was used throughout, but to satisfy airworthiness requirements a number of additions had to be incorporated including the fitting of VHF radio and an elementary elevator trimmer.

Apart from the fabric work, one of the largest jobs involving old skills was the manufacture of two new 10ft 6in dia-meter four-bladed wooden propellers. From glueing together the laminations which were then shaped with a small wood-chopper's axe to the final balancing when a cigarette paper placed on the tip of one blade caused the big propeller to turn was, fortunately, recorded on film. Making new honeycomb radiators with 9,000 thinwall copper tubes for each

Rolls-Royce Eagle engine was also a major task.

Everyone on the project was aiming for a first flight in May 1969 to give time to sort out the inevitable snags before the Paris Air Show in June; it would then be

The Vickers F.B.5 Gunbus replica, VAFA'S first project and the first VC10 for the RAF seen at Wisley. (*BAe*)

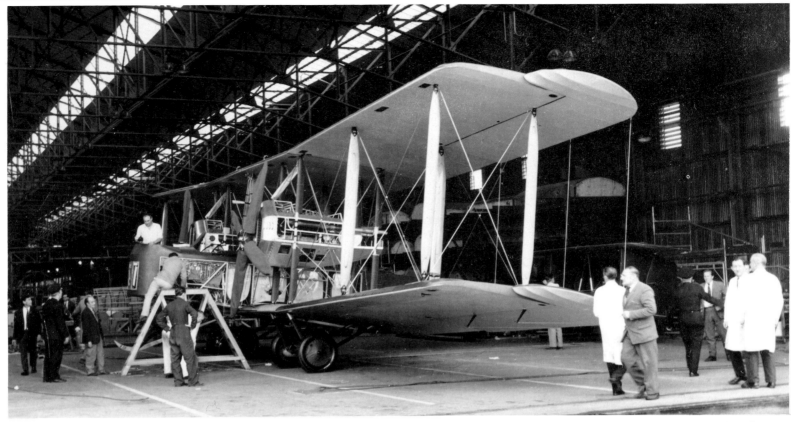

On the west side of the old Brooklands motor racing circuit, a few hundred metres from where many of the originals were built, a new Vickers Vimy, with original Rolls-Royce Eagle engines, nears completion. *(BAe)*

almost exactly fifty years since Alcock and Brown had first flown the Atlantic.

As May ended some of us began to think that the task was impossible, but Geoff Gregg and his team pulled out all the stops and, despite the last minute hitch of a broken tailskid, the Vimy was ready for flight on Tuesday 3 June.

It seemed that the whole workforce of BAC Weybridge turned out at lunchtime to see the first flight and most intended to stay on the aerodrome at Brooklands until the take-off. In no uncertain terms I was told that I would be held responsible for the salaries of the entire establishment if that happened, so it was decided to delay take-off until work finished for the day.

At about 1800hrs I climbed into the tiny cockpit, by now as familiar as the back of my hand, the magnificent old Eagles were cranked into life and run up, chocks waved away and the beautiful propellers moved another Vimy forwards at Brooklands for the first time in many, many years. But how to stop? Masses of people surrounded the machine, obviously unaware of the difficulties of manoeuvring an aeroplane with no brakes, so the engines were cut using the Edwardian electric light switches controlling the ignition. My heart ran a little faster for a minute or two – what if I broke the Vimy before flying it?

But there was no point in waiting any longer; the engines were fine and everyone wanted it to fly. With strict instructions for everyone to keep clear, and wing-walkers to help the taxi out to the take-off point, at 1840hrs the single throttle, with its odd twist knob on top, was pushed gently to fully open, control column central to avoid the nose skid digging in, and with a gentle hop and a skip the Vimy was airborne – I would have loved to have seen it!

The main problem taking off to the north at Brooklands is the height of the banking of the old motor racing circuit. This had been demolished on the southern side to allow jet bombers and airliners to take off in relative safety on the 3,600ft of paved runway, but the Vimy, with no brakes other than the tailskid, needed a grass runway into wind, hence the take-off to the north. Provided that both engines kept producing their total of 700bhp there would be no problem, but if one of them failed, could I ease it gently around the circuit? Those who had done the calculations just kept everything crossed until the Vimy passed the banking going up at a steady rate, if not exactly like a lift. It had, after all, been done before.

Turning south and climbing to about 1,500ft, the controls were checked for effectiveness and responded in the expected manner with the elevator light and effective, the rudder heavy but effective and the huge non-differential ailerons as heavy as lead, needing a bootful of rudder to counteract the further effect of the downgoing ailerons causing yaw and roll in the opposite direction to that required! Apart from the forces involved the harmonisation was similar to that of an early Spitfire.

Landing at Wisley some 30 minutes after take-off I was surprised to find no one to meet me – apparently the roads were busy and the ground crew took almost an hour to travel 3 miles. No one had seen my immaculate three-pointer.

In the evening of Wednesday 4 June, with full tanks and Ian Muir in the observer's seat on the left of the pilot, we took off to the east at Wisley. The wind was gusting from the north with considerable turbulence over the trees bounding the aerodrome on that side, and even before becoming airborne it was obvious that this was not the relatively gentle machine of the evening before, but a lumbering brute which seemed to be telling me that it did not want to fly in these conditions. As we barely cleared the trees, the physical and mental effort was intense – I dared not break it. It was obvious that test flying was impossible and I gasped to Ian that we would have to park it somewhere, and was incredibly relieved by his comment, 'Well, at this speed, we're not going to hurt ourselves!'

We could not land at Wisley, which runs east to west, with the wind across the runway and Brooklands was not a possibility due to heavy traffic crossing the centre of the runway, so Odiham was selected as a suitable parking area with sufficient grass into wind for us to make a safe landing. The RAF made us welcome. 'Well, it's not often that a Vimy drops in on us', said the station commander.

Thursday 5 June was spent making two fairly long test flights and a demonstration flight with John Carrodus, an Air Registration Board test pilot, on board to

see how I handled the aeroplane. The C of A was issued that evening together with the only flight manual ever made up for a Vimy.

On Friday 6 June at 0700hrs, we were driven to Odiham in fog – the final blow. My close friend Peter Hoar, also a pilot with BAC, who had occasionally flown the Gunbus, was to accompany me to Le Bourget to help with navigation and because he spoke fluent French. At about 0915hrs there seemed to be a slight improvement in visibility so we donned our cold weather flying gear. In my case this comprised a leather helmet and coat from the First World War, the coat reaching to my ankles and which would have stood up on its own if asked due to the amount of castor oil in which it was soaked. By 0930hrs there was a definite improvement in visibility so we decided to warm up the engines, and 15 minutes later, right on our deadline, pretended we could see far enough and aimed the Vimy into the murk. As it turned out the fog was fairly shallow and we climbed steadily to the east, the weather improving all the time, aiming to cross the English Channel at its narrowest point. Our tests had proved that following the loss of an engine, the serviceable engine merely enlarged the area in which a forced landing was inevitable – rather like a Mosquito on one engine with the undercarriage down.

The cold bit deeply and, despite wearing heavy gloves, our fingers and toes were numb long before we reached the Channel. Poor Peter, who had forgotten his goggles, could scarcely see out of eyes that looked like poached eggs!

The Vimy replica rolls out from the same shed which had seen the demise of TSR.2 production, the building of Concorde and one of the four final assembly lines for the Viscount. (*BAe*)

The test flights had shown that the trim was out of adjustment, heavy pressure being needed on the left rudder pedal and a lot of aileron applied through the beautiful and necessarily large four-spoke alloy steering wheel; fortunately, the elementary elevator trim took care of the longitudinal forces. It was a pity there had been insufficient time to re-rig the aeroplane, as was done originally.

Reaching the French coast we descended to about 1,500ft and bounced along under the cumulus clouds, perspiring now, instead of freezing, and worried that we might be a few minutes late at Le Bourget. The French were not going to be helpful – if we estimated landing after 1245hrs we must divert. Another fifty revolutions on the nacelle-mounted rev counters and a decision to have 'radio failure' if the order to divert had been confirmed appeared to be the only cures. To hell with approach procedures, a straight line to Le Bourget, and we landed

VAFA's then chairman, our contributor, 'Dizzy' Addicott, is ready to taxi. (*BAe*)

with about 30 seconds to spare. We needed a drink!

Hangarage was accepted, a decision which turned out to be a mistake.

Interviewed by a large number of press reporters the main question was, 'What's it like to fly – is it easier or more difficult than Concorde?' I could only reply that it was hard work and if it was not more

Walkers at the wingtips were very necessary at the first attempt. (*BAe*)

difficult than Concorde we must be making progress in the wrong direction – it was different. Hopefully, this had no bearing on the fact that on Saturday 10 June we were unable to take the Vimy out of the hangar. Skulduggery? Maybe, but the French flew their Blériot.

We flew on Sunday, but the big moment had been lost.

On Monday we flew back to Wisley where the Vimy was repainted in military colours and subsequently flown to Manchester where it was put on public display in an inflated tent. During this period we were asked to fly the aeroplane to London to take part in a celebration of fifty years of commercial flying. Despite an extensive search of Heathrow it was impossible to find an unobstructed grass area on which to land safely, apart from the impossibility of breaking our con-

Hello and goodbye. The replica's first flight from Brooklands on 3 June 1969 was also its last from this historic site. (BAe)

tract with Manchester, and it was slighty disconcerting to be accused of attending the function under false pretences.

After the exhibition at Ringway the aeroplane was to be flown to the Rolls-Royce test aerodrome, Hucknall. During the preparation for flight a small flame appeared on the port lower mainplane and, despite frantic efforts to extinguish it, spread rapidly along the wing, engulfing the upper wing so rapidly that the centre section tanks were almost unscathed. The airport fire service did all it could but the speed of the conflagration

Meeting at Manchester, a month before the fire which ended its flying career, the Vimy replica and the then daily BOAC Super VC10 flight to New York via Prestwick.

was such that from start to finish the fire took about 3 minutes.

The rebuild to static display standard was done – not by VAFA, the members had had enough – and since that time the Vimy has resided (a little forlornly, I feel) near the Gunbus in the RAF Museum, Hendon. It would have been nice to have been invited to the opening ceremony.

Triple First, she was called, in memory of the Atlantic, Australia and South Africa flights. Some of us had hoped that she would repeat all three. Maybe she just didn't want to. □

From Scapa Flow to Tokyo Bay: Grumman Aircraft in Wartime Service with the Fleet Air Arm

Dr René J Francillon

Over 3,200 Martlet/Wildcat, Tarpon/Avenger and Hellcat aircraft were delivered to serve with the Royal Navy. This account briefly examines their history and the units with which they served aboard aircraft carriers operating in the Atlantic, the Arctic, the Mediterranean, the Indian Ocean and the Pacific.

More than two out of every five carrierborne aircraft serving with the Fleet Air Arm (FAA) during the Second World War were built by the Grumman Aircraft Engineering Corporation in Bethpage, Long Island. The first squadron of Martlet I fighters was formed in the autumn of 1940, and nineteen squadrons of Wildcats, sixteen squadrons of Avengers and fifteen squadrons of Hellcats served aboard Royal Navy carriers or at shore stations at the time of the Japanese surrender. During the intervening years, carrierborne Grummans went on convoy duty in the Atlantic, the Mediterranean and the North Sea; took part in bitter operations against a for-

The Grumman Martlet, pictured here by Charles Brown, was the Royal Navy's first single-seat monoplane fighter when it entered service in late 1940. (*RAF Museum*)

mer ally during actions in North Africa and Madagascar; provided air cover during landings in North Africa, Italy and France; and avenged defeats in Hong Kong and Singapore by taking the British might to the Japanese capital. In the process, they established many claims to fame, including those for the destruction of the first and last enemy aircraft credited to American-built aircraft in British service.

The first of these claims was made on 25 December 1940 by two Martlet I pilots from No. 804 Squadron, Lieut L V Carter and Sub-Lieut A Parker, who forced down a Junkers Ju 88A attempting to bomb the Home Fleet at Scapa Flow. The last was credited to an Avenger crew from No. 820 Squadron which shot down a Japanese Zero fighter during a raid in the Tokyo area on 15

August 1945 (eight other enemy aircraft were claimed by Seafires from Nos. 887 and 894 Squadrons flying from HMS *Indefatigable*). Unfortunately, the Avenger was shot down almost immediately after claiming the Zero and thus became the last British flown aircraft lost in combat during the Second World War.

Arguably, it was in qualitative terms more than quantitative terms that contributions made by Grumman aircraft to Britain's naval aviation were significant. When the war started, the seven Royal Navy carriers embarked two types of biplane, the Fairey Swordfish torpedobomber and the Gloster Sea Gladiator single-seat fighter, and the Blackburn Skua dive-bomber monoplane. Types about to enter FAA service were the Fairey Albacore, another biplane torpedo-bomber, and Fairey Fulmar, a

Above: Martlet II AM997 07L of No. 888 Squadron ashore from HMS *Formidable* at Oran-La Senia, Algeria, in January 1943 following action during the North African landings. (*Courtesy R C Sturtivant*)

Below: An invasion-striped Wildcat V (8H of No. 896 Squadron) standing by at readiness on the flight deck of HMS *Pursuer* at the western end of the English Channel on D Day. (*Courtesy R C Sturtivant*)

two-seat monoplane fighter. Work had also been initiated on new torpedo-bombers and two-seat fighters of monoplane design, but the resulting Fairey Barracuda and Firefly were not to reach operational units until 1943.

The fastest aircraft in service with the FAA at the start of the war, the Sea Gladiator, had a top speed of 245mph at 10,000ft and was thus slower than most types of contemporary Axis bomber[1]. The performance of the Fulmar, which was still nine months from entering FAA service, was marginally better, with a top speed of 280mph at 9,000ft. By comparison, the Grumman G-36B ordered for the FAA by the British Purchasing Commission had a top speed of 315mph at 13,100ft.

Speed was not the only asset of the first Grumman aircraft to enter FAA service; the Martlet also carried a heavier armament (four 0.5in Colt-

Below: JV579 F, a conspicuously marked Wildcat V of No. 846 Squadron's Fighter Flight from HMS *Tracker* flyng near Limavady, Northern Ireland, in June 1944. Two months earlier, Avengers also equipping this squadron had shared with a Swordfish of No. 819 Squadron and with HMS *Beagle* in the sinking of two U-boats, U-288 and U-355. (*Courtesy R C Sturtivant*)

Browning guns for the Mk. I and six for the Mk. II versus four 0.303in Browning guns for the Sea Gladiator and the Roc). Above all, however, Martlets were superlative carrierborne aircraft. Most of these assets were retained by later types of Grumman aircraft operated by the FAA. There is no doubt that the Avenger and Hellcat were faster, sturdier and better suited to carrier operations than their British counterparts, the Fairey Barracuda and the Supermarine Seafire.

The operational careers of the three types of Grumman aircraft which were operated aboard Royal Navy carriers from August 1941 to August 1945 are summarised as follows:

Martlets and Wildcats

During the war, no fewer than 1,123 of Grumman's first monoplane fighters were obtained by the Royal Navy including 100 ordered by the United Kingdom, 127 taken over from French and Greek contracts, and 902 delivered under Lend-Lease. They were known as Martlets until 13 January 1944 when the name was changed to Wildcat to conform to US Navy nomenclature. The ninety-one Martlet Is were Grumman G-36A aircraft with non-folding wings which had been ordered by the French Purchasing Commission and were given British Service serials AL231/AL262, AX725/AX747, AX753/AX754, AX761, AX824/AX829, BJ554/BJ570 and BT447/BT456 after being taken over by Britain following the fall of France. One hundred Martlet IIs were Grumman G-36Bs (serials AM954/AM999 and AJ100/AJ153) ordered by the British Purchasing Commission, only the first ten of which were completed with non-folding wings. Thirty Martlet IIIs with non-folding wings were Grumman F4F-3As ordered by the US Navy but released for delivery to the Greek Air Force. Before this could take place,

1 For example, the top speeds of the Junkers Ju 88A-1, Savoia Marchetti S.M.79-II, and Heinkel He 111P-4, respectively, were 280mph at 18,050ft, 270mph at 12,000ft and 247mph at 16,400ft.

Greece was invaded by German forces and the aircraft were taken over by Britain and given serials BJ501/BJ530. Aircraft delivered under Lend-Lease included 220 Martlet IVs (Grumman F4F-4Bs, serials FN100/FN319), 312 Martlet Vs (Eastern-built FM-1s, serials JV325/JV636), and 370 Wildcat VIs (Eastern-built FM-1s, serials JV636/JV924, JW785/JW836 and JZ860/JZ889).

Most of these aircraft were taken on charge in the United Kingdom, where Lend-Lease Martlet IVs and Vs and Wildcat VIs were fitted with British equipment by Blackburn Aircraft Ltd. Others were taken on charge in Africa (notably seventeen Martlet IVs which had been shipped to Mombasa), in India (where fifty-four Martlet IIs had been shipped), in Australia (where a number of Wildcat VIs were shipped for embarkation aboard carriers operating in the Indian and Pacific Oceans) and in the United States (notably the Martlet IVs which carried US markings when going aboard HMS Victorious to take part in joint US-British operations in the

Martlet IV FN296 A all set and standing by on the catapult of HMS Archer around July 1943. During the spring and early summer of that year, HMS Archer embarked Martlets from No. 892 Squadron and Swordfish from No. 819 Squadron during three transatlantic convoys. (*Courtesy R C Sturtivant*)

Avengers of No. 852 Squadron based at Maydown, Northern Ireland, flying over a convoy in May 1944. (*Courtesy R C Sturtivant*)

Southwest Pacific, May to July 1943).

The first to reach the United Kingdom were the ex-French Martlet Is which entered FAA service with No. 804 Squadron at Hatston, Orkney, in October 1940. The first to go to sea aboard a carrier of the Royal Navy were a pair of Martlet IIs from B Flight, No. 802 Squadron, which in August 1941 went on convoy escort duty to the USSR aboard HMS Argus[2]. During the return voyage from Murmansk aboard HMS Victorious, these two Martlet IIs escorted Fairey Albacores during shipping strikes off the coast of Norway in September and October. From carriers and escort carriers of the Home Fleet, Martlets and Wildcats later provided air cover for Arctic convoys to Murmansk (Nos. 811, 813, 816, 819, 824, 825, 835, 842, 846 and 853 Squadrons), flew air cover and

2 It is worth noting that the initial deployment of Martlet IIs with folding wings took place three months before the US Navy took delivery of its first folding wing F4F-4s.

Avenger 835 from No. 857 Squadron flying over HMS Indomitable after returning from a strike on the Sakishima Gunto, 25 May 1945. Three weeks earlier, HMS Indomitable had been hit by a Kamikaze. (*Courtesy R C Sturtivant*)

anti-flak sorties during Operations Tungsten and Goodwood against the *Tirpitz* in April and August 1944 (Nos. 842, 846, 852, 881, 882, 896 and 898 Squadrons) and during anti-shipping strikes in Norwegian coastal waters from April 1944 to May 1945 (Nos. 813, 821, 824, 835, 842, 846, 852, 853, 856, 881, 882, 896 and 898 Squadrons). On 26 March 1945, during one of the last strikes of the war, Wildcats from No. 882 Squadron embarked aboard HMS *Searcher* shot down five Messerschmitt Bf 109Gs, thus proving that the obsolescent Grumman fighter was still very efficient.

In the Mediterranean and North African theatre of operations, Martlet IIIs were first assigned in September 1941 to No. 805 Squadron for operations over the Western Desert with the Naval Fighter Wing. In the Western Mediterranean, Martlet IIs were first operated by No. 806 Squadron aboard HMS *Indomitable* during Operation Pedestal in August 1942. Thereafter, carrier-based Martlets, and later Wildcats, took part in numerous operations in the Mediterranean and Aegean Seas and notably supported landings in Algeria during Operation Torch in November 1942 (Nos. 882, 888 and 893 Squadrons), in Sicily during Operation Husky in July 1943 (No. 888 Squadron), in southern Italy during Operation Avalanche in September 1943 (Nos. 878 and 890 Squadrons), and in southern France during Operation Dragon in August 1944 (No. 881 Squadron, which also took part in the mopping-up operations in the Aegean in September 1944).

In the Indian Ocean, Martlet IIs made their debut in May 1942 during Operation Ironclad when Britain feared that the Vichy Government would let Japan establish bases on the island of Mada-

On 20 December 1944, twenty-seven Avengers from No. 854 (HMS *Illustrious*) and No. 857 Squadron (HMS *Indomitable*) attempted to bomb the refinery at Pangkalan Brandon in Sumatra but were forced by bad weather to divert to the secondary target, the port of Belawan Deli. JZ594 W1A, an Avenger II from No. 857 Squadron, was photographed during that strike in the Netherlands East Indies. (*Courtesy R C Sturtivant*)

gascar. In three days of operations, Nos. 881 and 882 Squadrons operating from HMS *Illustrious* claimed six confirmed victories (four Morane M.S.406 fighters and two Potez 63.11 reconnaissance aircraft) and eleven probables for the loss of a single Martlet II. Following this painful operation, *Illustrious* and her Martlets remained in the Indian Ocean but saw little action until withdrawn in January 1943. Wildcats returned to this theatre of operations in 1944 with Nos. 832, 834, 851 and 890 Squadrons providing trade protection from the escort carriers *Begum*, *Battler*, *Shah* and *Atheling* beginning that summer.

Above all, however, it was in the Battle of the Atlantic that Martlets and Wildcats distinguished themselves while operating aboard escort carriers. The first to do so had been six Martlet IIs from No. 802 Squadron aboard HMS *Audacity* which shot down five Focke-Wulf Fw 200C patrol bombers and shared in the destruction of the submarine U-231 during two escort voyages to and from Gibraltar in September-November 1941. Unfortunately, *Audacity* was sunk on 21 December while returning to Britain. No other escort carriers were available until April 1943 when HMS *Biter* joined the fray with No. 811 Squadron. During the next sixteen months, until the Battle of the Atlantic was won, Martlets and Wildcats from Nos. 811, 813, 819, 824, 833, 842, 846, 881, 882, 892, 896 and 898 Squadrons hunted Heinkel He 177s, Focke-Wulf Fw 200s, and Junkers Ju 88s and Ju 290s (with at least one example of each falling to the guns of Wildcats) and strafed surfaced U-boats to distract flak gunners during rocket, bomb or depth charge attacks by strike aircraft.

Virtually none of the British-, French- or Greek-ordered Martlets survived the war (although a Martlet I, Al246, is now in the FAA Museum, Yeovilton) and the Lend-Lease Wildcats were dumped after the war ended as Britain had no need to purchase them and the USA already had too many aircraft to dispose of.

Tarpons and Avengers

Initially named Tarpon by the Royal Navy but renamed Avenger on 13 January 1944 to conform to the designation already adopted by the US Navy, the Grumman-designed torpedo-bomber became one of the most important types of carrier-based aircraft operated by the FAA during the last three years of the Second World War.

Between 1943 and 1945, the Royal Navy was allocated a total of 958 Tarpons and Avengers under Lend-Lease including 402 Grumman TBF-1Bs (Tarpon or Avenger T.R. Mk.Is, serials FN750/FN949, JT773 and JZ100/JZ300), 334 Eastern-built TBM-1Cs (Avenger T.R. Mk.IIs, serials JZ301/JZ634), and

AM958 in June 1941. This was one of the ten Martlet IIs delivered with non-folding wings. (*Air Ministry via National Air and Space Museum*)

222 Eastern-built TBM-3s and TBM-3Es (Avenger T.R. Mk.IIIs, serials JZ635/JZ746 and KE430/KE539). Most Avenger squadrons of the FAA operating in the Pacific formed up in the United States and were equipped with aircraft straight off the Grumman or Eastern lines or modified by Blackburn Aircraft at Roosevelt Field, New York. Squadrons operating in the Atlantic and Arctic were equipped with Avengers modified in the United Kingdom to meet British requirements. Major modifications included the installation of British gun sights, oxygen system and radio equipment; the substitution of an F.24 camera for the ventral gun; the relocation of the navigator closer behind the pilot; and the use of a folding radio mast rendered necessary by the lower ceiling of hangar decks aboard British carriers. One of the T.R. Mk.Is, JZ146, was modified even more extensively and served as a testbed for the Frazer Nash F.N.95 remotely controlled twin-gun barbette intended for the Fairey Spearfish.

No. 832 Squadron, the first FAA unit to be equipped with Tarpons, was formed in January 1943 at NAS Norfolk, Virginia, and trained in the United States before embarking aboard the USS *Saratoga* for operations in the Pacific. The squadron first went into action on 27 June 1943, when it provided support during the American landing on New Georgia in the Middle Solomons. Shortly after, No. 832 Squadron transferred to HMS *Victorious* before embarking successfully aboard HMS *Illustrious* and HMS *Begum* for operations in the Pacific and Indian Oceans (where Avengers of Nos. 832 and 851 Squadrons worked with HMS *Findhorn* and RLNDN *Godavari* to sink U-198 on 12 August 1944). Other Avenger squadrons which served aboard armoured and escort carriers of the East Indies Fleet and the British Pacific Fleet were Nos. 820, 828, 845, 848, 849, 851, 854, 857 and 885. Their greatest successes were achieved in 1945 during operations against Japanese oil refineries in the Dutch East Indies and against enemy bases on Formosa, Truk, the islands south of Japan, and the Japanese mainland.

In home waters, Avengers flew anti-submarine and anti-shipping patrols and mine-laying sorties from shore bases. Others operated from escort carriers on strikes against shore targets, coastal shipping and naval facilities in Norway (notably sinking U-711 and its depot ship at Kilbotn just four days before VE Day) and on convoy escort duty, mainly in the Arctic where they shared in the destruction of two U-boats (U-288 and U-355). Squadrons involved in these operations between the summer

A formation of Tarpons from an unidentified squadron. (*British Information Services via National Air and Space Museum*)

of 1943 and VE Day were Nos. 846, 850, 852, 853, 855 and 856. In addition, during the last war years, Avengers served along with other aircraft with an Anti-Submarine Warfare Development Unit (No. 703 Squadron at Thorney Island), a Tactical Trials Unit (No. 711 Squadron at Crail), the Operational Training Unit (No. 732 Squadron at Squantum), four Telegraphist/Air Gunner Training Units (Nos. 743, 744, 745 and 754 Squadrons), and a Target Towing Unit (No. 785 Squadron at Crail). An RAF target-towing and army co-operation squadron, No. 567, is also reported as having received at least one Avenger III. After the war, the Lend-Lease Avengers were either returned to the United States or, as they were no longer needed and their transport back to the United States was costly, were

JZ691, the fifty-seventh Avenger T.R. Mk. III, in US Navy glossy midnight blue finish, July 1945. (*Ministry of Aircraft Production via National Air and Space Museum*)

simply destroyed. The first career[3] of the Avenger in British service came to an end in June 1946 when No. 828 Squadron was disbanded.

Gannets and Hellcats

Next to the Vought Corsair, of which 1,977 were delivered to Britain under Lend-Lease, the Hellcat was the most numerous American aircraft operated by the FAA during the war years. The first 252 aircraft were F6F-3s which were to have been designated Gannet Is in British Naval service but were redesignated Hellcat Is (serials FN320-FN449 and JV100/JV221) before deliveries

[3] The second career of the Avenger in British naval service began in 1953 when a hundred TBM-3Es and TBM-3Ss were delivered. They received the serials XB296/XB332, XB355/XB404 and XB437/XB449 and were designated Avenger A.S. Mk. 4s (TBM-3Es operated as received from the US), Avenger A.S. Mk. 5s (TBM-3Ss fitted with British equipment including radar in a large radome beneath the starboard wing), and Avenger A.S. Mk. 6s (TBM-3Ss fitted with British equipment including radar in a radome beneath the forward section of the bomb bay). The ASW Avengers remained in service until 1962 when the last were operated by No. 831 Squadron from Culdrose.

began in May 1943. They were followed by 849 F6F-5s and 76 F6F-5Ns which, respectively, became Hellcat F.Mk.IIs (serials JV222/JV324, JW700/JW784, JW857/JW899, JX670/JX964, JX968/JX999, JZ775/JZ827, JZ912/JZ946, JZ960/JZ964, JZ968/JZ994, KD118/KD152, KD158/KD160, KE118/KE159, KE170/KE214 and KE220/KE265) and Hellcat NF.Mk.IIs (serials JX965/JX967, JZ890/JZ911, JZ947/JZ959, JZ965/JZ967, JZ995/JZ999, KD108/KD117, KD153/KD157, KE160/KE169 and KE215/KE219) with the FAA. Most Hellcats were operated as received from the United States, but Blackburn Aircraft fitted British rocket projectile launchers to a number of F6F-3s and modified some F6F-5s as Hellcat FR.Mk.II fighter-reconnaissance aircraft and others as Hellcat PR.Mk.II unarmed photographic-reconnaissance aircraft.

The first two squadrons to be equipped with Hellcat Is were Nos. 800 and 804 which began their training at RNAS Eglington, Londonderry, in the early summer of 1943. After first being deployed in December 1943 for North Atlantic convoy duty aboard HMS *Emperor*, the two squadrons went on the offensive during Operation Tungsten on 3 April 1944 when they provided cover for strikes against the German battleship *Tirpitz* anchored in Kaafjord in Norway. From then until the end of the war, Hellcats took part in several operations in Norwegian waters. Those from Nos. 800 and 804 Squadrons flew anti-shipping strikes in May 1944, with No. 800 claiming three enemy aircraft for the loss of two Hellcats during a dogfight against Bf 109Gs and Fw 190As on the 8th. Those from No. 1840 Squadron provided cover during two further strikes against *Tirpitz*: Operation Mascot on 17 July 1944 when they operated

Avenger I from No. 846 Squadron catapulted from the deck of HMS *Trumpeter*. Aircraft from this squadron sank the depot ship *Black Watch* and the submarine U-711 on 4 May 1945 at Kilbotn, Norway. (*Imperial War Museum via Alain Pelletier*)

from HMS *Furious*, and Operation Goodwood on 22-29 August when they were aboard HMS *Indefatigable*. The only other significant operations against German forces in which Hellcats took part were those which saw No. 800 Squadron provide cover for a convoy to Gibraltar, in June 1944, and air support for the landings in southern France, in August 1944, while operating from HMS *Emperor*.

In the war against Japan, British Hellcat squadrons went into action on 29 August 1944, when Nos. 1839 and 1844 aboard HMS *Indomitable* provided cover during strikes against a cement factory at Indaroeng and the Emmahaven harbour in the Dutch East Indies. These two squadrons scored their first victories against Japanese aircraft during operations over the Nicobar Islands

Aboard HMS *Empress*, July 1945, Hellcat F.Mk.IIs of No. 804 Squadron. (*Courtesy R C Sturtivant*)

in October 1944 and took part in strikes against Belawan Deli in December 1944, and Pladjoe and Soengi Gerong in January 1945. Remaining aboard *Indomitable* until May 1945, when this carrier was pulled out for refitting, Nos. 1839 and 1844 Squadrons provided Hellcat F.IIs and NF.IIs for operations with the British Pacific Fleet. Their NF.IIs then went aboard *Formidable* to keep providing the British Pacific Fleet with its only night fighter defence during the final operations of the war. Other Hellcat squadrons which saw service in the Indian and Pacific Oceans were Nos. 800 (aboard *Emperor* and *Shah*), 804 (*Ameer*, *Empress* and *Shah*), 808 (*Khedive*), 885 (*Ruler*), 888 (with Hellcat PR.IIs aboard *Indefatigable* and *Empress*), 896 (*Ameer* and *Empress*) and 898 (*Attacker* and *Pursuer*). FAA squadrons which formed up too late to see action were Nos. 881, 889, 891 and 1847. After the war, all airworthy Hellcats were returned to the United States. The last two squadrons to fly the type were Nos. 892 and 888 which disbanded in April and August 1946, respectively.

However, at least one Hellcat F. II (KE209) was retained in the United Kingdom and was airworthy as late as April 1953.

In addition to 3,264 Martlets, Wildcats, Tarpons, Avengers and Hellcats during the war, the FAA received fifteen Grumman J4F-2 Widgeon amphibians and requested the loan of two F7F-1 Tigercat twin-engined fighters. The name Gosling, initially given to the small twin-engined amphibians, was later changed to Widgeon as used in the United States. These fifteen aircraft (serials FP455/FP469) were primarily operated on communications duties in the West Indies between 1943 and 1945. The twelve surviving Lend-Lease Widgeons were returned to the United States after the war's end. The two Tigercats, TT348 and TT349, arrived in England after the war ended and were evaluated at Farnborough, Hampshire, in 1946 before being given back to the US Navy. □

Above: The 510th Hellcat for service with the FAA was this F.Mk.II JX696 seen on the lift aboard HMS Pursuer. (Courtesy R C Sturtivant)

Below: A Hellcat F.Mk.II being accelerated on HMS Battler, in September 1945, off the Ayrshire coast. (Courtesy R C Sturtivant)

Grumman aircraft data

	Martlet I 1,240hp Wright Cyclone G-205A	Martlet II 1,200hp Pratt & Whitney Twin Wasp S3C4-G	Wildcat IV 1,240hp Pratt & Whitney Twin Wasp R-1830-86	Avenger T.R. I 1,850hp Wright Cyclone R-2600-8	Avenger T.R. III 1,750hp Wright Cyclone R-2600-20	Hellcat II 2,250hp Pratt & Whitney Double Wasp R-2800-10W
Span	38ft	38ft	38ft	54ft 2in	54ft 2in	42ft 10in
Length	28ft 10in	28ft 10in	28ft 11in	40ft	40ft	33ft 7in
Wing area	260sq ft	260sq ft	260sq ft	490sq ft	490sq ft	334sq ft
Empty weight	5,000lb	5,544lb	5,598lb	10,627lb	10,700lb	9,414lb
Loaded weight	6,607lb	6,607lb	7,900lb	16,300lb	16,400lb	11,811/13,808lb
Max speed	304mph	300mph	298mph	259mph	262mph	392mph
Crusing speed	247mph	246mph	238mph	171mph	174mph	237mph
Service ceiling	30,800ft	29,000ft	28,600ft	23,000ft	25,000ft	36,500ft
Range	830 miles	830 miles	695 miles	1,020/1,910 miles	1,000/2,230 miles	Up to 1,115 miles
Climb	2,140ft/min	2,030ft/min	1,760ft/min	5,000ft in 4.3min	5,000ft in 3.8min	2,440 ft/min

No. 25 Squadron Royal Air Force

Half a Century of Developing Round-the-Clock Interception

John D R Rawlings

The 1937 Royal Air Force Display at Hendon was the last major occasion on which the RAF's Hawker Fury fighters were shown off to the public. For one youngster, our contributor, they made an unforgettable impression. Within months the Furies had gone, and No. 25 Squadron was beginning to learn how to develop skills which would lead eventually to twenty-four hour interception, virtually regardless of the weather.

No. 25 Squadron of the RAF shows off its Hawker Fury IIs in echelon stepped up.

The Furious Fighters

Saturday 26 June 1937 started like so many summer days, with a covering of radiation fog which lifted to form low stratus. We joined the crowds streaming out of Colindale Underground station and up Colindale Avenue that morning and entered Hendon Aerodrome by Gate No. 3 where we turned left to find Enclosure B which was for 5 shillings [25 pence today] ticket-holders only, and that was what we were.

The occasion was the eighteenth, and as it turned out, the final RAF Display at Hendon, billed as containing the greatest number of military aircraft ever seen together. There was an air of anticipation that here was something tremendous, enhanced by the august tones of the RAF Central Band, under its director of music, Sqn Ldr R P O'Donnell MVO, beaming out at us from the loudspeaker system (they weren't known as Tannoys in those days). As we settled ourselves on our benches we could see the rows of participating aircraft over to the left and a small, separate group to the right, the New Types Park.

The pessimists were already rumouring that this might be the last Hendon Display for a new era was dawning, an era of realism engendered by the increasing number of camouflaged monoplanes apparent amongst the preponderance of silver biplanes. That turbulent parliamentarian Winston Churchill was already scaring us that Germany meant business which could only be stopped by force, and even in the bright summer of 1937 there were grim forebodings in many thinking people's minds. As it turned out, these forebodings were correct and this was the termination of that grand inter-wars period of light-hearted RAF growth in which the silver biplanes epitomised the joy and wonder of life in the RAF.

Of all the biplanes from that carefree

era none was a finer example of the breed than the Hawker Fury. With its clean lines and rather tall undercarriage, it resembled a well-bred racehorse, while the forward-stagger of the wings gave it an eager look, like a sheepdog about to race after its flock. This classic appearance was coupled with superb handling qualities which made it a fine aerobatic mount. And even at this late stage in the 1930s it was the Fury which stole the show at Hendon that year. Only three squadrons had been equipped with Furies, the first being No. 43 at Tangmere in May 1931. Obviously, No. 43 believed themselves to be the best fighter squadron but this was soon challenged in 1932 when No. 1, also at Tangmere, and No. 25, along the coast at Hawkinge, received their Furies in February 1932.

The rivalry was intense, especially between the two Tangmere squadrons. No. 25, away at Hawkinge, was a little aloof from the other two but nonetheless very keen. Obviously, the accolade for the finest squadrons in the 1930s was to be chosen to take part in the Hendon display and over the years the rivalry to outdo each other climaxed each June. Aerobatics were the key and both individual and formation aerobatics became highly refined by the Fury squadrons. In 1933, No. 25 Squadron stole a march on the two Tangmere units by providing a flight of three Furies tied together by bungee cord. They took off and flew a full programme of air drill, then landed, still tied together. This became No. 25's trademark and the theme was developed by other squadrons as well.

So, on that 1937 display day, the Fury squadrons were still the attraction, despite the billing of a 'mass formation of 250 aircraft' and the appearance of Fairey Battles, Bristol Blenheims and Armstrong Whitworth Whitleys in squadron service for the first time. In fact, the Furies were in evidence from the start of the flying programme. The first event, the 'Headquarters Race', comprised a race around a circuit of 44 miles, for two entrants from each of the Home Commands, and the Air Ministry. Two Furies took part, one flown by No. 25's commanding officer, Wg Cdr C A Stevens MC and the other from Training Command where the type was in service at the RAF College, Cranwell. They came sixth and fifth in the race, respectively, having been heavily handicapped.

After this, at 12.45pm, came No. 43 Squadron's turn. From the ranks of participating aircraft taxied Fury I K3731, with silver and black checks denoting No. 43 Squadron and a yellow fin showing it to be B flight commander's aircraft. The pilot was P/O C B

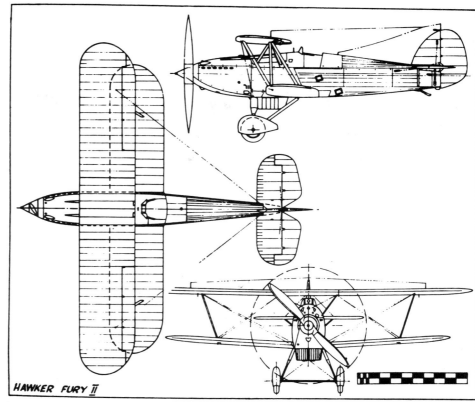

Hawker Fury Mk. II.

'Caesar' Hull who was later to lead the squadron in the dark days of August and September 1940 and whose prowess at Hendon was applauded by both *Flight* and the *Aeroplane*. A brisk succession of aerobatic manoeuvres, each leading into the next, entranced spectators.

No. 1 Squadron's turn did not come until four o'clock in the afternoon but this was to its advantage for by then the

sun had burned off the low stratus and it was a perfect summer's afternoon. Into the blue, sunny sky Flt Lt E M 'Teddy' Donaldson led three other squadron pilots into a scintillating formation aerobatic display predating the Red Arrows by three decades. The four silver Furies, emblazoned with No. 1's red bars, looped, rolled and dived, changing formation through the manoeuvres with the highest precision.

So what was left for No. 25 Squadron to do? The answer came immediately after No. 1 Squadron's Furies had landed and a flashback to 1914-18 had taken place with the flying of a Bristol Fighter and a Sopwith Triplane. From behind us and to the right came nine fighter biplanes, stacked up in echelon formation, the sun glinting on their highly polished engine cowlings. As No. 25's formation appeared it was apparent that these were Mk. II Furies, with glistening spats and Kestrel VI engines giving them nearly 20mph more speed and a

Refuelling a Fury I at Hawkinge in the early 1930s, the airmen wear the flat caps and cumbersome overalls characterising the period. (*via J D R Rawlings*)

Hawker Fury
Single-seat interceptor fighter. All-metal structure, fabric covered

Manufacturers	Hawker Aircraft Ltd. Mk. II subcontracted to General Aircraft Ltd
Powerplant	One 525hp Rolls-Royce Kestrel IIS (Mk. I). One 640hp Rolls-Royce Kestrel VI (Mk. II)
Dimensions	Span 30ft. Length 26ft 8in. Height 10ft 2in. Wing area 252sq ft
Weights	Empty 2,623lb. Loaded 3,490lb (Mk. I). Empty 2,743lb. Loaded 3,609lb (Mk. II)
Performance	Max speed 207mph at 14,000ft (Mk. I); 223mph at 16,400ft (Mk. II). Initial rate of climb 2,380ft/min (Mk. I); 3,200ft/min (Mk. II). Range 305 miles (Mk. I); 260 miles (Mk. II). Service ceiling 28,000ft (Mk. I); 29,500ft (Mk. II)
Armament	Two .303in Vickers machine guns (both Marks)

Mk. I equipped three squadrons, Mk. II equipped five squadrons rapid increase in rate of climb. Suddenly the leader, Sqd Ldr H H Down AFC, peeled off and with an ear-searing whine the eight Furies followed him down in a steep dive on a band of imaginary pirates in the middle of the aerodrome. Right down to ground level they came, then up into a steep climb, one after the other, and then a wing-over and down again, strafing the pirates into submission. Truly, No. 25 Squadron had administered the Fury's *coup de grâce* at Hendon.

After this glorious display, the Fury's day was all but over. Nos. 1 and 43 were in evidence in 1938 and Fury IIs equipped one or two other squadrons (Nos. 41, 73 and 87) but by 1938 the impetus had changed to the Hawker Hurricane and Supermarine Spitfire, then entering service, and from that time on the Hawker Fury began to join the dinosaur as a has been.

In fact, No. 25 Squadron's period of glory was brief: four months after its climactic arrival over Hendon, the squadron was stripped of its Furies and began a period of interim equipment, first with two-seat Hawker Demons and then with Gloster Gladiators, each type lasting only a few months.

Major change in Role

When No. 25 Squadron was first formed in September 1915 it was equipped with a variety of two-seaters and this multi-seat tradition was recreated when the squadron received the three-seat fighter version of the Bristol Blenheim Mk. I bomber.

Until this point, night-fighting appeared to have received scant attention at Air Ministry. On the ground, the evidence was everywhere of a nation preparing to be at the receiving end of air attack. Realistic anti-aircraft defences were openly discussed, citizens flocked to join the Air Raid Precautions (ARP) and be trained as emergency wardens and fire-fighters. Air Ministry knew, from night air exercises with Hurricane and Gladiator squadrons, that these single-seaters stood only an infinitesimal chance of finding what amounted to a straw-coloured needle in a haystack. A multi-seat fighter would at least provide one extra pair of searching eyes, a long endurance, and with a third man, the wireless operator, a better chance of getting down safely when the weather worsened on the ground.

Fortunately, not everything was quite what it appeared to be, for in the summer of 1937, with Furies disporting themselves overhead at Hendon, experiments, unique in the world, were taking place some 75 miles further east at Bawdsey Manor on the Suffolk coast.

Here Dr E G Bowen almost single-handedly was developing the world's first experimental airborne radar transmitter/receiver. When the proposal had first been mooted by Robert Watson Watt, the 'father' of British radars, in 1935, the task had looked near impossible as the equipment alone would weigh several tons.

In the following year, by producing a hybrid experiment where the transmitter was kept on the ground and the receiver carried on board a Handley Page Heyford, the concept of airborne radar was established. By September 1937, in the Home Defence exercises, the Bawdsey Avro Anson operating out of Martlesham Heath, with a civilian crew, received radar returns in the air from both ships and aircraft. Now, all became potentially possible and considerable competition would ensue as to where priorities should lie. The night-fighter requirement came first and development of suitable airborne interception (AI) radar would take a further three years before it became effective. However, at that time no one else in the world was even trying.

Meanwhile, as recounted, No. 25 Squadron received its Blenheims which were armed with a battery of four fixed under-fuselage .303in Browning machine guns. In the amidships turret, the gunner retained his single Vickers K gun. So set up, the three-man Blenheim fighter crews learned their new trade completely unaware of the highly secret electronic future.

Throughout the remaining months of peace, the squadron was working up operationally under the leadership of Sqn Ldr J R Hallings-Pott. On 22 August, the squadron moved to what would, within less than a fortnight, prove to be its first war station, North-olt. The first scramble came on 4 September, the day after Britain declared war on Germany, and proved to be a false alarm. On 15 September, the squadron moved to Filton, just north of Bristol, to provide fighter cover for the convoy of ships coming out of Avon-mouth and Cardiff, taking the British Expeditionary Force to France. Three weeks later the squadron moved back to Northolt.

In November 1939, history was made when the first AI radar-equipped Blenheims arrived on the squadron. At this stage, and for many months to come, the new equipment meant little. The range of an enemy it could give within limits; a bearing on the enemy it could not. Techniques, training and tactics all had still to be developed and with the early equipment, experiment rather than results was all that could reasonably be

expected. One flight, radar-equipped, was detached to Martlesham Heath for experimental night patrols over the North Sea.

The rest of the squadron explored the long-range possibilities of the Blenheim by making a fighter sortie against Bor-kum on 26 September. Unfortunately, they were unable to find the island. Two days later, in the company of No. 600 Squadron, they returned to shoot up the base and ships.

From January 1940, the squadron, operating out of North Weald, Kent, was engaged in convoy escort patrols and in continuing to work up with the new AI. Following the German invasion of the Low Countries and France on 10 May, No. 25 Squadron became involved in the evacuation from Dunkirk, flying daylight patrols along the Dutch and Belgian coasts and escorting damaged ships back to England.

In May, three single-seat Westland Whirlwind fighters arrived on the squadron, but two months later were taken away again while the squadron concentrated on its main task of practising interception following night 'scrambles'. Only on 3 September, at the height of the Battle of Britain, was No. 25 in action, and the first outcome was disastrous when two of the squadron's Blenheims were shot down by Hurricanes. The following night, fortunes improved when P/O Robe attacked three German bombers and damaged one. On the same night, P/O Herrick destroyed a Dornier Do 17 and a Heinkel He 111. Morale was restored and on the 13th, P/O Herrick scored again, bringing down another Heinkel He 111 and winning a DFC.

AI Mk.IV and Better Fighters

In September 1940, No. 25 Squadron was about to embark on a completely new dimension of effectiveness with the arrival of six of the new Bristol Beaufighters equipped with the first effective AI Mk. IV radar and four nose-mounted 20mm Hispano cannon, and after the fiftieth Beaufighter off the production line, six wing-mounted Browning machine guns as well.

With AI Mk. IV, for the first time, it became possible, after the co-ordination between pilot and radar observer had been learned, a frustrating and difficult business, to locate and then track high flying enemy bombers. There were still major problems to solve. For consistent results, a ground controlled interception (GCI) system based on a sufficiency of ground radar stations was still needed to put the fighter into proximity with the bomber. The second problem concerned the final moments of the interception

A Vic Three of No. 25 Squadron Fury Is turns low over the hangars at Hawkinge, displaying the squadron's distinctive black and white bar insignia. (*via J D R Rawlings*)

Wg Cdr C A Stevens, No. 25's commanding officer, shows off a Fury II in 1937. (*Wg Cdr E A Shipman via AST*)

itself when the radar returns from the enemy aircraft would disappear from the observer's screen before the target could be confirmed visually as being an enemy aircraft. This was necessary when en route British aircraft, which had defective identification friend or foe (IFF) transponders, or indeed had not switched them on, were encountered.

The first operational patrol with the Beaufighter was flown on 10 October; meanwhile Blenheim fighter operations continued. On 15 November, a Blenheim made a successful interception and the squadron was moved to Wittering, Northamptonshire, handing over its Beaufighters to No. 604 Squadron at Middle Wallop in Hampshire. Once established in the Midlands, No. 25 was re-equipped fully with Beaufighters achieving a probable kill on 15 January 1941. In April, two enemy aircraft fell to the devastating armament (the most powerful in the world at the time) of the

squadron's soot-black Beaufighters. Eleven more enemy aircraft followed in May, seven in June and one in July. By now Germany had become embroiled in its invasion of the Soviet Union and 'trade' tailed off. But now No. 25 Squadron had acquired both proficiency and confidence.

With the coming of autumn, the squadron added daylight coastal sweeps and convoy escorts to its stock-in-trade. The only shooting that took place was on 25 January 1942, when the Beaufighters were fired on by the convoy they were escorting. That month the squadron left Wittering after more than a year's residence and moved, first to Ballyhalbert, Ulster, and then in May to Church Fenton, Yorkshire, another prewar base with comfortable accommodation like Wittering. Also in January, F/O Picknett found and destroyed a Dornier

Fitters check out Beaufighter Mk.IF X7876 at dusk, at Church Fenton, Yorkshire, in 1942 before the night's work begins. (*via J D R Rawlings*)

Do 217 after what he described as 'a lively shooting match'. The summer of 1942 proved a busy one with an average of five victories a month.

Now an aircraft came along which promised more than the Beaufighter when in September the first of No. 25

Bristol Blenheim IF

Three-seat long-range patrol and night-fighter. All-metal stressed-skin construction

Manufacturers	The Bristol Aeroplane Co. Subcontracted to A V Roe & Rootes Securities
Powerplant	Two 840hp Bristol Mercury VIII
Dimensions	Span 56ft 4in. Length 39ft 9in. Height 9ft 10in. Wing area 469sq ft
Weights	Empty 8,839lb. Loaded 12,500lb
Performance	Max speed 260mph. Cruising speed 200mph. Initial rate of climb 1,540ft/min. Range 920 miles. Service ceiling 25,500ft
Armament	Five .303in Browning fixed machine guns firing forward; one .303in Vickers K gun in dorsal turret

Bristol Beaufighter IF

Two-seat night-fighter. All-metal stressed-skin construction

Manufacturers	The Bristol Aeroplane Co. Subcontracted Fairey Aviation
Powerplant	Two 1,400hp Bristol Hercules XI
Dimensions	Span 57ft 10in. Length 41ft 4in. Height 15ft 10in. Wing area 503sq ft
Weights	Empty 14,600lb. Loaded 21,600lb
Performance	Max speed 321mph. Initial rate of climb 1,351ft/min. Range 1,170 miles. Service ceiling 26,500ft
Armament	Four 20mm Hispano cannon, fuselage-mounted. Six .303in Browning machine guns wing-mounted
Radar	AI Mk. IV

Squadron's Mosquito F. Mk. IIs arrived armed with four 20mm cannon and four .303in Browning guns. The radar was similar to the Beaufighter's but the aircraft's performance was greatly enhanced. Working up with the new equipment began in October with the first operation following on 14 November, though the Beaufighters continued to predominate.

By New Year 1943, the squadron was equipped completely with Mosquitos, achieving its first victory when Flt Lt Singleton shot down a Dornier Do 217 later that month. Using Coltishall, Norfolk, as a forward base, squadron aircraft without AI radar began night intruder operations over northern France and the Low Countries looking for German aircraft flying over their own bases.

At first the air defence Mosquitos found they were losing many of the radar contacts made due to the Mosquito's unaccustomed turn of speed. This problem was soon overcome with experience. By April 1943, the squadron was claiming some railway locomotives destroyed as well as aircraft when Ranger sorties gradually became a main part of the squadron's activities. In June, a special C flight was formed and detached to Predannack on the Lizard peninsula, Cornwall. From there northwest France lay within easy reach and importantly the marauding Junkers Ju 88s, which had been achieving success against vulnerable RAF Coastal Command aircraft in the Bay of Biscay, were no longer safe.

In June 1943, C flight was re-equipped with the Mosquito F.B. Mk. VI for the intruder work over France, while the rest of the squadron retained F. Mk. IIs. Towards the autumn the squadron concentrated on bomber support work over Germany, operating in the RAF bomber streams and patrolling German night-fighter aerodromes to keep the enemy

A 'scanner nose' Mosquito N.F. Mk. 30, part of No. 25 Squadron's complement in the closing stages of the Second World War. (*J D R Rawlings*)

on the ground. These operations were reduced with the winter, the squadron being withdrawn for re-equipment with the Mosquito N.F. Mk. XVII with nose-mounted American centimetric AI radar, returning to operational status in February 1944. With the US SCR 720 AI, known in the RAF as AI Mk. X, No. 25 Squadron was able to play a key defence role in the Luftwaffe's 'Little Blitz', a German bomber offensive which was mounted against the United Kingdom initially with some 500 aircraft in the period 21 January to mid-May 1944. Altogether there were some two dozen raids in which the Luftwaffe lost over 300 aircraft.

The Luftwaffe, decidedly out of practice in operations against heavily defended United Kingdom airspace, never-

At West Malling, Kent, in 1949, No. 25 Squadron operates the ultimate service version of the Mosquito night-fighter, the N.F. Mk.36. Just visible on the fin is a tentative reintroduction of the squadron's black bars insignia. (*J D R Rawlings*)

De Havilland Mosquito

Two-seat night-fighter. Wooden construction

Manufacturers	The de Havilland Aircaft Co
Powerplant	Two 1,300hp Rolls-Royce Merlin 21, 22 or 23 (Mks. II, VI, XVII). Two 1,710hp Rolls-Royce Merlin 72 or 76 (Mk. N.F. 30). Two 1,700hp Rolls-Royce Merlin 113/114 (Mk. N.F. 36)
Dimensions	Span 54ft 2in. Length 40ft 6in (Mks. II, VI), 41ft 4in (Mks. XVII, 30, 36). Height 15ft 3in. Wing area 435sq ft
Weights	Empty 14,300lb (Mks. II, VI), 15,400lb (Mks. XVII, 30, 36). Loaded 22,300lb (Mks. II, VI), 21,600lb (Mks. XVII, 30, 36)
Performance	Max speeds 380mph (II), 358mph (VI), 378mph (XVII), 424mph (30), 404mph (36). Rate of climb 2,100ft/min (II, VI); 2,250ft/min (30, 36). Range 1,200 miles (II, VI); 1,000 miles (XVII, 30, 36). Service ceiling 36,000ft (all Mks.)
Armament	Four 20mm Hispano cannon in fuselage; Mks. II & VI also four .303in Browning machine guns in wings; Mk. VI also two 500lb bombs under wings
Radar	AI Mk. IV (II); AI Mk. X (XVII, 30, 36)

theless posed a difficult interception problem. To intercept the fast fighter-bombers, Messerschmitt Me 410s, required every bit of the Mosquito's speed, augmented by nitrous oxide injection when available, to prevent the quarry escaping over the French coast and safety, because the Mosquito, with its advanced AI, was not allowed to follow. British centimetric AI, the world's first, had started to come into use in RAF Fighter Command from the spring of 1942. Now, following the Brit-

The year 1951 sees the squadron equipped with jets, the Vampire N.F. Mk.10. No. 25 Squadron's identity is preserved in the black bars on the aircraft's tail booms. (*via J D R Rawlings*)

De Havilland Vampire N.F. Mk. 10
Two-seat night-fighter. All-metal stressed-skin construction

Manufacturer	The de Havilland Aircraft Co
Powerplant	One 3,350lb st de Havilland Goblin 3
Dimensions	Span 38ft. Length 34ft 7in. Height 6ft 7in. Wing area 262sq ft
Weights	Empty 6,984lb. Loaded 13,100lb
Performance	Max speed 538mph at 20,000ft. Initial rate of climb 4,500ft/min. Max range 1,220 miles
Armament	Four 20mm Hispano cannon fuselage-mounted
Radar	AI Mk. 10

ish gift of the cavity magnetron to the USA in August 1940, excellent United States production radar equipment was flooding back into Britain, and one good home for it was the Mosquito N.F. Mk. XVII. British centimetric radar-equipped Mosquitos were N.F. Mk. XIIs and both were greatly superior developments of the F. Mk. II, the original metric radar AI Mk. IV version.

The Luftwaffe's return to the bombardment of the United Kingdom in early 1944 had brought with it many new stratagems, some of the Germans' own devising and some borrowed from the RAF. The use of metallised foil (code-named Window) to deceive enemy radars had been kept secret by both sides for fear of retaliation. But after the RAF's use of Window from August the previous year in raids on Hamburg, the Luftwaffe could use its own version, Düppel. British GCI stations were duly swamped and confused but the crews of the centi-

metric radar equipped Mosquito night-fighters were not deceived, because the Düppel had not been cut to lengths to mimic its wavelength. Indeed the Germans had no idea that centimetric AI radar was being used against them. Based again at Coltishall, No. 25 Squadron shot down three raiders in February and nine in March; three of these had been despatched in a single sortie by Flt Lt Singleton and F/O Haslam on 20 February.

In April, the squadron was at the defending end of the intruder business as German night-fighters attacked RAF bombers returning to their bases in East

Re-equipment with Meteor N.F. Mk.12s followed in 1954. No. 25 Squadron's black bar insignia now has a silver infill, recalling the days of the silver-doped biplanes. (*via J D R Rawlings*)

Anglia. As summer approached, and with it D Day, the squadron's work was divided between these anti-intruder operations and its own intruding sorties which increased in importance with Allied troops back again in France.

Now commanding No. 25 Squadron was Wg Cdr C M Wight-Boycott, a leading night-fighter exponent of the Second World War and a pioneer of the concept of the all-weather fighter. The centimetric radar equipped night-fighter was the one defender, vectored by GCI, standing a chance of intercepting fast low-flying enemy fighter-bombers which were taking advantage of bad weather to make surprise attacks by day. The new commanding officer also led his squadron into new techniques for combating the V1 flying bombs crossing southern England by day and night. The V1's speed posed problems even for the Mosquito. As the launching sites in France were overrun, the squadron's task became simpler as the Luftwaffe began to air-launch V1s from Junkers

Gloster Meteor (Night-fighters)
Two-seat night-fighter. All-metal stressed-skin construction

Manufacturers	Sir W G Armstrong Whitworth Aircraft Ltd
Powerplant	Two 3,800lb st Rolls-Royce Derwent 9
Dimensions	Span 43ft. Length 48ft 6in (N.F.12), 51ft 4in (N.F. 14). Height 13ft 10in. Wing area 374 sq ft
Weights	Empty 12,500lb. Loaded 17,000lb (N.F. 12)
Performance	Max speed 541mph (N.F. 12), 576mph (N.F. 14) at 30,000ft. Initial rate of climb 4,800ft/min (N.F. 12), 5,797ft/min (N.F. 14). Range 950 miles
Armament	Four 20mm Hispano cannon wing-mounted
Radar	APS Mk. 21

Ju 188s and Heinkel He 111s. The technique was to deal with the carriers before they had an opportunity to launch.

To get nearer the source of the problem, the squadron moved to Castle Camps, Essex, at the end of October 1944, under the command of Wg Cdr L J C Mitchell DSO, and stayed there until the war in Europe ended the following May. In November 1944, the squadron was re-equipped with the Mosquito N.F. Mk. 30. These had the same British or American centimetric AI but had the more powerful 1,650hp Rolls-Royce Merlin 72 or 76, the latter with a cabin supercharger. With the new equipment the squadron continued to score, as the enemy's night-fighter defences began to run down due to lack of fuel and trained crews, while the Mosquitos continued to operate in the British bomber stream.

Peace brought relief but also uncertainty. Wartime bases such as Castle Camps were progressively closed down. For some time the squadron moved round East Anglia as its repeated new and temporary homes reverted to farm use. Meanwhile, on the squadron itself, personnel awaited their turn to come up on the points ladder of demobilisation. It was a time when, inevitably, a sense of purpose was lacking, the threat having vanished. However, in retrospect, surprisingly quickly by September 1946, the squadron had found a proper peacetime home at West Malling, Kent, along with two other Mosquito squadrons, Nos. 29 and 85. Now the ulitmate operational Mosquito night-fighter, the Mk. 36, with the even more powerful Merlin 113s, became standard equip-

Readiness at dusk. A No. 25 Squadron crew goes aboard a Meteor N.F. Mk.14 in the summer air exercises of 1955. (*via J D R Rawlings*)

ment, and the squadron settled down to a very different routine.

The Move to Jets

At last even the Mosquito began to look outdated, with the rest of Fighter Command jet-equipped and jet-bombers on the drawing board, whereas the late enemy had had them in the air. The race was now on to produce radar-equipped jet-fighters. Within the Hawker Siddeley Group, Gloster had handed over to Armstrong Whitworth the task of producing a night-fighter Meteor; meanwhile, de Havilland had produced its own private venture side-by-side seating Vampire with AI Mk. X radar in the nose. The Ministry of Supply ordered a small batch of the de Havilland product, the Vampire N.F. Mk. 10. In July 1951, the first of them came into service with No. 25 Squadron at West Malling.

Compared with the standard single-seat Vampire, the performance of the two-crew version complete with radar

Against an East Anglian coastal background, a section of Javelin FAW. 9s, led by Wg Cdr J H Walton, show off their armament of two Firestreak air-to-air missiles, pylon-mounted under the big delta wing. (*via J D R Rawlings*)

inevitably suffered. However, the new type enabled jet techniques to be developed for night-fighting; although the cockpit was rather cramped for both pilot and radar operator, the squadron quickly reached operational pitch. As No. 85 Squadron, also at West Malling, received Gloster Meteor N.F. Mk. 11s, a rivalry built up, each to be better than the other. As time went on the Meteor, being the bigger airframe, received more modern radar (the US APS21) in the Mk. 12 and Mk. 14.

In January 1954, No. 25 Squadron began to receive some Meteor Mk. 12s. By April the Vampire had gone and Meteor Mk. 14s were arriving as well. The later aircraft had the same radar as the Mk. 12 but a far better cockpit arrangement with a clear view canopy. With a mix of both types, No. 25 Squadron was soon showing its operational prowess in a succession of exercises. The RAF was now at the beginning of a long series of cuts in defence spending. Among the squadrons affected was No. 25, as West Malling became due for closure. On 30 September 1957, the squadron pulled out of its happy home, set in the Kentish hopfields, and went to RAF Tangmere, Sussex, another happy station with a great history.

There followed a change of role, as

Gloster Javelin FAW. 7/FAW. 9
Two-seat all-weather fighter. All-metal stressed-skin construction

Manufacturers	Gloster Aircraft Co. Subcontractor Armstrong Whitworth
Powerplant	Two 11,000lb st Bristol Siddeley Sapphire Sa. 7 (FAW. 7). Two 12,200lb Bristol Siddeley Sapphire Sa. 7R (FAW. 9)
Dimensions	Span 52ft. Length 56ft 9in. Height 16ft 3in. Wing area 928sq ft
Weight	Loaded 35,690lb (Mk. 7), 38,100lb (Mk. 9)
Performance	Max speed 616kt at sea level (FAW. 7), 610kt at sea level (FAW. 9). Service ceiling 52,800ft
Armament	Two 30mm Aden cannon wing-mounted; four Firestreak AAMs on wing pylons
Radar	AI Mk. 17

No. 25 Squadron became the one all-weather fighter squadron on a station shared with two day-fighter squadrons, Nos. 1 and 34, with Hawker Hunter F.5s. A few months later, the 1958 Defence White Paper announced the disbandment of the squadron with a cut-off date of 1 July. For the first time since the 1920s there was no No. 25 Squadron.

In a strange piece of bureaucracy, the Ministry of Defence, having perhaps realised its mistake, on the very same day, resurrected No. 25 Squadron by renumbering No. 153 Squadron at RAF Waterbeach, just north of Cambridge. Inevitably, it took time before No. 25 Squadron's mantle and considerable traditions felt comfortable on No. 153, a Meteor night-fighter squadron.

However, in the following year, 1959, there came a big stride forward when No. 25 Squadron was re-equipped with the massive Gloster Javelin all-weather fighter. Specifically, the variant was the Javelin FAW 7 which carried AI Mk. 17 radar and was equipped with four de Havilland Firestreak air-to-air missiles with infra-red guidance, the first to be used by the squadron.

This version of the Javelin had two Armstrong Siddeley Sapphire 7 engines each of 11,000lb static thrust, giving it a maximum speed of 620mph at 40,000ft. At last here was a round-the-clock interceptor able to take on fast jet-bombers. The squadron was soon operational and taking its share of quick reaction alert (QRA) duties.

At the end of 1959, the FAW 9 version of the Javelin began to augment the Mk. 7s. The new mark had reheat on the Sapphires, which further augmented the performance. For the next two years the squadron maintained its operational role at Waterbeach before that base too was closed and given over to the Royal Engineers. Now No. 25 Squadron was

Still in the business of round-the-clock interception, No. 25 Squadron at Wildenrath in the 1970s with a detachment of Bloodhound Mk. 2 surface-to-air missiles. (*via J D R Rawlings*)

directed north of the Scottish border for the first time since it had been formed at Montrose in 1915. At Leuchars, on the coast of Fife, No. 25 was in company

---

Bloodhound Mk. 2 (British Aerospace)
Surface-to-air missile

Powerplant	Two Bristol Thor ramjets; four jettisonable solid propellant boosters
Guidance	Semi-active homing. A receiver in the nose of Bloodhound picks up the reflected radiation from the target which has itself been illuminated by continuous wave ground radar
Warhead	High explosive with proximity fusing
Dimensions	Length with boosters 27ft 9in. Max body diameter 1ft 9½in. Wing span 9ft 3½in
Range	In excess of 50 miles

---

with two squadrons of English Electric Lightnings. And here it continued to oppose a possible Soviet threat from across the North Sea for another year. The time of phasing out the Javelin was approaching, the squadron flying its last one on 30 November 1962, the day the squadron itself was disbanded as an all-weather fighter squadron.

Surface-to-Air Missiles

There are those who will say, 'that was the end of No. 25 Squadron' but, of course, they would be wrong. On 1 October 1963, the squadron was born again and in the same role, in air defence of the United Kingdom. This time the place was North Coates, Lincolnshire, the surface-to-air (SAM) headquarters of the RAF. Since then the squadron's equipment has comprised Bloodhound ground-to-air missiles developed initially by the Bristol Aeroplane Company in the years following the Second World War. Having learned the missile business, in 1970 there came another new step. Deployed since 1920 in the direct defence of the United Kingdom, much of the time in the protection of its cities, an operational detachment was now sent to Laarbruch in Federal Germany for the defence of the RAF's 2nd Tactical Air Force based there. Unlike a squadron of aircraft, once a SAM unit fires its rounds they cannot be recalled and the force is spent.

Unsurprisingly, little is heard of the Bloodhound squadrons whose presence is observed as a battery of missiles on the edge of an RAF base, pointing in the direction of a likely threat. However, the

---

missile units co-operate in all the air exercises and tactical evaluations, keeping them up to pitch.

In August 1970, the squadron moved its headquarters to RAF Brüggen to consolidate its defensive role there, putting another detachment at RAF Wildenwrath in February 1971. During the 1980s, No. 25 Squadron was withdrawn to the United Kingdom to take up its traditional role of home defence, using updated Bloodhound Mk. 2 missiles. Its

---

No. 25 Squadron's pre-Fury History

Formed at Montrose on 25 September 1915, No. 25 Squadron RFC served with a miscellany of two-seaters before going to France where it flew F.E.2bs on so-called fighter-reconnaissance until early 1917 when it transferred to bomber duties. That July it was re-equipped with the Airco D.H.4 and flew long-range, high-altitude bombing raids until the war ended. Disbanded after the war, it was re-formed as an interceptor fighter squadron on 20 April 1920, and flew a succession of single-seaters (Sopwith Snipes – which it took to Constantinople for a year, Gloster Grebes and Armstrong Whitworth Siskins) from its home base at Hawkinge before the Hawker Furies arrived.

---

headquarters and one flight are at RAF Wyton with detachments at Barkston Heath and Wattisham.

Quietly, unobtrusively, No. 25 Squadron, whose Fury IIs took Hendon by storm on a summer's day more than half a century ago, is still fulfilling the role in which it has served Britain for nearly seventy years, the defence of this realm.   □

Since this article was prepared the Ministry of Defence (RAF) has announced that No. 25 Squadron is re-forming at mid-year with Tornado F.3s.

By 1990, the squadron will be fully operational in its traditional role of round-the-clock long-range interception, using the most advanced equipment available to search out and destroy intruders in war, and backed by in-flight refuelling as already described in issue No 1.

Building Britain's Best Heavy Bomber

The Editor provides a background commentary on the Lancaster and on some of the key men who made it possible, as an introduction to a unique series of early production photographs.

Who was fooling whom? Roy Chadwick holds the wind-tunnel model of the Lancaster with the wing set high and a boxcar body – in other words, an Avro York. To the censors of the day, the aircraft had to be a Lancaster. However, on 14 July 1942, Col J J Llewellin, then Minister of Aircraft Production, in reply to a parliamentary question, disclosed, 'We have tried conversion of one of our bombers into a transport plane, and the House will be glad to hear that it is flying today'.

By 17 December 1942, the forceful Robert Perkins, MP for Stroud and a former airline pilot, was pleading in the House of Commons for twenty Yorks and ten Sunderlands to be made available to BOAC so that the Corporation would be better able to operate alongside Pan American. The British public was not let into the secret, even then, that the York's parentage was out of Lancaster by Avro. It could have been a Halifax conversion or even a Stirling, though this had been eliminated by the replies to other questions. Creating further confusion, Capt Harold Balfour (later Lord Balfour of Inchrye), Britain's long-serving and highly capable Under Secretary of State for Air, said of the York in a December 1942 debate, 'As soon as this machine has completed its trials it will be put into production and deliveries in considerable numbers will

take place during next year'. First pictures of the York were released only eleven months later.

In July 1942, Col Llewellin had confirmed in the House, referring to the production of transport aeroplanes in the USA, that it was better for Britain to order 'ready made' rather than to jig and tool in the United Kingdom.

RAF Transport Command was not formed until 1943 and the York became the mainstay of that Command in the early postwar years, with seven squadrons taking part in the Berlin Airlift.

Few Yorks were in service during the war. The first prototype flew on 5 July 1942, only five months from go-ahead, and three more prototypes followed, the most famous wartime York being Mr Churchill's *Ascalon*. There were others, but though the York was certainly the best transport aircraft produced in Britain up to that time, its short-term future, in which it would have shone, had to be sacrificed to the overriding need to produce the maximum number of Lancasters. By the time the York was available in quantity it had been technically outclassed in terms of both speed and payload/range. However, Yorks would continue to operate unspectacularly and profitably for many years. As the saying went, 'Nobody ever heard of an operator running Yorks who went bust!'

To rediscover the pictures published here forty-seven years after they were taken is to re-enter a new, old world. Taken by a news feature cameraman Ruben Saidman for the popular British weekly *Illustrated*, they were originally published on 15 August 1942.

In the strict sense, these pictures do not tell us much about how the Lancaster was put together, for the photographer did not, for instance, point his camera at the key major assembly – on the Lancaster the wing centre section married to the roof section of the 33ft long bomb bay, together with the main undercarriage, forming the structural heart of the design. However, the cameraman and fine grain film captured much that is of interest. After all, millions of words exist already on the Lancaster design per se and still more are believed to be on the way.

First, the period February/March 1942, as we know now but did not realise at the time, was significant. On 22 February Air Marshal Sir Arthur Harris had been appointed as Air Officer Commanding in Chief of RAF Bomber Command. He inherited a mixed bag of bombers, 221 Wellington, 112 Hampden, 54 Whitley, 29 Stirling, 29 Halifax and just four Lancasters. At first glance, after almost thirty months of war, this looked meagre. Germany, however, had its problems too. Hitler's Blitzkrieg tactics against the Soviet Union, the summer before, having failed, the German economy was long overdue for putting onto a proper long-term war footing. Within days of the Harris appointment, in Germany Albert Speer had been made armaments minister, there being no connection between the two appointments it should be added. Coincidentally, the first officers, of what would become the vanguard of the US 8th Air Force, were arriving in Britain.

With the USA coming into the war just over three months before, the scale, tempo and geographic extent of operations was to become quite different. But the capability of the Allies to strike back massively, rather than selectively, would still take years rather than months. In the overall armoury, the singular effectiveness of the Lancaster stands high.

With these pictures, taken in March 1942 and with publication withheld by the censor until that August, we have a privileged viewpoint virtually at the

beginning of the Lancaster quantity production story. Avro, for long a member company of the Hawker Siddeley combine, had demonstrated a great release of talent under the forceful leadership of R H (later Sir Roy) Dobson, its managing director, known universally in the industry even then as 'Dobbie'. Roy Chadwick, by then design director, had been with the company for over twenty years and had been responsible for the design of many aeroplanes including the unspectacular and elderly, but still thoroughly excellent Anson, which then formed the backbone for navigator training in the Commonwealth Air Training Plan; and, universally, for every fetching and carrying task within its capabilities. A third member of the team was Vickers-trained S D Davies who had later been project engineer on the Hurricane with the Hawker company before joining Avro in 1938.

Early in the war, Stuart Davies was moved over to the works side of the company to organise the experimental

In March 1942, the Avro team responsible for the Lancaster meets, for the benefit of the news cameraman and, as it happens, posterity, under the chairmanship of Roy Dobson (later Sir Roy), seated centre, the managing director. On his right is the design director Roy Chadwick. On Dobson's left is the works director C E Fielding. Also represented in the picture are the other departments in the management team – production, inspection, contracts, subcontracts, employment, the works engineers and finance. When this picture was first published, the caption began, 'The Lancaster takes form – on paper and in a little wooden model on the managing director's desk'. Thus the Avro York became one of the war's less well kept secrets.
(*All photographs by courtesy of Popperfoto*)

department as a self-contained manufacturing unit. So it was that an Avro Manchester was taken apart on the shop floor, with Roy Dobson directing and, so it is said, with Roy Chadwick hacksaw in hand, and a back-up team of draughtsmen with their drawing boards, in the shop with the Manchester then in pieces. However apocryphal parts of this delightful story may be, it is a fact that the considerable thought given to Lan-

caster detail design, ab initio, was to bring the aircraft into production and effective operational service with far less anguish than Short's Stirling or Handley Page's Halifax, which had to experience the learning curve which Avro had already gone through with the Manchester.

In his then role, Stuart Davies does

In this posed version of reality, the design director Roy Chadwick makes a random descent on a draughtsman (a not quickly forgotten experience) to check his work while the chief draughtsman, Jimmy Turner, centre, looks on.

not appear in these essentially production pictures. But without his prototype, the Lancaster could not have been brought into production in the timescale it was. And had that not happened, the Lancaster might well not have appeared at all. In fairness, once the prototype had flown, the Air Ministry became converted to the need for the aircraft but was then doubtful of Avro's ability to make it quickly in quantity.

At this point enter C E Fielding, the newly appointed works director seen standing on Dobson's left in the conference picture. Teddy Fielding had joined Roy Dobson at A V Roe and Company in 1918. Both had subsequently risen through the company on the works management side of the business. With Dobson and then Fielding charged successively with the task of getting aircraft built on time and to price, the realities of the importance of the production versus design debate had been made apparent daily to Roy Chadwick, the chief designer.

There was healthy respect by each side for the other, and at the most critical period in Britain's history for almost a thousand years, the balance with the Lancaster came out right. Not only did it prove to be the best weight lifter by far and the highest flyer of the

York in the wind tunnel, not yet with a central fin.

heavies, it was also marginally faster. The pictures give no clue to the quality and resourcefulness of the production planning where every stage-completion

In the tracing department, the master drawings coming out of the design office are traced in ink ready for reproduction in quantity.

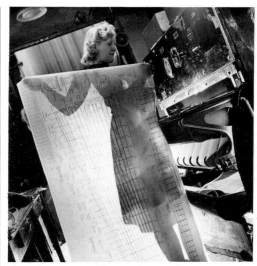

The print room, with a tracing ready to go into the machine for reproduction in quantity, for use throughout the Lancaster production group. Modifications to aircraft, calling for amendments to drawings, keep the print room one of the busiest places in the organisation.

A corner of the 'steel stores'. Half a million manufacturing operations later, a Lancaster will emerge. Engines, gun turrets, airscrews, pumps and the myriad items bought out or Air Ministry supplied have their own production schedules and are not included.

Above: Fuel cells before testing and inspection, and self-sealing proofing.

Below: Loaded from two sides, the 6,500 ton hydraulic press is a key production tool. From aluminium alloy sheet it forms a huge range of pressed items, needed in the manufacture of wing ribs, fuselage formers and many other parts.

had to be met before production was allowed to proceed. For example, all riveting operations had to be completed before the five major sections of fuselage could join the final assembly line. Each section had been prepared beforehand to a high standard with operational systems, including turrets, in position; aircraft systems with hydraulics, electrics and instruments were all installed. Only then could these major sections join the production line. The system was applied throughout the thirty-six major sections which made up the Lancaster.

Initially, Avro's own assessment of a peak rate for Lancaster production was eighty aircraft a month. In fact, following the creation of a Lancaster production group, of which Teddy Fielding was chairman, Lancaster production in August 1944 would reach 293 aircraft in the month, together with more in the form of spares.

All told, 7,366 Lancasters were built. We now know that Avro itself made 3,670 of them at a peak rate of 150 a month at Manchester and Yeadon. From Armstrong Whitworth at Coventry came

In the view room (inspection), each item of detail receives a part number and, after viewing, a works inspection stamp.

These light alloy castings, after further machining, will be main undercarriage beams.

1,329. Vickers-Armstrongs in shadow factories at Castle Bromwich, alongside Spitfires, and at Chester, beside Wellingtons, built 535. Metropolitan-Vickers at Manchester made 1,080 and the Austin Motor Company in Birmingham 330. Finally, Victory Aircraft in Canada manufactured 422.

Altogether, scores of thousands of men and women were to become involved in the factories of the Lancaster group supported directly by the workforces of a further 600 direct subcontractors and suppliers. In 1942, with volume production beginning to get under way, Peter Masefield (now Sir Peter), then technical editor of the *Aeroplane*, reported that 10 per cent of the workforce were skilled tradesmen and about 40 per cent were women, with an average working week of 61 hours.

There is little more to say to enhance the value of the pictures, these many years later, other than to give a diary of the Lancaster's critical beginnings.

Diary of Difficult Days

It is difficult now, virtually half a century later, to imagine the pressure on those making the decisions in the summer of 1940. In aircraft procurement, Lord Beaverbrook had been made Minister of Aircraft Production, a new

Making up the early stages of a web for a wing spar.

Riveting up a rib in the wing main structure between the spars.

Riveting wing skins.

A corner of the machine shop.

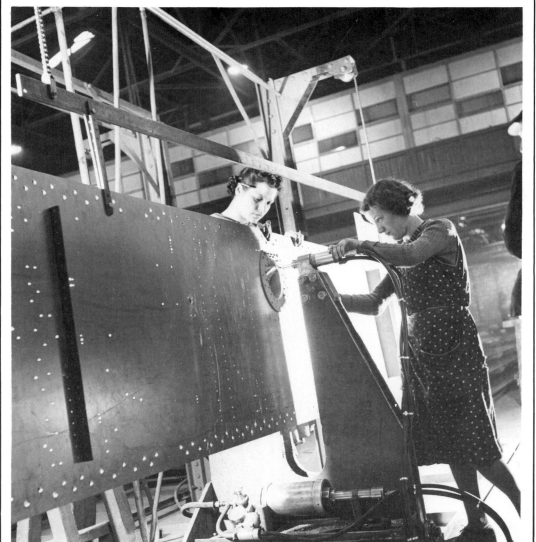

The D2 section for the roof of the bomb bay in its vertical building jig; together with the wing centre section, to which it will be attached, this will form the structural heart of the complete aircraft.

Avro works inspector, on station, in the bomb aimer's position.

The final assembly line at Woodford, Cheshire, the first of many.

appointment, in May, by the incoming Prime Minister Winston Churchill. Immediately, Beaverbrook set out to galvanise the industry, to solve the problems of the day, and to get more aircraft into the squadrons immediately. Any decision designed to produce results in a month's time or longer was disallowed. A supreme effort for the time being was to be concentrated totally on the types already in full production, namely the Hurricane, Spitfire, Wellington, Whitley and Blenheim.

Despite such directions, vital forward decisions *were* taken, none more important than that by Roy Dobson and Roy Chadwick to come to terms with reality.

Man-handling a main wheel. Only at about that time were bomber airfields becoming equipped with runways.

An outer wing receives its spray camouflage and roundel.

Below: The 33ft long bomb bay with bomb carriers in position.

Their twin-engined Avro Manchester bomber, to the difficult P13/36 specification of 1936, had been flying in prototype form since July 1939 but had crashed; a second prototype had flown at the end of May. Overall, the programme was in trouble, as was the Heinkel He 177 in Germany, at the same period, and for a related reason. In both cases the powerplant held the key. Both were unsuitable. In Britain there was sufficient resolution to cast the problem aside and start again. In Germany there was not.

The powerplant of the Manchester comprised two Rolls-Royce Vultures, each of a nominal 1,760hp. This engine, big for its day, resulted from taking two of the later twelve-cylinder liquid-cooled vee Kestrels (late 1920s technology) and setting them in an X arrangement, driving the airscrew through a common crankshaft via a reduction gear. The Vulture was new and undeveloped. Had it ever become reliable after the expenditure of much effort and time, it would still not have looked attractive. The big bombers now envisaged would need much more power.

For Avro the problem was serious and considerable. If its major product fell short of the mark, and it did, then unquestionably under 'the exigencies of war', a favourite phrase, the company

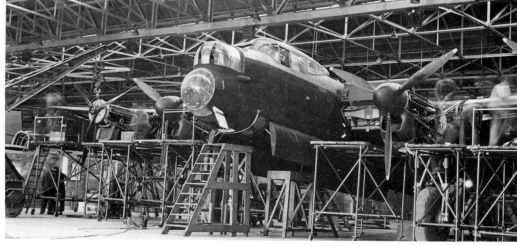

Nearing the head of the line. In final assembly at Woodford.

Completed Lancaster R5493 under tow from the flight sheds to the flight line for production flight testing. The aircraft marshaller is Walter Holland and the tractor driver Alf Beacon.

Cleared for flight. Avro's chief test pilot, Capt H A 'Sam' Brown, and his deputy, Bill Thorne, prepare to go aboard L7577.

Bill Thorne checks the flight engineer's instrument panel as 'Sam' Brown prepares for take-off.

would be ordered to build the Halifax. To hitch the Avro waggon to the star of the Rolls-Royce Merlin was easily devised, but less easy to get agreed and built. Fortunately, Merlin production no longer depended on the parent factory at Derby alone. There was a new Rolls-Royce managed shadow factory at Crewe and another at Glasgow almost ready. The Ford Motor Company in Britain was being asked to come in, too. It would do so, building a new factory at Manchester, equipped with purpose-built tooling for making Merlins. Ford in neutral USA was not so willing, but Packard was, gutting completely one of its car plants to start afresh on Merlin production. A Manchester with four Merlins, a nominal 5,000hp, would be an infinitely better proposition than the Manchester as existing, with a problematic 3,500hp.

As so often on other occasions, Air Marshal Sir Wilfrid Freeman, the member of the Air Council responsible for aircraft production in the period 1938-40, together with Air Marshal Tedder (later Lord Tedder), then holding the appointment of Director General of Research and Development, were receptive to the Avro proposals but imposed tough conditions. The prototype four-engined Manchester had to fly by the end of the year.

Snag clearance on No. 3 engine. Rolls-Royce Merlin XX of 1,280hp driving a de Havilland hydromatic, fully-feathering, constant-speed airscrew.

Chocks away as 'Sam' Brown gives the thumbs-up signal.

To get the revised four-Merlin prototype into the air quickly, standard Manchester undercarriage beams would be used on the same wing centre section, though the new version would have the potential to operate at much higher weights. Fortunately, Merlin XX engines were available as built-up power-plants, the first to come out of Rolls-Royce for the Bristol Beaufighter Mk. II. At that time the Ministry wisely provided insurance engine alternatives against the risk of enemy bombing of single sources, in this case of factories producing the Beaufighter's standard

A rare wartime opportunity. Flight test and flight shed crews gather for a posed photograph.

Lancaster operations

Midsummer 1940	Decision taken to build a prototype Lancaster
2 Jan 1941	Adapted Manchester flown with four Merlins. It was already named Lancaster
24/25 Feb 1941	Manchesters first used operationally. Target, German warships at Brest
Oct 1941	First production Lancaster flown. Manchester taken to pieces on the floor of the experimental department, resulting in new production drawings, new components and an entirely re-evaluated production aeroplane
24 Dec 1941	No. 44 Squadron RAF, the Rhodesia Squadron, received its first Lancasters for training
22 Feb 1942	Bomber Command's new AOC-in-C took over
Night of 3/4 Mar 1942	First operations by Lancasters – minelaying off northwest German coast
Night of 24/25 Mar 1942	First Lancaster lost on operations, from No. 44 Squadron, minelaying off Lorient
17 Apr 1942	12 Lancasters of Nos. 44 and 97 Squadrons penetrated deeply into southern Germany, at low level, in daylight to raid the MAN diesel engine manufacturing plant (engines for U-boats) at Augsburg. Seven shot down. Sqn Ldr J D Nettleton awarded the Victoria Cross
Jun 1942	Last operations flown by Manchesters
11 Jul 1942	44 Lancasters flew 850 miles in daylight to Danzig to attack the U-boat yards. 24 bombed, at dusk. Two lost
17 Oct 1942	94 Lancasters made a dusk attack on the Schneider works at Le Creusot and a transformer station nearby. One lost
24 Oct 1942	88 Lancasters attacked targets in Milan. Four lost

Further operations by Lancasters in daylight during 1942 involved small numbers being despatched to Essen in the Ruhr on at least four occasions, and to other German targets. These raids appear to have been early attempts to bomb through cloud using Gee, a task for which the equipment was proved to be unsuited.
□

An Air Transport Auxiliary (ATA) crew will deliver the completed aircraft either direct to a squadron or to a maintenance unit. The tail of the ferry Anson which has brought them is to the right of the picture.

Formal handover, informal setting. ATA accepts the Lancaster from an Avro works inspector, on this day probably on Home Guard duty after work. The RAF sergeant was attached to Avro at Woodford.

powerplant, the air-cooled Bristol Hercules. The Rolls-Royce power eggs were arranged to pick up on the aircraft's structure at only three points, with all the associated pipes and controls and electrical connections between airframe and engine being made at the airframe's fireproof bulkhead. For the inboard engines, some modification was required to make the slimmer Merlins occupy the wider nacelles designed for Vultures.

The other major item, which could have wrecked any plans to produce a prototype quickly, concerned the wing. The Merlin Manchester would require a bigger wing. Fortunately, two sets of specially extended wing spar extrusions were already in the factory, to meet a proposal to extend the wing of the Vulture-powered Manchester.

It is worth stressing that all the innovation, despite the rule-by-threat practised by Beaverbrook's Ministry of Aircraft Production, was going ahead long before the Manchester entered service. The dates speak for themselves, and it is sobering to recall that the girls engaged on detail manufacture, in the pictures, were almost certainly building the very Lancasters that were to be engaged in the heroic daylight actions of that faraway summer.

Acknowledgements

My thanks are due to Harry Holmes of British Aerospace, Commercial Aircraft Ltd, Woodford, Cheshire, and to Wg Cdr L R Byram RAF retd, for advice and checking of particular aspects of this article.

VC10/Super VC10: Small Fleet Operators

Dr Norman Barfield

Detailed here are operations by VC10s and Super VC10s when functioning in small fleets or even singly, often in mixed-traffic configuration. The second operator of the VC10 was British United Airways (BUA) which put the type into service in November 1964, six months after the British Overseas Airways Corporation (BOAC).

During 1960, significant changes had been taking place in British aviation, both in manufacturing and airline operations. Vickers-Armstrongs (Aircraft)'s parent, Vickers, together with English Electric and the Bristol Aeroplane Company, had formed the British Aircraft Corporation (BAC). Meanwhile, the independent air operator Hunting-Clan and the old-established Airwork, founded in 1928, had merged to form BUA. The managing director appointed to tie the operation together and make it prosper was F A Laker, much later to become better known, after his Atlantic Skytrain venture and a well deserved knighthood, as Sir Freddie Laker.

By the following year, BUA became stronger when shipping and insurance interests formed Air Holdings as the airline's financial parent. A new spirit was abroad in Britain as, in May, BAC announced its new short-haul jet, the BAC One-Eleven, on the back of a launch order from BUA. In September the airline followed up with an order for two Standard VC10s.

BUA's basic requirement was for a Britannia replacement, capable of carrying economic payloads from the same kind of difficult airports from which BOAC operated. It saw the VC10 as having much more appropriate airfield performance than the Comet 4, yet requiring fewer passengers than the Boeing 707 to break even. However, the diverse nature of the airline's payloads led to the specification of a large freight-loading door in the forward fuselage.

Three years later, in September 1964, the ebullient Freddie Laker took delivery of the first at the Farnborough Air Show, Hampshire, where the world could watch it setting off, with the latest Rolls-Royce car on board, for Central Africa on a pre-service demonstration tour.

BUA, with just two VC10s, each with the big side freight door, pioneered markets to South America and Africa. BOAC, losing £1¼ million a year on the admittedly economically difficult routes to South America, using Comet 4s, had given them up. BUA stepped in, without sales offices and without traffic rights in Brazil, on a two services a week basis from November 1964. At the end of its

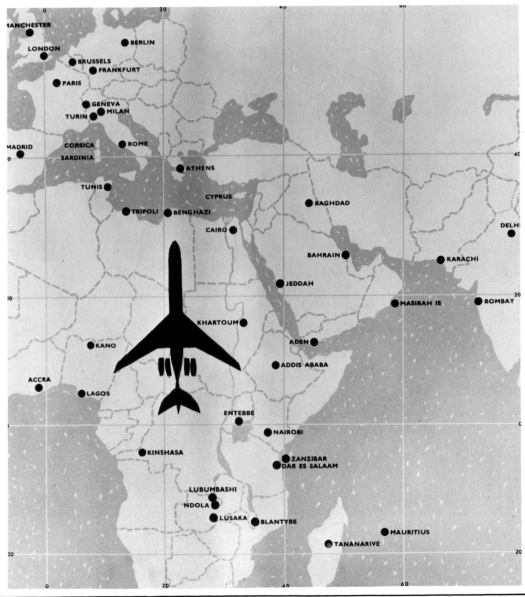

'Big jet for Africa'. Points on the East African Airways Network. EAA was the joint airline for Kenya, Tanzania and Uganda. (EAA)

first year, against a budgeted loss of £500,000, it had done substantially better. Fitted out to carry ninety-three economy-class passengers, sixteen first class and with capacity for up to 5 tons of cargo, the two aircraft made ninety-nine trips to South America in their first year.

In the same period, the two VC10s were also providing services to East and Central African destinations. For a month, further services were run to Freetown, West Africa, until the VC10 service had to be withdrawn while the runway was strengthened, Britannias being substituted. For the African runs the VC10s were arranged as 84-seaters plus cargo, and after eight months a third VC10 was chartered from Ghana Air-

BUA began to operate the VC10 from the autumn of 1964.

ways. Time was also found to operate trooping services to the Middle East for the British Government and some inclusive tour flights as well.

Each of BUA's own two VC10s flew just over 3,000 hours in the first year and, as Freddie Laker reported at the year-end press conference, in making 390 out-and-back trips altogether to overseas destinations, the VC10 had carried the equivalent of 140 passengers and baggage and 4½ tons of cargo on every round trip. BUA, in short, was in no doubt about the technical and economic merits of the mixed-traffic VC10 over the routes for which it had been initially designed. Full payload, 4,000 mile sectors out of high-altitude airports on the equator, with take-off in the middle of the day, direct to London Gatwick was something no other jet could do at the time.

The late Capt P A Mackenzie, BUA's chief pilot, speaking at a Royal Aeronautical Society branch occasion eleven months after its VC10s entered service, said that the typical in-flight introduction by one of its captains over the public address was, 'Ladies and gentlemen, it is my *privilege* to command a BUA VC10'. He went on to refer to the challenge presented to the pilot by being 25ft ahead of the main wheels when taxiing at small airports, adding that pilots who had come straight off DC-3s had flown the VC10 after five hours.

BUA later increased its VC10 fleet to four by purchasing the third aircraft of the Ghana Airways order and buying the prototype VC10 (built to production standards – Type 1109) which had been bought initially by Laker Airways (Laker had, by that time, left BUA to form his own airline). The ex-Laker VC10,

G-ARTA, had been leased previously to Middle East Airlines in Lebanon. In turn, in 1970, BUA merged with Caledonian Airways. The VC10s flew on, first in a Caledonian/BUA livery and later as British Caledonian. Caledonian, which before the merger had decided to buy Boeing 707-320s for its direct services to California, eventually ran down its ex-BUA VC10 fleet and sold them.

One British Caledonian VC10 went to the Royal Aircraft Establishment (RAE), Bedford, for instrument test flying; one to Air Malawi in Central Africa (which continued to fly between London and Blantyre until the serious fuel crisis of 1979 and the lack of passengers forced the airline to discontinue its long-haul international services); and one to the Sultan of Oman's Royal Flight. G-ARTA was eventually broken up.

The next airline to 'go VC10' was Ghana Airways of West Africa. Ordering three Standard VC10s (Type 1102) in

G-ATDJ at London Gatwick in a revised livery following the merger of BUA and Caledonian Airways in November 1970. The airline was known as Caledonian/BUA until September the following year when the name British Caledonian was adopted. (*BAe*)

1961 (one to the all-passenger specification and the subsequent two to have BUA-type freight doors), the two aircraft delivered in 1965 were under-used, especially as it had not proved possible to secure new route licences, hence the airline's third delivery position was made over to BUA. In order to increase use of the second aircraft, it was leased in the winter to Middle East Airlines. Unfortunately, this leased aircraft was destroyed in the Israeli attack on Beirut International Airport on 28 December 1968. Ghana's first VC10 continued in operation for a further two years.

Having ordered two Standard VC10s and subsequently cancelled them before they were built, Nigeria Airways nevertheless found that the competition it faced from the BUA VC10 operation forced it to lease BOAC Standard VC10s. These were used in a joint operation, painted in Nigerian green and white colour scheme, on flights to Europe, mainly London, and to East, North and Central Africa. BOAC also operated a number of flights jointly with associated airlines, Gulf Air in the Middle East and Air Ceylon.

With the widespread introduction of the Boeing 747, compounded by the fall-off in demand due to the fare increases that came in the wake of the oil crisis in 1973, BOAC decided to withdraw its Standard VC10s. This gave Gulf Air the opportunity to buy five of them to build up its international network, pending delivery of the Lockheed TriStar. Initially acquiring two aircraft for around £½ million each, then gradually taking up the other three, this virile little airline used the VC10 on routes to Europe and the East as well as to complement its local Middle Eastern services. One Gulf VC10 was also leased to Air Ceylon. After three years' operation, and as the TriStar fleet arrived, Gulf Air's VC10s were phased out, eventually being bought for the Royal Air Force, via an aircraft broker, for conversion into air-to-

air refuelling tankers.

East African Airways Corporation (EAAC) then became the only airline other than BOAC to purchase the enlarged Super VC10, its first service being operated from London to Nairobi on 7 December 1966. Being the consortium airline of the three states of Kenya, Tanzania and Uganda, EAAC's colourful livery included a combined tri-national 'sunburst' on the aircraft fin, which was rotated so that the appropriate symbol appeared at the top of the insignia on each of the three nationally owned aircraft. The EAAC Super VC10s (Type 1154) also flew to South and West Africa, the Middle East and to other European centres, from its three national capitals – Nairobi, Dar-es-Salaam and Kampala (using Entebbe Airport). Ultimately building up its fleet to five aircraft (and in 1970 taking delivery of the last of the VC10 family to be built), the airline continued to operate the aircraft for the next seven years.

However, mounting political tensions in the area, compounded by the resulting financial difficulties, ultimately caused the corporation to go out of business. This resulted in British Aerospace repossessing the remaining four aircraft (one had been written off in a runway accident at Addis Ababa due to faulty brake maintenance) which were returned to Filton (Bristol) in the United Kingdom where, like the five ex-Gulf Air Standard VC10s, they were converted to aerial tankers for the RAF.

The overall airline safety record of the VC10, given the operationally difficult jobs it had to perform throughout the world, was exemplary as far as the aircraft itself was concerned.

Apart from a BUA incident and the East African accident already noted, an ex-BOAC Standard VC10 being operated by Nigeria Airways in 1969 was lost on the approach to Lagos from Kano due

East African Airways Super VC10, in its natural habitat. Seen here at Nairobi (Embakasi) Airport, now Jomo Kenyatta International, elevation 5,327ft, the aircraft could fly a commercial payload direct to London Heathrow, a 4,250 mile sector. (*EAA*)

Above: In happier days, a Middle East Airlines VC10, 9G-ABP, on charter from Ghana Airways, arrives at Beirut International Airport. (*MEA*)

Below: A Ghana Airways VC10 on test from Wisley in November 1964. (*BAe*)

to crew error in height observation and in the absence of suitable instrument landing system (ILS) equipment. However, this accident, together with others around the world, served to focus attention on the development of a ground proximity warning system (GPWS) which was subsequently fitted to all VC10s as a result of the US Federal Aviation Administration (FAA) requirements for airlines operating into and within the United States. Two BOAC/British Airways Super VC10s also fell victim to hijacking incidents by Middle Eastern factions and were written off.

Having served with BUA and Middle East Airlines, an ignominious end befell the VC10 prototype. During an empty return to London Gatwick on an exceptionally blustery night in January 1972, the aircraft made an extremely heavy landing, bouncing three times and severely buckling the fuselage beyond economic repair. However, some good also came from the incident. Later broken up at Gatwick, the scrap metal was cut into

VC10-shaped plaques by apprentices in aid of an associated children's trust.

Significantly, after airline service, the VC10 was chosen for personal use by several Middle Eastern heads of state and potentates during its latter years of civil operation. An ex-BOAC Standard VC10 was sold to Sheikh Zayed of Abu Dhabi, President of the United Arab Emirates (UAE), who used the aircraft for seven years in VIP configuration with the UAE title in English and Arabic along the fuselage side and sporting a huge falcon emblem on the fin. The Ruler of Qatar used a similar aircraft on his private and diplomatic missions, albeit still in the ownership of, and crewed by, British Airways, until July 1981.

After ten years of airline service, an ex-BUA/British Caledonian VC10 was bought in 1974 by the Sultan of Oman. After custom furnishing by BAC at Hurn (Bournemouth), this superbly appointed aircraft was widely used by the Sultan's Royal Flight for the next thirteen years. Replacing it with a much larger Boeing 747SP, the Sultan graciously donated the VC10, complete with all accessories, to the newly formed Brooklands Museum Trust, fittingly the original home of the aircraft. Arriving at Brooklands via London Heathrow in July 1987 (a week after the twenty-fifth anniversary of the maiden flight from the same short 'backyard' runway by G-ARTA), this aircraft is now not only the museum's largest and grandest exhibit but also a permanent tribute to the last and biggest of the five-decade Vickers aircraft lineage to emerge from Weybridge. □

Regional Airport: Hamburg

Gerhard Katzsch

Hamburg, old trading seaport of the Hanseatic League, has been equally enterprising in the air age. Today's thriving travel, cargo and engineering activities, centred at Fuhlsbüttel, rest on foundations laid in 1910 when the airport site was prepared to serve the world's first air passenger carriers, the airships made by the Zeppelin company.

Airport: Hamburg-Fuhlsbüttel

Latitude: 53°37'55"N
Longitude: 09°59'22"E

Relation to city centre: 12km north
Population resident within 50km: 4 million
Runways: 05/23 3,250m; 15/33 3,665m
Operation: 0600–2300hrs, special permits required for take-offs/landings at night
Customs: yes
Principal aids: ILS on 3 of the 4 approach directions; on runway 23 landings to ICAO CAT IIIb conditions in poor visibility
Owner: Flughafen Hamburg GmbH which has the following shareholders: City of Hamburg 64%, Federal Republic of Germany 26%, Federal State of Schleswig-Holstein 10%

Airport directors: Dr Claus Lau, commercial; Dipl-Ing Richard Schleicher, traffic and technical
Aircraft movements/year (1988): 120,000
Passengers/year (1988): 6 million
Principal operators serving: Lufthansa British Airways, Air France, SAS, KLM, Pan American and 20 others
Scheduled service destinations: principal German domestic airports, Amsterdam, Gothenburg, Copenhagen, London, Milan, Paris, Helsinki, New York. Total 49 international, 11 domestic destinations
Holiday destinations include: Fuerteventura, Las Palmas, Palma de Mallorca, Colombo, Monastir, Istanbul. 66 destinations are served in 24 countries

Principal operators based at airport: Lufthansa Engineering Base, Aero Technik Klaus Siemers GmbH, Air Service Flugbetrieb u Luftfahrttechnik KG, LFE Avionic Services, Luftfahrtelektronik GmbH
How to get there: coach service from central railway station direct to airport with stops at major hotels in the city centre; public bus services 4 lines; shuttle bus every 10 min to nearest suburban railway station (Ohlsdorf); daily scheduled coach services to the neighbouring cities of Kiel (110km) and Lübeck (50km) 9 and 5 times per day, respectively, reduced schedules at weekends
Car parks: long-term 3,700 cars, short-term 300 cars

Being the largest city in Federal Germany and with a tradition dating back many centuries, Hamburg has become the country's largest trading, media and manufacturing centre. The importance international trade has for the city is underlined by the fact that there are no fewer then seventy-seven consulates in Hamburg – in these terms it ránks second worldwide after New York – and the city has also become an international banking centre. Sixteen of the largest German manufacturing companies are located at Hamburg, among them microchip and medical equipment manufacturers as well as the Hamburg branch of Messerschmitt-Bölkow-Blohm (MBB), Germany's leading aerospace company which, employing about 20,000 people, manufactures *inter alia* Airbus components and fits out the 'green' Airbuses arriving from the assembly line at Toulouse.

There are facilities in Hamburg for large-scale congresses and important trade fairs and exhibitions. All this needs an efficient airport and Hamburg has certainly lived up to that challenge.

Being northern Germany's principal gateway airport, Hamburg has expanded rapidly in recent years and gained its full share of the general expansion air transport enjoyed in the last two decades. Aircraft movements numbered 92,534 in 1977 and the 1988 estimate puts that figure at 120,000. Likewise, the number of passengers who used the airport grew from 3.95 million in 1977, to an estimated 6 million in 1988. The corresponding figures for cargo are 46,616 tonnes in 1977 and 75,000 in 1988, for mail 10,039 tonnes in 1977 and 17,000 in 1988. The 1988

The Lufthansa Engineering Base. A lot of maintenance work is done for airlines other than Lufthansa. (*Lufthansa*)

figures are again estimates.

Among the twelve large German airports, Hamburg ranks fourth in terms of passengers, says airport director commercial Dr Claus Lau. In international services the emphasis is on Europe. All large economic and cultural centres are served from Hamburg. Because of the traditionally close economic relations with the Scandinavian countries, that part of the world is served particularly well. No other German airport offers more direct services to Scandinavia.

Check-in for domestic departures. This terminal was opened on 5 March 1980. (*Flughafen Hamburg*)

Intercontinental services from Hamburg reflect the economic importance of the city. They link Hamburg with New York and Miami, with Tokyo and Osaka. Dr Lau particularly points to the rapid growth of air cargo which has made Hamburg the largest cargo airport in northern Germany. As far as imports are concerned, Hamburg ranks third worldwide after Frankfurt and New York in the Lufthansa network.

Asia is very important, too. In Hamburg the share of goods received from Asia is over 45 per cent, Dr Lau emphasises. The importance of the Far East cargo link is highlighted by a Lufthansa Boeing 747 freighter service which links Hong Kong, Bangkok and Dubai with Hamburg once a week. Lufthansa also operates a direct freighter service between Hamburg and New York.

Regional and commuter air services have experienced a particularly rapid growth in Germany in recent years. Initially, Hamburg did not profit much from this new line of development which was closely associated with a change in German legislation. However, having made a remarkable marketing effort in recent years, the airport is now served by seven regional airlines which provide scheduled services to five destinations in Germany and five in Belgium, the Netherlands, Switzerland and Sweden. 'It is principally our business community which profits from these services', Dr Lau comments.

The airport director is also proud of the development of holiday flights from Hamburg in recent years. No fewer than thirty-seven airlines are operating charter services from Hamburg to holiday destinations in Europe and overseas in the 1989 summer season. In the summer

of 1988, the Düsseldorf-based LTU, among West Germany's largest and most efficient of those airlines specialising in charter services, branched out to Africa (Mombasa), the Maldive Islands and Sri Lanka (Colombo). In the 1988/89 winter season, LTU added Bangkok, the Thai island of Phuket and Miami to its network out of Hamburg.

In the summer of 1989, charter services are operating (previous year in brackets) to sixty-six (thirty-three) destinations in nineteen (twelve) countries. The principal destination country continues to be Spain with fifty-three (thirty-nine) flights per week. Like the scheduled airlines, the charter operators are using increasingly modern low-noise equipment such as the Airbus A300, the Boeing 757 and the Boeing 737-300.

The rapid growth of the number of holiday flights out of Hamburg shows how considerable is the potential in northern Germany for this type of operation if properly tapped. A contributary factor is the congestion of German airspace at centres such as Frankfurt, Munich and Düsseldorf with the associated air traffic control and slot problems. The allocation to airlines of convenient slots is not a subject of discussion in Hamburg.

'Many of the holiday flights from Hamburg', says Dr Lau, 'take away aircraft movements from the large focal points. The regional air services also make a great contribution to what I might call the straightening out of schedules by staying away from congested centres. The fact that airlines can get in Hamburg the departure and arrival times they want is a potential for the development of the airport.'

In line with the metropolitan function

Lufthansa operates scheduled freighter services to the Far East and North America. The latest handling equipment makes loading and unloading a smooth operation.

The Airbus A310 is a frequent visitor at Hamburg Airport. (*Lufthansa*)

of Hamburg in trade and industry, the non-commercial 'general aviation' is an important factor at Hamburg Airport. The majority of non-commercial operations are executive flights which carry more than 60,000 passengers annually. Fifty leading companies, among them Daimler-Benz; Nixdorf, the leading German computer manufacturer; Quelle, the large mail-order house; and Springer, one of the country's leading publishing groups, have stationed their executive aircraft at Hamburg Airport.

For many years, Hamburg's airport has been investing a lot of money to make it a good neighbour. 'An airport', Dr Lau says, 'is like a sewage works. Everybody needs it, but no-one wants to live near it.' Hamburg follows the widely adopted practice of giving an up to 30 per cent rebate on landing fees for quiet aircraft meeting the ICAO Annex 16 conditions. In addition and co-operating with the City of Hamburg, several pro-

The new control tower with the latest ATC equipment took over in September 1988. Its DERD-MC system (display of extracted radar data – mini-computer controlled) converts radar data into a form suitable for computers. (*AEG*)

grammes for soundproofing houses in the neighbouring residential areas have been completed. By 1987, voluntary contributions to such measures amounting to DM38 million (£12 million) had been made. More than half of this expenditure had been borne by the airport company. Another such programme worth DM18 million (£5.7 million) will be launched during 1989.

Employing a workforce of about 13,000 people of whom about 8,000 work at the Lufthansa Engineering Base and 5,000 in services, Hamburg's airport is an important growth centre in the region. About 180 companies are associated with the airport and more than 1,000 additional jobs were created during the last five years.

'Like any major airport throughout the world, Hamburg is a large building site', says airport director Dipl-Ing

Entrance to the international terminal used for services operated by twenty-six international airlines. It incorporates parts of the first terminal opened in 1929. (*Flughafen Hamburg*)

Richard Schleicher who is in change of the traffic department and everything technical at the airport which includes the planning of airport facilities to meet the challenges of the future. To enable the airport to handle the volume of traffic expected in the coming years, a comprehensive expansion programme is in hand. It involves investments of DM700 million (£222 million) by 1997.

One half of this amount will be spent on a new terminal for scheduled services, a passenger access unit with eleven traffic fingers and a parking garage. This will increase the passenger handling capacity to 7.5 million a year. Apart from increasing the airport capacity, an important aim is to improve the service and comfort available to passengers. Departures and arrivals will be handled on two different levels, a concept which has already proved sound in Hamburg and elsewhere. Generous shopping and catering facilities, conference rooms and VIP lounges will make the new terminal versatile and attractive to passengers and visitors alike.

At the air side of the building, a pier will link all the passenger facilities like a spine, including the existing terminals. This pier will accommodate the departure lounges and shops including the

Artist's impressions of the new terminal buildings to be erected during 1989–92/93 at a cost of DM350 million (£111 million). On the air side a pier runs along the entire terminal building with aircraft docking at finger units. (*Flughafen Hamburg*)

duty free shop. From the departure lounges, the eleven fingers will provide under-cover access to aircraft ranging from the Boeing 747 to the Fokker 100.

A circular parking garage adjacent to the new terminal will accommodate 800 cars and opposite the terminal another 2,000 cars can be parked. 'This is in line with the general philosophy of the development plan', comments Herr Schleicher, 'and that is to keep distances short from the points at which passengers leave their cars and coaches to check in.' The public access roads will meet this philosophy and their general layout will make it easy for large numbers of arriving passengers to leave the airport smoothly in a short time.

Cargo was the sector in recent years which saw the most rapid expansion. The cargo handling facilities will therefore be enlarged to an annual capacity of 120,000 tons. Forward planning for an even larger capacity is under way. More than seventy forwarders have their offices at Hamburg Airport and one of

their aims is to develop new types of co-operation between the city's seaport and airport.

'General aviation is an important feature of the activites at our airport', Herr Schleicher emphasises, and explains that a separate terminal and aircraft hangars with workshops and office facilities will be built for general aviation. This new terminal, an important feature for executive flying, will be equipped with everything needed, *viz* flight planning advice, air control service, customs, immigration and security facilities, a duty free shop, catering and telecommunication facilities and, of course, a large departure lounge. At present about 140 private and executive aircraft are stationed at Hamburg Airport. Additional parking facilities can be provided should their number increase.

Hamburg Airport is situated close to a motorway but so far has no direct access to it. After many years of discussion with representatives of the resident population and local government, agreement has now been reached that the airport will have a four-lane access road to the motorway.

Marketing efforts for the future, Dr Lau explains, will be principally directed towards strengthening the links with Scandinavia, for which Hamburg is already a hub, and the United Kingdom as well as Eastern and Southeastern Europe.

For intercontinental services, marketing will particularly aim at areas of importance for Hamburg's overseas trade. These are the markets in the Pacific Area, Southeast Asia and China.

Receipt from 1910 for a minimum subscription of 1,000 Marks to the Hamburger Luftschiffhallen-Gesellschaft (Hamburg Airship Hall Company). (*FHG*)

In recent years, Hamburg has become the West European Centre in trade with China and is, for instance, home of the Chinese Trade Centre which covers the whole of Europe. Hence in passenger traffic with Singapore and China, Hamburg ranks second after Frankfurt and ahead of Munich and Düsseldorf. Another growth market for Hamburg air transport is the United States.

History

Looking back to the origins of Hamburg Airport we find that the city had an airport before air transport had started as we know it today, ie with machines that are heavier than air. In Germany, the era of air transport began with craft which were lighter than air, the legendary Zeppelin airships.

The Deutsche Luftschiffahrts AG (Delag), the world's first airline, was founded on 16 October 1909 and by August 1914 total airship travelling time was 3,139 hours and 10,088 paying passengers had been carried without a single injury – but these were not scheduled services.

Men with a vision in Hamburg were quick to grasp the importance of this new mode of transport. Leading businessmen, among them Albert Ballin, chairman of the Hamburg-based HAPAG, then the world's largest shipping company, saw to it that funds were subscribed to both the Delag headquartered at Frankfurt and a port facility for airships in Hamburg. This resulted in the construction of a large shed (hall in German) to accommodate an airship like a garage, which was built on the site of the present Fuhlsbüttel Airport.

Measuring 160m long, 45m wide and 25m high, it could take two airships. It was used by three passenger airships and one navy airship. When the hangar was commissioned in January 1912, the scene seemed to be set for a brilliant future in air transport by means of airships. In those days they were considered in Germany, and elsewhere, to be the only craft suitable for passenger transport by air.

However, the first men with their incredible flying machines – heavier than air, made of wood and fabric and kept together by glue and wire – appeared at the brand new airport before the First World War stopped all civil flying. No one knew it at the time, but this ended the airship era at Hamburg-Fuhlsbüttel.

In December 1917, AEG, the General Electric Company equivalent in Germany which had been manufacturing aeroplanes since 1910, together again with HAPAG and the Zeppelin transport company, formed an airline, the Deutsche Luft-Reederei GmbH. It started flying in February 1919 and a month later the first scheduled service was inaugurated between Hamburg and Berlin. The heavier-than-air era had begun at Fuhlsbüttel. By the end of 1919 a total of 233 passengers had been carried, quite an achievement if one considers the turbulent end-of-war period in Germany, the state of the 'airport' and types of aircraft available. In 1920, even a bucket-and-spade service was opened to the North Sea resort island of Westerland, a service which has not lost its popularity up to the present day.

The first international services to Copenhagen, Amsterdam and Rotterdam were introduced very soon, and on weekends flying displays attracted thousands of visitors.

The terms of the Peace Treaty which Germany had to sign at Versailles involved the destruction of the airship hangar which was blown up in 1921. The airport was in a desolate state. In 1921, the city made available funds for the airport and before the end of the year work had started. Buildings were repaired, an apron, an underground petrol storage tank and other facilities provided. A beacon with ninety-six bulbs was installed at the top of a tall chimney and it could be seen from a distance of 80km. There was even an early form of air traffic control. Nineteen petroleum lamps lit the landing strip at night. Fuhlsbüttel was an all-weather airport by 1922.

The airship shed was opened in 1912. In those days airships were the only conceivable vehicles for air transport. However, bold men with their flying machines (the picture shows a Rumpler Taube (dove)) were an attraction at the brand new airport. (*FHG*)

In 1935, the airport area was extended from 236 to 551 acres. The flock of sheep used for 'maintenance' had to be enlarged. The terminal in the background is still part of the present group of terminal buildings. (*FHG*)

Fuhlsbüttel was the first German airport to receive a wireless station, which was installed in 1923. It could make contact with places such as London, Zürich and Warsaw, a brilliant achievement in those days.

The airport ground was very swampy and hence it was important to provide good drainage which involved hard work for several years. The soil was stabilised also in adjacent areas which could then be included in the airport boundaries.

Air transport developed very rapidly in the mid-1920s. One had to book one or two weeks in advance to get a seat. In 1925, the number of passengers handled at Fuhlsbüttel grew from 2,318 to 12,499, or by 439 per cent, the most substantial growth rate ever achieved.

The first large terminal was opened in 1925, a second one a year later, and a completely new terminal which replaced the two earlier ones was opened in 1929. This was the building that gave the airport its appearance and survived in essence to the present day. By the end of the 1920s, 5,427 aircraft movements per annum were recorded at Fuhlsbüttel and 17,973 passengers and over 600 tons of cargo handled. Air transport was a flourishing business.

The economic crisis in the early 1930s dealt a severe blow to air transport. Between 1930 and 1933, Fuhlsbüttel lost 35 per cent of its volume of cargo and about 10 per cent of passenger traffic. However, from 1934 air transport grew again rapidly. The prewar peak was achieved in 1937 with over 10,000 aircraft movements for the first time, 57,194 scheduled service passengers and almost 1,300 tons of cargo. The number of routes originating in Hamburg had increased to twenty-two. Among them was the 4,050km Hamburg–Belgrade–Athens–Rhodes–Damascus–Baghdad route, which Lufthansa aircraft covered in under twenty-four hours.

In August 1939, Fuhlsbüttel was closed to commercial aviation. It was requisitioned by the Luftwaffe for military operations. Whether it was the camouflaging which prevented the airport's destruction during the Second World War is unknown, but the Royal Air Force was able to take over a working airport in May 1945. Fuhlsbüttel was now known – in English – as Hamburg Airport.

There was no German aviation in the postwar years but commercial services were begun in 1946 by British European Airways (BEA) which from 1 September operated a twice daily London–Amsterdam–Hamburg–Berlin service.

There were no paved runways in those days; only a grass field was available. The swampy ground had been a problem in prewar days but now, when aircraft weights exceeded 40 tons and approached 60 tons, something had to be done. A primitive sort of runway was provided by steel plates over a length of 1,500m. However, air transport grew rapidly and hence there was a need for concrete runways. Work on runway No. 1 started in April 1948. In the summer of that year the Berlin Airlift began. The utmost was done to complete the concrete runway rapidly. Hamburg was one of the airports from which supplies were flown to Berlin. Managed by BEA, aircraft were operating round the clock.

The first of October 1950 was an important date, when German airport administration took over from British military authorities. In addition to BEA, other foreign airlines served Fuhlsbüttel Airport. The number of passengers us-

ing the airport grew rapidly but air transport really began in postwar Germany when Lufthansa inaugurated its first postwar scheduled service on 1 April 1955, when a Convair CV-340 took off from Hamburg for Munich.

Before that happened an important act was performed that proved to be vital for the future of Hamburg Airport. In December 1953, Luftag, the Lufthansa predecessor company, signed a long-term lease for an existing large hangar with the airport administration. This was the nucleus of the present Lufthansa Engineering Base which now provides work for more than 8,000 people and is Hamburg's second biggest employer.

The dual runway system already existed but as aircraft grew larger and heavier and more powerful, with the advent of the jet era, the runways had to be lengthened and strengthened in 1956, 1961 and 1964 to their present lengths. The terminal facilities had to be extended continually. The first extension of the passenger terminal was opened in November 1961. As the holiday traffic experienced a rapid expansion in the 1970s, handling facilities for these passengers had to be extended by stages. The present large charter terminal was opened in September 1982. Cargo also required a large-scale extension of storage room, lorry parking spaces and office accommodation which was completed in 1976.

The hundreds of millions of Deutschmarks previously invested in the extension and modernisation of passenger terminals provided buildings that clustered around the old airport nucleus of 1929. The expansion programme now under way is for the first time a completely new concept which sets the scene for an up-to-date airport fit for the challenge of the twenty-first century. □

Don't Forget the Forum

Although it is up to the readers to choose the subjects for the forum pages, while we are so new it may be helpful to suggest a possibility that comes to mind after reading this issue.

Any reader with strong views on over-ocean airline operations with twin-engined aircraft anywhere in the world, bearing in mind the 120-minute rule and the possibility of this being extended to 180 minutes, has the opportunity to make points.

We hope we are going to get a good response on the subjects of your choice: the more succinct, the greater the chance of being quoted fully, and the better our *Review* will be.

BOOK REVIEWS

Blackburn Aircraft Since 1909
A J Jackson

Putnam Aeronautical, London EC4Y 8DR.
Hardbound, 145 x 223mm, 584 pp, illustrated, £24.00.

Sadly, since this book first appeared, in 1968, Jack Jackson has died, so he was not able to revise this second edition of a classic company history. For some years, *Blackburn Aircraft* has been sought after by collectors of the celebrated Putnam company histories, but now it is readily available again, with some revision by David Dorrell. In the main, this has consisted of updating the sections on the Beverley and Buccaneer and adding a three-page appendix on the work of the Brough factory up to 1988. This contains references to a pair of interesting military trainer projects to RAF requirement AST.412, the jet-powered P.164-108 and the propeller turbine-powered P.164-109 which, superficially at least, bears a marked resemblance to the Pilatus PC-9 which British Aerospace offered in competition with the Embraer Tucano, offered by Shorts.

Perhaps of greater interest, however, is new appendix G, which largely concerns the identity of the Blackburn Single-Seat Monoplane of 1912, now preserved and flown as part of the Shuttleworth Collection. M G K Byrne, previously sales and publicity manager for Hawker Siddeley at Brough, sent the author some details from an early sales book. Both Byrne's notes and A J Jackson's reply are provided for our edification, and it is revealed that this aircraft was, in fact, a new machine built to replace the one damaged by Harold Blackburn, and was not the one owned by Cyril Foggin. A reference to this appendix has thoughtfully been added to the relevant section of the main text.

One point that has not been picked up concerns the Antoinette monoplane which Robert Blackburn found in a Colwyn Bay garage in 1916 and bought for £60. This aircraft, which is now displayed in the Science Museum, London, was thought to have been the one flown by Hubert Latham at the Blackpool Flying Meeting in 1909 (*not* 1910, as this book states), but it was certainly not Latham's aircraft, and John Bagley of the Science Museum now thinks it possible that it was the machine bought for Vivian Hewitt in 1909 by a wealthy uncle. As Hewitt lived in Bodfari, not far from Colwyn Bay, it seems much more likely that this is the machine which Blackburn bought.

Apart from this small point, and the fact that some of the pictures have suffered somewhat in 're-reproduction', this is still an excellent and unsurpassed history of one of Britain's great aircraft manufacturers. Don't miss it second time around.　　**PJ**

My Part of the Sky
Roland Beamont

Patrick Stephens, Thorsons Publishing Group, Wellingborough NN8 2RQ, England.
Hardbound, 145 x 223mm, 208pp, illustrated, £10.95.

Roland Beamont's flying career was exceptional. He gained his wings at an RAF Flying Training School shortly after the outbreak of war, and after proving himself to be an exceptional fighter pilot he became an exceptional test pilot, nursing types such as the Canberra, Lightning, Jaguar and Tornado through their test programmes.

With the exception of two short breaks, 'Bee' served with fighter squadrons continuously from October 1939 to October 1944, when he force-landed his Hawker Tempest behind enemy lines and became a prisoner of war. Sub-titled 'A fighter pilot's firsthand experiences, 1939-45', this volume recounts the author's experiences during those momentous years, based on 'narrative diaries' which he kept when he could find the time. These were typed up in 1945, and are published as written, 'with only minor editing and some explanatory linking passages'.

Throughout the war, Bee flew Hawker fighters, starting with Hurricanes and progressing to Typhoons and Tempests. He was with No.87 Squadron with the BEF in France and during the Battle of Britain, and later served with Nos. 7, 56 and 609 Squadrons. He then served as a wing commander, commanding Nos.150 and 122 Wings.

The style of the book makes it lively reading. The reader is spared the day-to-day repetition of a conventional diary, yet the immediacy is still there, and the author conveys vivid impressions of squadron life and aerial combat.

The testimonials and letters in the appendices were scarcely necessary, and one wonders whether they might not have been reduced to make room for an index, which is conspicuously lacking. Nonetheless, this is an absorbing book, well illustrated by six four-page sections of black and white photographs, reproduced on glossy paper.　　**PJ**

The Aviation Gouaches of Thijs Postma
Introduction by Wim Kroese

Edu' Actief, Postbox 56, 7940 AB MEPPEL, Netherlands.
Hardbound, 290 x 230mm, 100pp, illustrated, £10.95.

The aviation writing and art of Thijs Postma has become increasingly familiar to enthusiasts all over Europe, but I doubt whether the scope of his pictures was known to more than a few. This nicely produced book, in landscape format, presents the reader with forty-seven of the artist's finely worked gouaches, printed on one side only of the high-gloss pages so that items can be removed for framing if desired.

The nature of the technique, which is described in the introduction, means that, often, the backgrounds are of a rather ethereal nature, though in Postma's later pictures this has improved. The aeroplanes themselves are well rendered and the colours are impressive. I particularly liked those depicting Martin and Osa Johnson's Sikorsky S-38 and S-39 amphibians, the US Coast Guard Fokker FLB, and the dam-busting Avro Lancaster, but all tastes and periods are catered for – right up to modern jets.

While I greatly enjoy aviation art, books such as this leave me in a quandary. They are a delight to look at, but once the pleasure wears off, what can you *do* with them? Few people have sufficient wall space to hang forty-seven plates, and it seems a shame that such splendid art should remain buried between boards on a dusty bookshelf. If art for art's sake is your pleasure, however, you will want this one.　　**PJ**

World Encyclopaedia of Aero Engines
Bill Gunston

Patrick Stephens, Thorsons Publishing Group, Wellingborough, Northamptonshire NN8 2RQ. Hardbound, 200 x 242mm, 192pp, illustrated, £19.95.

It is strange that, while there is a plethora of books on aeroplanes of all types and ages, books on the engines that enabled them to fly are thin on the ground. Bill Gunston's encyclopaedia is one of the few, so it is no surprise that the publisher has seen fit to produce a fully revised second edition only three years after the first edition appeared. This edition has eight more pages than its predecessor and many additional illustrations, as well as entries for newly established engine manufacturers such as CFE and Eurojet.

This volume is really far too thin to do justice to so vast a subject, but it is a very useful source of reference for those needing information quickly. The illustrative material is a bit weak in some of the earlier engines (the three-cylinder Anzani is in a Deperdussin, *not* a Blériot) and in some cases pictures have been pushed into the bottom bleed and trimmed rather drastically to make space. There is no mention of Clerget's early in-line engine, but as the book concentrates on 'significant' engines this is hardly surprising.

It would have been nice to have some tabular data, either in each company entry or in an appendix, but what figures there are have to be gleaned from the text. Perhaps this could be considered for the third edition. Unless another publisher produces something bigger, this is going to be the standard basic volume on aero engines, and it is meeting a very long-felt need. **PJ**

British Carrier Aviation
Norman Friedman

Conway Maritime Press, London. Hardbound, 295 x 248mm, 384pp, illustrated, £35.00.

This is a history of the British aircraft carrier told by the leading authority on US naval aviation and very flattering it is too. Dr Friedman begins by pointing out that the carrier was a British invention; between the wars we introduced the armoured hangar and, later, the angled deck, mirror landing sight and the steam catapult. He also reminds us of the operational 'firsts' such as the first shore strike in 1914 and Taranto, and the first effective torpedo-bomber operation.

He shows how the carrier evolved during and after the First World War to meet two imperatives. Aircraft were needed at sea to find the enemy battle fleet, slow it down with torpedoes and spot for the battleships' guns in the decisive encounter. Surface raiders in distant seas were seen as the major threat to trade, and carrier aviation as the appropriate counter. The author does not accept the view that Boards of Admiralty in this era were insufficiently 'air-minded'; he maintains that they did the best possible with limited resources and dying industries.

The requirement for long-range search and strike, combined with low take-off and landing speeds, led to a very large wing area and light weight. In the interwar years, these aspects could only be provided by the biplane; the Swordfish was a result of the lack of a powerful aero engine rather than the Board living in the past. In turn, this vicious circle led to a reluctance to expose the flimsy aircraft to the rigours of a deck park and hence to the closed hangar.

The book follows the evolution of the carrier, from early trials before the Second World War to *Invincible*, using large numbers of British official documents and the many reports sent to the US Navy in both wars. The logical development is made clear by details of the many studies, not built for one reason or another, but still forming part of the evolutionary chain. Many of the missing links are shown in clear drawings, never before shown to the public. There are chapters describing the development of aircraft and of operational policy showing how they interacted with ship design.

The book is extremely well illustrated with many little-known photographs and original drawings. The comprehensive text is supplemented by a large number of footnotes on each page, much easier to use than when relegated to the back.

It is hard to fault; I would have liked a little more on the flexible flight deck and there are more spelling errors than usual. Overall, it is a most important study which is very readable and can be enjoyed by specialist and non-specialist alike. **DKB**

A line of Sea Hawks starts up, using cartridge starters. A photograph from *British Carrier Aviation*

THE PUTNAM AERONAUTICAL REVIEW

ISSUE NUMBER THREE **OCTOBER 1989**

Editor: John Motum

Consultant Editor: John Stroud

Assistant Editor: Linda Jones

Design: Swanston Graphics Ltd

Managing Editor: Julian Mannering

Publisher: William R Blackmore

© Conway Maritime Press Ltd

ISBN 0 85177 526 8
ISSN 0955-7822

Published quarterly by:
Conway Maritime Press Ltd
24 Bride Lane, Fleet Street
London EC4Y 8DR

Telephone: 01 583 2412

**Annual subscription for
four issues:**
£16.00 UK, £18.00 overseas,
$33.00 USA and $39.00 Canada.
All post paid.

Typeset by Swanston Graphics Ltd, Derby.
Printed and bound in Great Britain
by Page Bros (Norwich) Ltd.

CONTRIBUTORS

Dr Norman Barfield. No stranger to readers of the *Review* or to Brooklands, where the well-known Sopwith aircraft types and all the Hawker aircraft, up to and including the Hurricane, were flight tested, Dr Barfield in this issue turns to the life and times of Sir Thomas Sopwith.

Derek Dempster, a former pilot with the RAF and the RAuxAF, has been writing about aviation since 1950. His books include *The Tale of the Comet* and *The Narrow Margin* – co-author Derek Wood – which was awarded the C P Robertson Trophy and filmed under the title *Battle of Britain*. His writing for TV includes the prize-winning documentary series *Command*. He created *BOAC News* and founded the magazine *Airports International*.

Guy Roberty has approaching forty years' professional aviation experience, including five years with the Belgian Air Force and the remainder with an international airline, with service in London, Johannesburg, Stockholm and, currently, Montreal. He has held a PPL for twenty years, collects aviation books, writes occasionally, preferably on aviation history. Claims to be a total aviation person.

Johann-Friedrich (Hanfried) Schliephake trained as a radio operator and served principally on instruction duties in the wartime Luftwaffe. From 1944 he was a radio navigator with the No.1 Minensuch-Staffel of minesweeping Junkers Ju 52/3ms. He rejoined the new Luftwaffe in 1956 serving until retirement. He is an aviation author and historian with a special interest in aircraft armament.

Sqn Ldr William Simpson OBE DFC **Croix De Guerre** joined the RAF in 1935; as a pilot he went to France with the AASF in September 1939. Wounded and badly burned he narrowly survived the German Blitzkrieg of the following May. In 1944 he was invalided from the RAF, joining the *Sunday Express* as its air correspondent. Subsequently, he held a succession of professional PR appointments and consultancies to the airline and aerospace industries and professional bodies. He has had two books published.

John Stroud is an aviation writer, author and historian with a lifetime of involvement in or close to the airline industry, joining Imperial Airways in 1933, and subsequently recording many of its developments as they occurred. In this issue he turns back the clock to the very birth of air transport.

Hugh J Yea, consultant and writer on commercial and military aviation history, joined Imperial Airways in 1935. Apart from RAFVR service from 1939 to 1943, he served successively with IAL, BOAC and British Airways Overseas Division. A third of his forty-one years' service was spent abroad in West Africa, the Middle and Far East, Bermuda, the USA and Canada.

Front cover: A Handley Page O/400 used on cross-Channel services in 1919.
Back cover: Pioneering passengers in a Deutsche Luft-Reederei A.E.G. G V in 1919.

Editorial

In this issue our contributors discuss aspects of the role of some of the earlier aeroplanes, in peace, war and politics, in shaping world history and indeed reshaping the lives of virtually every individual. We pause to look at the life of one man, born almost sixteen years before the Wright brothers made their first flight in December 1903 and who died only this year.

Just over a hundred years ago, during the lifetime of Otto Lilienthal, Dr Octave Chanute, Percy S Pilcher and F H Wenham, on 17 January 1888, Thomas Octave Murdoch Sopwith was born, the eighth child but eldest son of Thomas Sopwith, a Scottish civil engineer. By that time Sir George Cayley, 'Father of British Aeronautics', had lived and died and been temporarily forgotten, as had many other air pioneers in many lands. The Australian pioneer Laurence Hargrave (1850-1915) who developed the box kite which later led to the cellular biplanes by Santos Dumont and the Voisin brothers, was one of those at work when Sopwith was born. Gottlieb Daimler's internal combustion engine was then in its infancy, as was Karl Benz's first motor vehicle.

The subsequent Sopwith century (the lifespan of one man, Sir Thomas Sopwith CBE who died on 27 January 1989) was to be unprecedented. By the time Sopwith came of age and was bitten by the flying bug, aviation was about to become a group activity and he was to become a prime mover of aero-industrial projects which in little over a generation would bear the stresses of two world wars.

Our contributors explore what happened in Europe in the autumn of 1939 after the deterrent of Britain's Expansion Air Force had failed to deter. Although useful aircraft were still in adolescence, giant strides had been made to the point at which a European dictator, Adolf Hitler, believed that air power would be able to do for Germany what he considered sea power had done for Great Britain in earlier times. The aeroplane was about to become a powerful tool.

Today, when aircraft help to sustain influence worldwide, military and commercial, they are often seen, particularly in aviation periodicals, as ikons to be revered by the enthusiastic for their own sake. Our contributors seek to dig rather deeper than that. And overall we look for reader response on the Forum page, to increase our knowledge.

Returning to 1939, we contrast the shock of Blitzkrieg, giving quick results within days, with the measured response by Britain and France whose governments were well aware that once war came, for them it could only be a long one. Industrial capacity would need to be created, allies sought, men mustered from across the world. The democracies had, in fact, underestimated the effectiveness of Blitzkrieg which in continental Europe was to be all-consuming within months. The action by Germany against Poland and the early preparation by France and Britain were to result in a conflict on a scale unimaginable when Britain was deploying its Fairey Battles to France and its airliners to southwest England. In nearly every battle yet to be fought, however, the control of the air would prove decisive to the result. By 1939 Sir Thomas Sopwith had barely reached his personal half-way point.

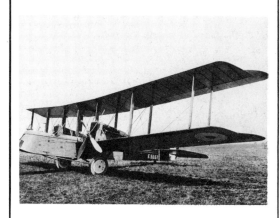

On 14 May 1919, a de Havilland D.H.10 (picture shows a prototype) became the first aeroplane to operate a night mail flight, from Hawkinge, direct to Cologne in three and a quarter hours.

Civil aviation, as opposed to using aeroplanes for war, began in the wake of a return to some kind of normality following the end of the First World War. John Stroud tells us what happened. But what was the motivation? In a world which, following the end of the Second World War, had learned to become accustomed to the dominant position held by United States air transport, how had Europe come out on top the first time around?

The answer almost certainly lies in requirement and geography. The war had provided aircraft with engines of some reliability (witness the Atlantic flights) and a capability to lift useful loads. There was no shortage of pilots to operate them or aircraft manufacturers to try their hands at the transition from war to peace. Indeed should they prove unable to make that change, their businesses would almost certainly have to be wound up. National capitals separated by distances of only 200 to 300 miles provided a challenge which the aeroplane could meet, the sponsors believed. There was an increasing demand to re-establish trading links broken by war, and first there were special needs to satisfy in connection with the army of occupation in Germany and the demands of the Peace Conference in Paris.

Not surprisingly, an organisation in being, the Royal Air Force, was the first to start when on 1 March 1919 No.120 Squadron began an international air mail service from its base at Hawkinge, Kent, to Maisoncelle in northern France and then on to Cologne. The squadron's aircraft at the time were two-seat single-engined D.H.9s, three, flying in formation, being needed to deliver the twenty-three mail bags of the first service.

Re-equipment with the D.H.10 followed. This rare twin-engined bomber which would have found a role in Trenchard's independent bomber force, had the war continued, was now to find its own niche in history when on the night of 14/15 May a No.120 Squadron aircraft flew the first night mail non-stop from Hawkinge to Cologne, a distance of some 300 miles, in just over three hours.

By 23 August 1919, the military requirement to carry the mail had ceased. Altogether the forces air mail service had carried some 90 tons of mail in 7,164 bags without loss. And No.120 Squadron had carried the bulk, being involved in 130 of the 173 mail flights operated.

Why bother to mention all this now? There are several reasons. Today No.120 Squadron continues to serve with the Royal Air Force, though between 1919 and 1941 it was disbanded. It still operates what started out as a de Havilland product, the maritime Nimrod, which has a long and successful lineage out of Comet. The writer could scarcely let John Stroud, who now lives and works in Scotland, get away with writing about the real beginnings of commercial transport without reminding him that just a few miles down the road at Kinloss, Morayshire, there were others with a small claim to fame. And finally, the writer had to declare his own interest for he served with the squadron for six months in 1945 before it was disbanded yet again, though not for long.

All this and more, which the *Review* will be exploring, spells out much more than history – it is heritage, described by the *Oxford English Dictionary* as 'that which has been or may be inherited'.   □

September 1939: In Poland the Luftwaffe Strikes and Wins

Johann-Friedrich Schliephake

As Supreme Military Commander Hitler issued orders concerning the objective, preparation and sequence of all combined operations by means of Directives for the Conduct of War. For the attack on Poland, code-named Case White, Hitler had given Directive No.1 as early as 3 April 1939. This directive was the basis for the plans to be made by the Military Supreme Command (OKW) and the High Commands of the three fighting forces, viz army, air force and navy. It specified that the armed forces had to begin the fighting 'with surprise, powerful blows and to destroy the Polish armed forces'.

Hitler/Stalin Pact

On 23 August 1939 the German/Soviet talks which had started on 20 May resulted in the signing of a pact. Among other items it contained a non-aggression treaty limited to ten years and a secret extra protocol in which the division of Poland between Germany and the Soviet Union was agreed.

Army and Luftwaffe Attack Plans

On 25 August, less than two days after the conclusion of this pact, the German forces, ie the army with two army groups and two air fleets, were already deployed in a widespread arc along the Polish border. There was Army Group North under the command of Generaloberst Fedor von Bock with No.4 Army (Gen der Artillerie Günther von Kluge) in Pomerania and No.3 Army (Gen der Artillerie Georg von Küchler) in East Prussia; Army Group South under the command of Generaloberst Gerd von Rundstedt comprised No.8 Army (Gen der Infanterie Johannes Blaskowitz) in Lower Silesia, No.10 Army (Gen der Artillerie Walter von Reichenau) in Upper Silesia, and No.14 Army (Generaloberst Wilhelm List) in Moravia. These forces amounted to a total of fifty-four divisions with 3,195 light and medium armoured vehicles.

Hitler's plan provided for the two army groups to encircle the majority of the Polish army in a surprise pincer movement to prevent its retreat in good order behind the rivers Vistula, Narew and San where enemy formations would have been out of the reach of the German armour. The main thrust in the direction of Warsaw was the task of No.10 Army whose flanks were to be covered in the north by No.8 and in the south by No.14 Army and also the majority of Nos.3 and 4 Armies with Gen Heinz Guderian's XIX Armoured Corps in the vanguard. The weaker parts of the Army Group North were to join in the Polish Corridor (the part of Poland between the German Reich and the province of East Prussia which, as a result of border lines drawn after the First World War, was isolated from the Reich) to prevent the retreat of the Polish North Army to the south. For this plan to succeed there needed to be German air superiority over Poland right from the start.

Ground and air forces now had to co-operate in war for the first time on a large scale. Their actions would be based on the evaluation of battle experience which the Condor Legion had acquired in the Spanish Civil War and it was decided to tailor the disposition of the air force directly to the deployment of the ground forces.

No.1 Air Fleet Command and No.4 Air Fleet Command, the latter having been formed from the Austrian Air Force Command immediately after the occupation of the rest of Czecho-Slovakia in the previous March, were tasked to the air war against Poland under the command of the Luftwaffe Supreme Commander.

No.1 Air Fleet 'East' under Gen Albert Kesselring had to rely on co-operation with the Army Group North, and No.2 Air Fleet 'Southeast' under Gen Alexander Löhr on co-operation with Army Group South. Gen Kesselring (Chief of Staff Oberst Speidel) had under him

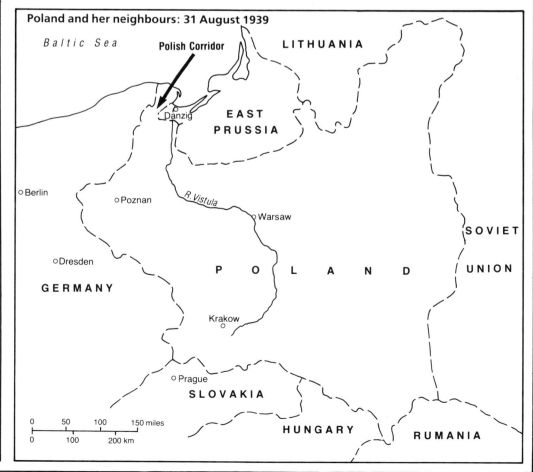

Poland and her neighbours: 31 August 1939

German Assault: Sept 1939

No.1 Fleet East

Baltic Sea

LATVIA
LITHUANIA
EAST PRUSSIA
Putzig
Vilna
4th Army
3rd Army
Torun
Grodno
Poznan
Plock
Warsaw
Brest
Lodz
Deblin
Radom
Golab
8th Army
Tomaszow
POLAND
10th Army
Lublin
Lutsk
Katowice
Krakow
Lemberg
No.4 Air Fleet South
14th Army
GERMANY
R. Vistula
SOVIET UNION
SLOVAKIA
HUNGARY
RUMANIA

German Army
Polish army
Polish Airforce

Poland overwhelmed: post Soviet attacks of 17 Sept

Baltic Sea

LATVIA
LITHUANIA
Tilsit
Kovno
Kolberg
Danzig
Königsberg
EAST PRUSSIA
4th Army
3rd Army
Stargard
Minsk
POLAND
Slutsk
Liegnitz
8th Army
10th Army
R. Vistula
Tomaszow
Novograd-Volynskiy
GERMANY
14th Army
Lemberg
SOVIET UNION
SLOVAKIA
Kamenets-Podolskiy
HUNGARY
RUMANIA

German army
Russian attacks

Some of the Combatants

	Type	Crew	First flight	Powerplant	Loaded weight	Max speed	Cruising speed	Service ceiling	Range	Armament	Max bomb load
Fighters	Messerschmitt Bf 109E-3	1	1938	1,100hp Daimler-Benz DB 601A-1	2,610kg	560km/hr	485km/hr	11,000m	560km	Two 20mm cannon Two 7.92mm MG	
	P.Z.L. P.11C	1	1936	645hp P.Z.L. (Bristol) Skoda Mercury VI-S2	1,796kg	390km/hr	320km/hr	11,000m	800km	Two 7.7mm MG plus two more in wings some aircraft	Two 12kg
Zerstörer (destroyer)	Messerschmitt Bf 110C-1	2	1938	Two 1,100hp Daimler-Benz DB 601A-1	6,028kg	475km/hr	350km/hr	9,800m	980+km	Two 20mm cannon Four 7.92mm MG One 7.92mm MG on flexible mounting	
Dive-bomber	Junkers Ju 87B-1	2	1938	1,200hp Junkers Jumo 211Da	4,250kg	345km/hr	280km/hr	6,000m	600km with bombs	Two 7.92mm MG One 7.92mm MG on flexible mounting	500kg (one 250kg, four 50kg)
Medium bombers	Dornier Do 17M-1	3	1937	Two 900hp Bramo Fafnir 323A-1	8,185kg	410km/hr	350km/hr	6,700m	1,375km	One 7.92mm MG Two 7.92mm MG on flexible mountings	1,000kg
	Heinkel He 111P-1	4	1937	Two 1,175hp Daimler-Benz DB 601Aa	13,300kg	425km/hr	360km/hr	7,400m	2,000km	Three 7.92mm MG all on flexible mountings	2,000kg (eight 250kg)
	P.Z.L. P.37B Loś B (Elk)	4	1938	Two 918hp P.Z.L. Bristol Pegasus XX	8,560kg	410km/hr	350km/hr	6,000m	1,500km	Three 7.7mm MG all on flexible mountings	2,580kg
Ground attack	Henschel Hs 123A-1	1	1935	660hp BMW 132A-2	2,175kg	309km/hr	270km/hr	6,800m	530–860km	Two 7.92mm MG	300kg (four 50kg, ten 10kg)
Reconnaissance/ light bomber	P.Z.L. P.23B Karaś (Crucian-Carp)	2	1936	680hp P.Z.L. Bristol Pegasus VIII	3,526kg	319km/hr	274km/hr	7,300m	600–1,500km	One 7.7mm MG Two 7.7mm MG on flexible mountings	700kg

NOTES: a) In general the most up-to-date variants of the types in service have been chosen for this table. b) All armament quoted comprises fixed forward-firing machine guns or cannon except where stated.

At a base near the German/Polish border, Junkers Ju 52/3m transport crews wait for the next sortie.

Above right: In the fog on the morning of 1 September 1939, Junkers Ju 87 dive-bombers of the Geschwader staff of St.G. 2 wait for take-off.

No.1 Air Division (Generalleutnant Grauert) in Pomerania and the Luftwaffe Command East Prussia (Generalleutnant Wimmer); Gen Löhr (Chief of Staff Oberst Korten) commanded the No.2 Air Division (Generalmajor Loerzer) and the units of the Air Commander for Special Purposes Generalmajor Freiherr von Richthofen.

A battle area had been assigned to air force units based in East Prussia which extended from the eastern bank of the River Vistula (Wista) to Warsaw, Brest-Litovsk, and Vilna with a 500km long front. The remaining units of No.1 Air Fleet and No.4 Air Fleet shared the areas west of the Vistula and south of Warsaw, an area which extended over a front of up to 700km.

At the beginning of the war the two air fleets were to make a surprise attack to destroy the Polish air force on the ground by successive strikes; only when this had been accomplished were they to be used for direct support of the ground forces.

A 250kg bomb is attached to the underfuselage crutch of a Junkers Ju 87.

A total of 2,093 modern operational aircraft was available to undertake these tasks. The twin-engined bombers totalling 810 aircraft comprised Heinkel He 111P and the Dornier Do 17 E, M and Z. A further 406 single-engined ground attack and dive-bombers were available, the Henschel Hs 123A and Junkers

Junkers Ju 87s of 1./St.G.1 each loaded with one 250kg and 50kg bombs over Poland.

Ju 87B respectively. There were 456 fighters on strength, a combination of Messerschmitt Bf 109Ds and Es and the twin-engined two-seat Bf 100C Zerstörer (destroyer). For short-range reconnaissance there was a total of 288 Heinkel He 45Cs, Heinkel He 46Cs and Henschel Hs 126As on strength. Long-range reconnaissance was the task of the 133 Dornier Do 17Fs and Ps which were available. In addition there were liaison and transport aircraft, viz the Fieseler Fi 156C Storch and Junkers Ju 52/3ms and a wide variety of seaplane types of various coastal units for reconnaissance over the Baltic and along the Polish coastline.

Polish Forces on the Ground and in the Air

The opposing Polish forces comprised thirty-eight divisions and thirteen brigades of which no fewer than eleven were cavalry brigades. Between them they had 1,134 mostly obsolescent light and small tanks. The six Polish air force divisions, reorganised only in July, essentially comprised one fighter brigade with five squadrons for the air defence of Warsaw, one bomber brigade comprising nine squadrons and in addition some reconnaissance and liaison squadrons.

Altogether the Polish Supreme Commander, Marshal Edward Rydz-Smigly, had at his disposal 463 operational front line aircraft. They comprised the following types: 158 P.Z.L. P.7 fighters, 154 P.Z.L. P.37 Łoś, P.23 Karaś and Potez 25 medium and light bomber aircraft, 84 Lublin R-XIII, P.W.S.19 and P.Z.L. P.23 reconnaissance aircraft and a further 66 miscellaneous aircraft. With the exception of the 36 P.Z.L. P.37 Łoś bombers all the other aircraft types were obsolete and poorly armed.

This enemy could never constitute any danger to the ultra-modern Luftwaffe, its crews believed. Keenly motivated by massive propaganda, they waited impatiently during the last few days of August

for the order to attack which came on 25 August shortly after 1830hrs when the two air fleets received a radio message from the Luftwaffe Supreme Commander, Generalfeldmarschall Hermann Göring: 'Ostmarkflug 26 August, 0430hrs', this being the code signal for the attack on Poland by the Luftwaffe. A few hours later Hitler withdrew his order to attack as Mussolini had suddenly said that Italy was not yet ready for war, and Britain had once more unmistakeably reaffirmed her guarantee of assistance to Poland.

Last Days of Peace

During the next few days Hitler tried in vain to upset the coalition of Britain, France and Poland. He did not succeed and this resulted in the ultimate failure of the talks with Poland about the city of Danzig and the Polish Corridor which initially had taken a positive course. On 31 August at 1240hrs, Hitler gave the final order to attack Poland under the code name Case White. Day of attack: 1 September 1939; time of attack 0445 hrs.

First Day of War

Elbing airfield, home base in East Prussia of St.G.1 (Dive Bomber Group), was covered by dense fog in the early hours of 1 September 1939, when three Junkers Ju 87Bs of No.3 Staffel were readied for take-off for the first strike on Polish territory. The officer leading the flight, Oberleutnant Bruno Dilley and his two wingmen, Leutnant Horst Schiller and Unteroffizier Gerhard Grenzel, had special orders to prevent the Poles blowing up a strategically important railway bridge near Tczew on the Vistula, which carried German supplies across the Polish Corridor to East Prussia. It was known that the Poles had their explosive charges in place and would blow up the bridge from a remotely controlled firing point set up in a shed at Tczew railway

Returning to base after a mission.

station. The Luftwaffe's task was to destroy that shed.

Nineteen minutes before the outbreak of war, at 0426hrs, the three dive-bombers took off, each loaded with a 250kg bomb under the fuselage and four 50kg bombs under the wings. In an ultra low-level flight below a 50m thick layer of stratus cloud the three aircraft were dashing at tree-top height towards their target which appeared in front of them eight minutes later.

The strike was successful; the bomb-charge ignitor points for the bridge were destroyed with their connecting leads. Even so the Polish pioneers succeeded in piecing the cables together under a hail

Immediately following a mission a Junkers Ju 87 is refuelled and rearmed.

of bombs dropped during an attack flown one hour later by No.III group of KG 3 which had taken off from Heiligenbeil with its Do 17Z aircraft. Before the arrival of a German armoured train which was to occupy the bridge by a *coup de main*, part of it was blown up by the Poles at 0630hrs.

At 0445hrs the old battleship *Schleswig Holstein*, which had been commissioned in 1908, fired the first shells of the Second World War in the taking of the Westernplatte near the city of Danzig and in turn had been fired upon. At precisely the same time, in the south in the No.4 Air Fleet region, where the weather was better, twenty-one Ju 87Bs of No.I Group of St.G.2 *Immelmann* led by its commander Maj Oscar Dinort took off from Nieder-Ellguth am Steinberg airfield near Oppeln. Their target was the airport at Kraków. During the flight back to base Oberleutnant Frank Neubert became the first man to win an air victory in the war by shooting down a P.Z.L. P.11 flown by the officer commanding No.122 Fighter Squadron, Capt Medwecki.

2nd-Lieut Wladyslav Gnys who accompanied Capt Medwecki could

A Junkers Ju 87 Staffel rests.

A Henschel Hs 123 ground attack aircraft.

escape the German machine gun fire only by taking violent evasive action by banking sharply in his highly manoeuvrable P.11. Shortly afterwards, at about 0530hrs, Gnys succeeded in shooting down the first German aircraft, a Ju 87 of No.I Group of St.G.2. At 0600hrs No.I Group of KG 1 *Hindenburg* (Battle Group) bombed the base of the Polish naval air division at Puck with He 111P aircraft. Among the dead was the commander of the naval air division, Lt-Cdr Szystowski, who became the first naval officer to be killed in action in the Second World War.

In the south only one close support and two dive-bomber units were available on this morning to the Air Commander for Special Duties Freiherr von Richthofen. Coming from his operational base at Altsiedel Hauptmann Otto Weiß, Captain of No.1 Staffel of II.(Schlacht)/LG 2 (Close Support Training Group), crossed the Polish border shortly after sunrise to initiate the close support operations in front of the spearhead of XVI Armoured Corps.

II.(Schlacht)/LG 2 had available on that morning more than thirty-six operational Henschel Hs 123s. These unpretentious, robust biplanes which withstood enemy fire extraordinarily well were equipped with two rigidly mounted MG17 machine guns and underwing stores which could be four 50kg bombs. The attack by No.1 Squadron was followed by No.2 Squadron led by Oberleutnant Adolf Galland.

These close support operations were to prove what had been tried out already during the Spanish Civil War: von Richthofen's forward aircraft control officers operated in direct contact with the army's spearhead units. Should Polish forces slow down the advance of German troops, the forward control officer concerned requested close support assistance by radio directly from von Richthofen, and without any loss of time he could deploy the units under his command against enemy resistance.

On account of the poor weather Göring, one hour after the start of the war, had cancelled a massive air raid on Warsaw with all available battle groups. This was known as Operation Water Edge and had been planned for the afternoon of 1 September. Even so, a first strike was flown on Warsaw-Okęcie Airport. It was operated by He 111P aircraft of No.II Group of LG 1 (Training Group) from Powunden/East Prussia, accompanied by Bf 110C-1s of No.I (destroyer) Group of LG 1 from Jesau/East Prussia. During their return flight the destroyers were engaged in a brief aerial combat by eight P.Z.L. P.11 fighters, one of them being shot down.

In the north the poor weather continued. No.I Group of JG 21 (Fighter Group) led by Maj Mettig, based at Gutenfeld in East Prussia, was one of only six groups in No.1 Air Fleet which was able to operate. In his Bf 109D Leutnant Rödel of the 21st Fighter Group scored his first victory.

In accordance with their training the Luftwaffe units had orders to be on the offensive. Their first strike had to hit the enemy air force hard on the ground and in the air in order to reduce and possibly eliminate the enemy air threat. In accordance with the aerial warfare doctrine of the Italian Gen Douhet this meant there would be control of the air before there was a decision on the ground. This was the Luftwaffe's guiding principle during the campaign.

As cloud had now broken up over Kraków this airfield was hit again. At about 1250hrs Oberstleutnant Maier's No.I and Maj Evers' No.III Groups of KG 4 *General Wever* took off from their base at Langenau, north of Breslau in Upper Silesia, for their first sortie over enemy territory. Almost 100 tons of high-explosive bombs of various sizes, dropped from sixty Heinkel He 111Ps, exploded on the airfield, on hangars and living quarters and between parked aircraft. The escorting Messerschmitt destroyers of No.I Group of ZG 76 (Destroyer Group – *Hauptmann Reinecke*) which had taken off from the Ohlau base did not make contact with the enemy. After this the dive-bombers of No.I Group of St.G.2 under Maj Dinort flew their second mission that day to continue the destruction. Dinort's dive-bombers were followed by Nos.I and II Groups of KG 77 which were based at Brieg with their Dornier Do 17s. From these units No.III Group under Oberst von Stutterheim dropped thickwalled splinter bombs on the airfield from such a low altitude that many aircraft suffered damage from their own bomb fragments.

KG 76 flew missions from Breslau-Schöngarten and Zipers-Neudorf to Radom, Łódź, Skierniewice, Tmaszow, Kielce and Czenstochowa airfields. No.II Group of KG 77 bombed the Krosno and Moderowka airfields, while one group of St.G.77 made its first target the line of defensive bunkers in front of Lubliniec. There followed further attacks on the airfields of Katowice and Wadowice. Finally, Lvov airfield was attacked and heavily hit by No.II Group of KG 4 *General Wever* under Oberstleutnant Erdmann from their base at Oels. During this mission the crew of Unteroffizier Hölscher of No.5 Squadron was shot down and taken prisoner.

While No.I Group of St.G.76 (Hauptmann Siegel) in the morning attacked fortified positions in the township of Wieluń where Oberleutnant Peltz with No.1 Squadron had the market square as his target, a dreadful inferno started to develop in this area as noon approached.

Reconnaissance aircraft reported heavy concentrations of Polish cavalry near Wieluń, especially in front of the left wing of XVI Army Corps, as well as considerable troop movements in the same area. By 1250hrs No.I Group of St.G.2 (Maj Dinort) took off for the second time from its base at Nieder-Ellguth to attack the unprotected Polish cavalry brigade with 250kg and 50kg fragmentation bombs. This unequal fight, dive-bombers against cavalry, was

Armourers load a Henschel Hs 123 with SC50 bombs.

A Dornier Do 17 long-range reconnaissance aircraft being prepared for a mission.

continued by thirty more Ju 87s of No. I Group of St.G. 77 (Oberst Schwarzkopf), and after reconnaissance aircraft had reported more troop concentrations near Wieluń No. I Group of KG 77 (Maj Balk), rearmed after its attack on Kraków, took off again and finished off the cavalry brigade. Abandoned by their own air force, only a few survivors managed to escape. This Luftwaffe mission for direct support of ground troops succeeded in breaking up the assembly of Polish troops north of Czestochowa in front of the spearhead of the German troops.

As the weather improved the wave of destructive strikes of the Luftwaffe extended to the airfields in the north and those situated in the Polish Corridor were especially affected. Poznań Ławica was the target of No. II Group of KG 26

In the shade of a Dornier Do 17Z crews of I./KG 76 relax awaiting their next sortie. (*Archiv KG 76 Rehm*)

while No. I Group of KG 152 (II./KG 1 from 18 September), also equipped with the Heinkel He 111P, dropped high-explosive bombs on Torun Airport, the most modern in the north of Poland. Torun was again attacked in the late afternoon, this time by No. I Group of KG 1 *Hindenburg*.

No. I Group of KG 53 *Legion Condor* (Maj Mehnert) had orders to attack the airfields of Gniezno (1234hrs), Wrzesnia (1258hrs) and Sroda (1320hrs). Its Heinkel He 111s were loaded with ninety-six 50kg thickwalled splinter bombs and 160 cylindrical splinter bombs also weighing 50kg each. The airfields of Grudziądz and Bydgoszcz and an ammunition store were the targets of No. II Group of KG 3. At 1646hrs the 4th Squadron of KG 2 took off from Schippenheim in East Prussia with Do 17s to attack Brest-Litovsk. Further targets of KG 2 were Plock and Lida.

Dive-bombers of No. 1 Air Division to which were subordinated Nos. II and III Groups of St.G. 2, No. IV Group of (Stuka)/LG 1 and No. 4 dive-bomber squadron of 186 (T) Coastal Unit which had been earmarked for the aircraft

Twenty 50kg bombs under the fuselage of a Dornier Do 17Z of KG 76 waiting to be loaded. (*Archiv KG 76 Rehm*)

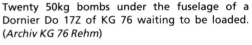

carrier *Graf Zeppelin* flew missions against the Polish navy and harbour facilities. Altogether there were 120 Ju 87Bs supported by six modern navy destroyers.

With an earsplitting howl the dive-bombers bore down on Oxhöft and the naval port of Gdynia where the small diving workshop ship *Nurek* and the torpedo boat *Mazur* were sunk. The air strikes on the port of Hel aimed in

A Dornier Do 17Z of I./KG 76 dropping bombs on Piotrków railway station on 3 September 1939. (*Archiv KG 76 Rehm*)

Messerschmitt Bf 110C Zerstörers (destroyers).

particular at the two largest Polish warships, the destroyer *Wicher* and the minelayer *Gryf*. Both ships suffered heavy damage despite defensive fire by ship-borne and coastal anti-aircraft batteries.

However, the last large-scale air strike of this first day had yet to be made; the target was the Polish capital. Although Operation Water Edge had been cancelled by Göring in the early hours of the morning because of poor weather conditions this mission had by no means been given up, and the right conditions came in the late afternoon; before KG 27 (Oberst Behrendts) had arrived from its home bases Hannover-Langenhagen, Wunstorf and Delmenhorst, which were 750km away, No. II Group of LG 1 (Hauptmann Allolio) flew its second mission from Powunden with orders to attack the Warsaw airfields of Okęcie, Goclaw and Mokotów. Meanwhile, No. I Group of St.G. 1 (Hauptmann Hozzel) dived onto the radio stations of Babice and Lacy. This was Dilley's second mission that day. However, the transmitter aerial masts withstood the attack and the strike did not yield the results that had been expected.

Shortly after 1730hrs the three Groups of KG 27 arrived over Warsaw with their ninety He 111Ps. Cover was provided by No. I Group of (Z)/LG 1 (Hauptmann Schleif). Over the very centre of Warsaw the bomber formations were engaged by two Polish fighter squadrons led by Col S

Head-on the Messerschmitt Bf 110, destroyer, lives up to its name.

Pawlikowski comprising about thirty P.11Cs. The destroyers thus became engaged in the first of the larger air battles of this war, five Polish fighters being shot down within a few minutes.

By nightfall dense fog came down again, forcing the Luftwaffe to discontinue its activities for the time being. On the first day of the war against the Polish air force there had been thirty group attacks, seventeen of them directed against ground installations (airfields) which were either totally or partially paralysed. Eight missions were flown in direct support of the ground forces and five against naval targets. Thirty Polish aircraft had been destroyed on the ground and nine shot down in air battles. Nowhere had the Polish defences succeeded in repelling or weakening the German strikes. Meanwhile, the Luftwaffe had lost fourteen of its own aircraft.

After this stocktaking the Luftwaffe High Command could be satisfied that the Polish air force had been destroyed. However, although the installations at the peacetime airfields and the reserve and training aircraft, as well as those aircraft that were under repair, had been almost completely destroyed, not a single front line aircraft had been hit. Most of the fighter, bomber and reconnaissance aircraft had been transferred to unprepared airfields within the country two days before the start of the war, the main purpose of this move being to have them available for the defence of Warsaw.

With the Vistula below, a Heinkel He 111P returns to base.

A view over the observer's shoulder in a Heinkel He 111 flying at low level over a Polish township.

Even so, the Luftwaffe had won air supremacy right from the first hour owing to qualitative and quantitative superiority.

Destruction Continues – Second Day

When at dawn on 2 September cloud and fog formed again, conditions similar to those on the first day, the Luftwaffe units took off again to continue the work of destroying the enemy. The targets were Warsaw, Poznań and Dęblin. KG4 bombed the three airfields at Dęblin. The accompanying twelve Messerschmitt Bf 110 destroyers of No. I Group of ZG 76, amongst them Leutnant Lent, destroyed by gunfire eleven aircraft parked on the ground.

Von Richthofen's dive-bombers cleared the way ahead of No. 1 Armoured Division, and enemy positions near Działoszyn and afterwards troop concentrations southeast of Częstochowa were battered and dispersed. Forty dive-bombers of No. I Group of St.G.2 and

A Heinkel He 111 of KG 1 dropping incendiary bombs on Warsaw.

No. I Group St.G.77 surprised a Polish infantry division when it detrained at Piotrkow station causing heavy casualties. St.G.77 attacked marching troop formations near Radomsko. At about 0935hrs the Polish submarine *Rys* was attacked by naval aircraft and damaged. At Hel harbour the destroyer *Wicher* and the minelayer *Gryf* were again the targets of dive-bombers.

First Polish Strikes

Success on the other hand did not attend the six Polish P.Z.L. P.23B Karaś B light bombers of No. 24 Eskadry when they flew their first mission against the Ger-

man XVI Armoured Corps north of Częstochowa. Eighteen further Karaś bombers of No. VI Group (Dyon) flew missions also in the Częstochowa area against advancing German tank columns; when after having dropped their bombs and disobeying their orders they embarked on a low-level machine gun attack five aircraft were shot down.

Lieut Malinowski of No. 41 Polish Reconnaissance Squadron at about 1100hrs dropped eight 50kg bombs during a reconnaissance flight in his Karaś on the railway sidings full of military transports at Neidenburg in East Prussia. This was to remain the only Polish bomb attack on German territory during the September campaign. The planned air attack on the city of Königsberg in East Prussia with all available P.Z.L. P.37B Łoś medium bombers was rescinded at the last moment. The Polish fighter aircraft did not succeed either, despite repeated German air strikes on Warsaw, in getting near German aircraft.

During a search and strike operation over Łódź Nos. 1 and 2 Squadrons of ZG 76 became involved in fierce dogfights with P.Z.L. P.11C fighters and Leutnant Lent and Oberleutnant Nagel succeeded in shooting down one P.Z.L. each, three Bf 110s being lost in this combat. No. I Group of ZG 2 (Hauptmann Gentzen) flew from its operational base at Groß-Stein near Oppeln, Upper Silesia, with its Messerschmitt Bf 109Ds

A Polish armoured train near Lochow is destroyed by a dive-bomber attack.

to the Radom area, destroying during the mission sixteen Polish aircraft in the air and on the ground.

By the end of the second day of the war the destruction of all important peacetime airfields of the Polish air force had been completed. Targets for air strikes also included hubs in the ground transport system to dislocate troop movements in the hinterland; communications installations of every kind were also attacked.

From the Third Day – Luftwaffe Intervenes Increasingly in Ground Fighting

Whereas Luftwaffe operations were largely of a strategic character during the first two days, there was an obvious shift to tactical missions from 3 September. The close co-operation with the ground forces was exemplified by more and more heavy bombardments of pockets of resistance, artillery positions and troop concentrations.

However, the Polish airmen did not remain inactive. Encouraged by the Declaration of War on Germany by Britain and France that morning, eighteen Karaś bombers attacked Nos. 1 and 4 Armoured Divisions near Radomsko with 50kg fragmentation bombs, inflicting heavy losses upon the Germans. Another mission flown by six Karaś bombers of No. 24 Eskadry against XVI Armoured Corps north of Częstochowa did not meet with success.

The Luftwaffe success, on the other hand, continued. In the north eleven

Warsaw-Okęcie Airport shortly after a bombing attack. The airfield, the adjacent hangars and the government aircraft factory have all been hit.

dive-bombers of No. 4 dive-bomber Staffel of 186 (T) Coastal Unit descended again onto the Polish navy base of Hel with Hauptmann Blattner's bombs hitting the stern of the minelayer *Gryf*. At about 1500hrs the ship was attacked once more and sunk by Hauptmann Stein's aircraft. Oberleutnant Rummel and Leutnant Lion, both in 4./186 (T), sank the destroyer *Wicher* by a direct hit. This meant that the two most powerful ships of the Polish navy had been eliminated in this attack, one Ju 87 being shot down.

After an attack by the two dive-bomber groups of Oberst Baier's training group on the area ahead of XI Army Corps, the city of Działoszyn fell into German hands without any further resistance. Having done that both dive-bomber groups, together with the Hs 123 close support unit, repeatedly attacked the 7th Polish Division which was encircled in the area southeast of Częstochowa. Totally demoralised by the powerful attacks from the air, the division admitted defeat the following morning.

At about 1430hrs battle and dive-bomber groups, accompanied by destroyers of LG 1, appeared once more over Warsaw-Okęcie and bombed the most important centre of the Polish aircraft industry, the Państwowe Zaklady Lotnicze (P.Z.L.) Works. In the course of that mission thirty P.Z.L. fighters became engaged in a fierce fight with Messerschmitt destroyers, the result of which was again that five Polish fighters were shot down with only one German loss. That evening the German High Command (OKW) announced that the uninterrupted missions of close support aircraft and dive-bombers had made essential contributions to the rapid success of the advancing troops.

During the night of 3/4 September the No. 100 Air Communications Unit (later KG 100) flew the first beam bombing mission from Köthen in the Province of Saxony to a military target in the Polish city of Palmiry using the X system. On the following nights more targets in Poland were attacked by the X method by No. 1 Squadron (Ju 52/3ms) and No. 2 Squadron (He 111s) of No. 100 Air Communications Unit. These were the first night bombing missions of the Second World War as well as the first bombing sorties ever to make use of VHF radio beams, both as track guides and for bomb release.

Fourth Day – Polish Defence Collapses

On 4 September the Polish defence effort started to collapse. Continuous attacks from the air had caused terrible disorder, Gen Przedrzymirski's Modlin Army hav-

An earlier photograph of Warsaw-Okęcie shows a P.Z.L. P.37B Łoś medium bomber and P.Z.L. P.11C single-seat fighters.

ing been hit particularly badly. Thus Warsaw lost its protection against attacks which could now be expected from the north.

In the early morning hours twenty-seven of the modern P.Z.L. P.37 Łoś Polish medium bombers of Nos. X and XV Dyon received permission to take off for the first time and flew several operations against XVI Armoured Corps in the Radomsko-Piotrków area. In the afternoon five Łoś bombers of No. X Dyon repeated attacks on motorised German columns and artillery positions west of Wieluń, while five Łoś bombers of No. XV Dyon attacked advancing German columns near Kamiensk. Seven bombers did not return from these missions and two of No. X Dyon were destroyed on the ground at their operational base airfield at Kuciny.

Polish fighters also succeeded in shooting down German short-range reconnaissance aircraft on several occasions. Over Łódź a Bf 109D Staffel of No. I Group of ZG 2 led by Oberleutnant von Roon became engaged in a fierce

fight with the fighter Eskadry of the Łódź Army in which the P.Z.L. P.11C fighters were greatly inferior to the Bf 109D. Eleven P.Z.L. fighters were shot down in that engagement.

Shortly before nightfall the Karaś bombers of No. VI Dyon, also, took off to attack motorised units between Ciechanow and Pultusk, two aircraft being lost. Unexpectedly, the twenty-five P.Z.L. fighters of Brygada Poscigowa still operational were transferred to different unprepared airfields. Some were discovered by a couple of Bf 109Ds of No. I Group of ZG 2, three P.11s of the Łódź Army being set on fire by fighter gunfire on the airfield at Widzew to which they had been relocated. Throughout the day the attacks of German strike aircraft were aimed increasingly at marching columns, railway transport and river crossing operations to prevent the orderly retreat of the Polish army.

Fifth to Ninth Days

On 5 September six Łoś bombers of No. XV Dyon mounted an attack against XVI Armoured Corps which, after the destruction of the Polish defence forces in the Piotrków Trybunalski region, was able to advance towards Warsaw almost without any enemy resistance.

Owing to the increasing losses of obsolescent Polish reconnaissance aircraft, the remaining few fighters and bombers were now also used increasingly for reconnaissance work.

The Luftwaffe continued to destroy the lines of communication of the Polish armies, successfully bombing the railway stations of Zduńska-Wola, Skarzysko, Tarnów and Wrzesnia. In the early hours of 6 September six Łoś bombers of No. XV Dyon again attacked German armoured units near Maków Mazowiecki where three of the bomber aircraft were shot down by Messerschmitt fighters.

So far the losses which the Polish air force had suffered amounted to over 60 per cent of its original operational strength. For this reason the remainder of the Brygada Pościgowa was moved to the Lublin area which meant that the Polish capital lay bare, without fighter cover an easy prey to Luftwaffe attacks.

In the meantime the Luftwaffe flew uninterrupted attacks against railway installations and bridges. Its area of action was now that between the River Bug and the central reaches of the Vistula. No. 4 Air Fleet moved the largest part of its units, which were to provide tactical support to the ground forces from Silesia into Poland, to Kielce and Radom. After the Polish Prusy Army had been defeated XV and XVI Armoured Corps could advance without resistance

to Kielce in the direction of Warsaw.

After No. 4 Armoured Division, assisted by Maj Spielvogel's Hs 123 close support aircraft, had seized the Warsaw Okęcie airport early in the morning of 8 September, about 120,000 Polish soldiers barricaded themselves in the streets of Warsaw to defend their capital. While Nos. 8 and 10 Armies embarked on a large-scale pincer movement around Warsaw, two moderately strong Polish armies joined together in the rear of these two German armies in the Kutno area between the Vistula in the north and its tributary, the Bzura, in the south. Maj Gen Kutrzeba's Poznań Army and Gen Bortnowski's Pomorze Army had managed to break away to the east from the Polish Corridor and the western part of Poland in night marches, seeking shelter in the forests in daylight, so that their movements escaped the Luftwaffe's attention. The battle which broke out by the River Bzura on 9 September was to be the largest of the Poland campaign.

Battle of Ilza

At the same time strong Polish units concentrated in front of the right-hand wing of No. 10 Army south of Radom and north of the Łysa Góra mountain ranges with emphasis on the sprawling forests near Ilza. It was here that the first large-scale battle of encirclement of this war took place between 8 and 13 September. During the course of the battle the German anti-aircraft (AA) artillery, Maj Weißer's No. I Group of No. 22 AA Regiment, first proved their efficiency in firing at ground-based targets.

A shot-down P.Z.L. P.23 Karaś light bomber.

It was not only Nos. I and II Groups of KG 55 and No. I Group of KG 77 which participated in the fighting, but also about 150 dive-bombers (the first Groups of St.G. 1, 2 and 77, the latter subordinated to No. III Group of St.G. 51, as well as the Hs 123s of No. II Group of (Schlacht)/LG 2. Fighter aircraft and destroyers under the command of Freiherr von Richthofen supported the operations of the ground forces and played a decisive role in the outcome of the encirclement. At its end 60,000 soldiers (six divisions) of the Polish Prusy Army led by Gen Dab-Biernacki were taken prisoner.

Battle of the Bzura – 9 to 20 September

Meanwhile, Gen Kutrzeba succeeded in achieving deep breaches into the German lines which affected in particular No. 30 Infantry Division. This serious situation prompted the G.O.C. Army Group South, Generaloberst von Rundstedt, to request for the first time powerful air support in the Kutno area. This brought into action with immediate effect at the Bzura several Kampfgeschwader (KG 3, KG 4 and KG 55) which in the previous days had flown attacks on Warsaw as well as industry and railway targets east of the Vistula and also von Richthofen's close support dive-bomber units. Here again the 50kg fragmentation bombs took a terrible toll of the Polish troops as they had in the Radom battle of encirclement.

No. 100 Air Communications Unit continued its night attacks by the X method, its targets being industrial installations in Warsaw. During the night of 10/11 September on one of these missions, the crew of Oberleutnant Korthals succeeded in hitting an ammunition depot from a height of 26,000ft with their last high-explosive bomb. The air pressure of the exploding ammunition had such a powerful effect even at that height that reportedly several rivets on the aircraft 'popped' while the explosions shattered window panes up to 35km away.

The share the Luftwaffe had in the battle of the Bzura grew more important from day to day. It succeeded in repelling the Polish attempts to break through, not only near Lęczyca but also near Lowicz between 12 and 14 September and near Sochaczew on 15 and 16 September. As it had to muster all the aircraft available to fight the Polish troops encircled at the Bzura, the Luftwaffe had to stop almost all other operations except reconnaissance flights over the area east of the sweeping bend of the Vistula. The transport command flights were not affected and could continue without interruption. They gave support to the Luftwaffe operational units when they were moved to forward bases and carried large quantities of ammunition and provisions to the front line.

During the battle of the Bzura, too, the most effective missions were flown by von Richthofen's close support and dive-bomber units. No activity or even intention of the enemy escaped the attention of his Do 17F reconnaissance aircraft of No. 1 Staffel (F)/124 (long-range reconnaissance group), their reports being followed immediately by appropriate action at focal points. From dawn to dusk up to ten missions were flown by the Hs 123s of No. II Group of (Schlacht)/LG 2 under Hauptmann Otto Weiß after Maj

Spielvogel failed to return from a reconnaissance flight in his Fieseler Fi 156 Storch on 13 September. No. I Group of (Z)/LG 1, which so far had won thirty victories in the air, operated with their Messerschmitt destroyers together with the Henschel and Junkers close support aircraft and dive-bombers on low-level strikes, firing with airborne weapons at everything that moved in the area of continuously reducing encirclement.

In the afternoon of 20 September the last resistance of the Poznań and Pomorze armies died down at the Bzura. Afterwards, thousands of vehicles and guns battered by gunfire and bombing, horsemen killed in action and dead horses lay scattered over the pinewoods, ridges and emergency bridges over the river, Gen Bortnowksi and 170,000 men being taken prisoner. The scattered units

remaining managed to escape the scene of this inferno and to fight their way through Kampinos heath to the Warsaw and Modlin fortresses. After the 60,000 men of the Lublin Army had capitulated on 20 September, there was no front left in the central part of the operational area.

Greatly impressed by the successes his Luftwaffe had achieved, Göring said when visiting a Kampfgeschwader, '... and if you perform in the west as you did against Poland, then the British, too, will learn to run'. Gen Wimmer, equally impressed by the results of their large-scale operations, said to his units, 'Gentlemen, your commitment has come to an end at this point; and now – eyes left – to the west!'

Luftwaffe Begins to Transfer Units to the West

The bulk of the Kampfgeschwader that had seen action was moved back to its home bases between 19 and 22 September to be readied for impending use in the west. No. I Group of JG 1 had been transferred back already, after only eight days of operational flying, No. I Group of KG 53 following on 12 September.

As the Bzura battle of annihilation neared its climax the encirclement of Warsaw and Modlin fortresses had been completed. Meanwhile, Łoś bombers made occasional attacks on the advancing German troops on 12, 14 and 16 September.

Soviet Union Intervenes

On 17 September seven large Soviet units fought their way across the border into the eastern part of Poland as was provided for in the Hitler-Stalin Pact. This move surprised the rest of the Polish air force which had been withdrawn to bases in the east of the country. However, the Poles shot down two Soviet SB-2 bombers and five I-16 fighters while Soviet troops reported the shooting down of seven Polish fighters and three bombers. The Polish air force units received the order to evacuate all aircraft still operational to Romania. Within the following twenty-four hours over a hundred Polish military aircraft crossed the border en route to Romania, among them the last thirty P.Z.L. P.37 Łoś bombers.

At this stage of the fighting the strongest fortress in eastern Poland, Brest-Litovsk, was captured by XIX Armoured Corps (Gen Guderian). On 18 September, after the Polish army had been smashed and the Polish government had fled to Romania, the war had ended for all practical purposes. Even so the fighting had not stopped completely.

A Messerschmitt Bf 109E-1 being rearmed.

Warsaw

After the 120,000 defenders under Gen Rómmel had refused to surrender Warsaw and had transformed the city into a very well defended fortress, twelve He 111s of No.I Group of KG 4 on 16 September for the first time dropped large quantities of leaflets urging the Polish civilians 'to leave Warsaw to the east within twelve hours if the city did not surrender'.

More leaflets were dropped without any result on 18 and 19 September. This was followed in the early morning of 20 September by German attacks on Polish positions east of the Warsaw suburb of Praga. These the Poles managed to repel for the time being. On 21 and 22 September 178 members of the Corps Diplomatique and 1,200 foreigners were allowed to leave Warsaw.

On 22 and 24 September leaflets were again dropped over Warsaw but the Poles still refused to surrender. Now Göring gave orders to Wolfram Freiherr von Richthofen to break Polish resistance under any conditions in order to prevent a prolongation of the campaign.

At dawn on 25 September a barrage of fire from about a thousand guns and about 400 aircraft began the assault on Warsaw. As the bulk of the He 111s and Do 17Zs had already been moved to the western front in view of the perceived threat of an Anglo-French attack, dive-bombers were primarily used. Available were 240 Ju 87Bs in eight groups assisted by about thirty Hs 123 close support aircraft, about a hundred Do 17E horizontal bombers of KG 77 and thirty Junkers Ju 52/3ms from a Kampfgeschwader for Special Duties.

On that day 486 tons of high-explosive bombs and seventy-two tons of incendiary bombs were dropped on Warsaw in 1,776 sorties. As the He 111s had already

Messerschmitt Bf 109Ds lined up at a Luftwaffe airfield in East Prussia.

been moved to the west, the incendiary bombs had to be dropped from the slow Junkers transport aircraft. This was done by two soldiers who dropped the small 1.3kg and 2kg incendiary bombs by salvo release from the half-open loading hatch in the starboard side of the fuselage.

Carried away by the wind, some of the incendiaries fell into the advanced positions of a German infantry unit and its commander requested that the bombing be stopped immediately. However, Hitler, who had observed the attack from a command post, refused this categorically. Two of these transport aircraft were shot down by Polish AA artillery.

The city's situation quickly deteriorated beyond hope. Fires could no longer be extinguished; the bombs had destroyed the water mains and the streets were blocked by the debris of destroyed houses. On the following night, 26/27

September, a few Polish aircraft are believed to have taken off from the bomb-stricken Mokotov airfield, among them two gliders and in a P.11 fighter Lt-Col M Izycki as a passenger in place of the communications equipment in the radio compartment. Izycki was later to become the commander of the Polish air force that fought as an integral part of the Royal Air Force in Britain.

On 26 September attacks on Warsaw and the encircled Modlin fortress continued and negotiations for surrender started, to put a quick end to more bloodshed in this now hopeless fight. The following day Warsaw surrendered. Two days later the brave defenders of Modlin fortress gave up, and at the same time all Luftwaffe operational flying over Poland ceased. The remainder of the Polish navy cut off on Hel peninsula surrendered on 2 October, and on 6

Messerschmitt Bf 109Es of I/JG20 are ready for take-off. (*MBB*)

October the Polish campaign ended when 16,857 Polish soldiers capitulated near Kock, east of Dęblin.

Losses

The Luftwaffe casualties amounted to 759 men, the aircrew toll of this total being 189 dead, 126 wounded and 224 missing. Of the front line aircraft 285 were a total loss. Most of them had fallen victim to Polish AA artillery during their otherwise very successful low-level attacks. A further 279 aircraft had suffered damage at the rate of over 10 per cent and therefore were also deemed lost.

The losses of the Polish air force were 61 dead, 63 wounded and 110 missing, while the aircraft losses amounted to 366 machines, viz 36 bombers, 112 light bombers, 116 fighters, 81 reconnaissance aircraft and 21 other categories.

Striking the Balance

Although the Luftwaffe had not in August 1939 attained its final stage of armament efficiency, it nevertheless had been able in the Polish campaign to fulfil the tasks entrusted to it. These were the destruction of the enemy air combat

Preparing to scramble, a Messerschmitt Bf 109E of I./JG20 (later JG51) at Brandenburg-Briest. (*MBB*)

forces by surprise bomber strikes on airfields, close support of the German ground forces and air defence of the home territory. In a surprisingly short time the Luftwaffe succeeded in eliminating the enemy air forces so that it was capable of using its own personnel and equipment for effective close support of the ground forces. Continuous dive-bomber and Kampfgeschwader attacks destroyed the enemy ground forces, prevented reserve units from coming into action and disrupted lines of communication and transport. Air superiority had been almost one hundred per cent.

Even so some shortcomings were evident. For attacks on enemy ground troops there was a shortage of small-sized fragmentation bombs. There was no radio telephone intercommunication between the fighters and other units as no corresponding airborne equipment was available. This resulted in unbearable delays as communication had to be channelled through ground stations. In addition there was no suitable equipment for the quick tidying up of bombed enemy airfields for the Luftwaffe's own use, after they had been overrun by ground forces.

Despite these shortcomings, the newly

created service, the Luftwaffe, had not only come out of its baptism of fire with flying colours but had also made an essential contribution to the quick end of the Polish campaign. The ground forces were spared heavy losses by the continuous close support missions under the direct command of and co-ordinated by the Air Commander for Special Duties. The principles of operational command and the guidelines for training, too, had proved to be adequate.

However, the end of the Polish campaign had not resulted in ending the state of war as the Western Powers were not prepared any longer to tolerate Hitler's policy of annexation by force and rejected his vague offers of peace. ⊏

All photographs via the author except where otherwise credited.

September 1939: The Paper Tiger Goes to War

Derek Dempster

The author looks at the facts and expresses views about key British attitudes towards air power in the first four decades of the twentieth century. By 1 September 1939, Britain's bomber force was proved to be insufficiently strong to deter aggression. The paper tiger had been called to account.

When, at 0426hrs on 1 September 1939, Luftwaffe Oberleutnant Bruno Dilley took off from a rough little airfield near Elbing in East Prussia, and set course for Dirschau on the Vistula at the head of a Schwarm of Stuka dive-bombers to strike the first blow of the Second World War, the three Junkers Ju 87s formed part of a total bomber force of 1,552 aircraft ready for war.

In Britain at the time, Air Chief Marshal Sir Edgar Ludlow-Hewitt, Air Officer Commanding-in-Chief, Bomber Command, had less than a third of that number with which to mount an offensive. His 536 bombers were distributed among fifty-five squadrons. Thirty-three of these, with 480 aircraft, were front line units; the rest, except for two, were kept back for training and to cover initial war wastage.

By the afternoon of 2 September, ten of the front-line squadrons had landed in France as part of the Advanced Air Striking Force. They were armed with the single-engined Fairey Battle which, though advanced in its day, was now slow, short-ranged, poorly defended and completely incapable of bombing Germany from England.

The twenty-three squadrons that remained in Britain were generally better off. Six of them had the twin-engined Bristol Blenheim IV, the RAF's fastest bomber. The rest were equipped with Vickers Wellingtons, Armstrong Whitworth Whitleys and Handley Page Hampdens which, though slower than the Blenheim, were capable of carrying heavier bomb loads over considerably longer ranges.

Why Britain was so inadequately prepared was undoubtedly due to the traditionally cautious, unimaginative and deep-rooted conservatism of the nation's rulers coupled to the parsimony of the Treasury – regardless of the colour of the government in power. History can show

Viscount Trenchard, pictured in 1940, was the first Chief of Air Staff of an independent air force. When the RAF's simple, and later to be derided, biplanes took over the League of Nations Mandate for the air control of Iraq from October 1922, the political justification for the continuance of the RAF was assured. In ten years to the end of 1929, when Trenchard gave up his stewardship, the RAF had secured just 11½ per cent of the money in the total defence vote, the navy having taken 46¼ per cent and the army 42¼ per cent.

Today, Trenchard should be remembered particularly for the brilliance with which he used his meagre funds to give the RAF the best training in the world at Cranwell, Henlow and Halton; for the links he forged within the community (akin to the tradition of the county regiments) through the creation of the Auxiliary Air Force, the University Air Squadrons and the Short Service Commission; finally, but in sequence first, for the creation of the RAF Staff College in the winter of 1922/23. Among the students on the first course were Sholto Douglas, Keith Park, Richard Peirse and Charles Portal. In Germany, the Luftwaffe staff college was opened only in 1936, suggesting that in training and leadership, the twin essentials when waging war under difficulties, the RAF, thanks initially to Trenchard, got it right.

many examples of both propensities, and when you read of the loose thinking and obliquity of judgement that permeated every attempt to take British military aviation forward, it is surprising the United Kingdom was ever able to muster an air force at all, let alone create the world's most powerful one by the end of the First World War, and later rebuild another from a virtual scrap heap into the powerful and effective weapon *force majeur* compelled it to grow into during the Second World War.

By way of illustration, it is worth a brief look at the record from 1902. In that year, Col J L B Templer, Superintendent of the Balloon Factory at Farnborough, visited the Brazilian aviation pioneer, Santos Dumont, then making headlines in Paris, and on his return recommended that experiments with rigid dirigibles be started immediately. Instead of encouraging him, the War Office told him dryly that the estimates for his balloon depot – listed at £12,000 per annum — were to be cut by half.

In 1904, the War Office sent Col J E Capper, Templer's successor, to the United States to persuade the Wright brothers to pursue their experiments in Britain. The Wrights were willing, however the Treasury would not sanction the expenditure.

In April 1909, the Army Council decided to stop experimenting with aeroplanes because the cost had proved too great – £2,500. Over the same period German Government investment encouraging aviation had reached £400,000. Less than four months later, Louis Blériot flew across the English Channel, exposing Britain's maritime invulnerability.

How much Blériot's flight influenced the Secretary for War, Richard Haldane, has never been made clear, but soon after, aviation in Britain received formal recognition. However, it was not until 1911 that Prime Minister Asquith asked the Committee of Imperial Defence (CID) to '... consider the future development of aerial navigation for naval and military purposes ...'.

The army and the navy could by then collectively muster eleven airmen, and for advice the CID turned to a small and informal group of them. One was Frederick Sykes. A reconnaissance expert who had fought in the Boer War, he had learned to fly in 1910. He had also been to Libya in the following year to observe the Italian colonisation campaign and about which he wrote: 'What specially rivetted my attention was the use which the Italian military authorities were making of the aeroplane ... [They] confirmed me in the opinion that the aeroplane was going to play a decisive role in the coming struggle, but I could find hardly anyone to share this view'.

'The opposition by senior officers to air experiments amounted almost to mania,' he added. 'General Nicholson, the Chief of the Imperial General Staff, was of the opinion that aviation was a useless and expensive fad advocated by a few cranks whose ideas were unworthy of attention.'

Writing to an acquaintance about the matter, Field Marshal Lord Haig told his correspondent to '... tell Sykes he is wasting his time; flying can never be of any use to the army'.

The general attitude to flying machines was prejudiced by an ingrained and aristocratic attachment to the horse. Aeroplanes frightened horses. It was an attitude that permeated much of each government's thinking and it was the unconscious cause of official conservatism, parsimony and reluctance to encourage technical advance, except under duress, even after Hitler had come to power. That aviation made any progress at all was due principally to dedicated enthusiasts who understood its potential and wanted to realise it.

The recommendations Sykes and his colleagues drafted hurriedly for the CID led to the creation of the Royal Flying Corps (RFC) in 1912. By then, the Aircraft Factory at Farnborough had begun to produce some reasonable aeroplanes, but when war broke out in 1914, the RFC could still muster only 113

Viscount Swinton, Secretary of State for Air for three years from mid-1935, presided over the tripling of the strength of the RAF in that period. Aided by the Scottish engineer Viscount Weir who had held equivalent appointments in the First World War, enormous strides in air rearmament were made in what was still a peacetime economy, where the threat of economic collapse through financial imprudence was viewed by the Treasury with at least equal concern as the posturing of dictators. But by the spring of 1938 the parliamentary mood had changed. Rearmament was perceived not to be proceeding quickly enough and Swinton, the architect of the Expansion Air Force, was made the scapegoat and forced into resignation.

aeroplanes. In contrast, Germany started the war with 384 aeroplanes and a fleet of thirty airships.

Although the navy had looked into the possibilities of delivering torpedoes by air, no one appears to have given serious thought to arming aeroplanes for combat. Not one British aeroplane carried any defensive armour. Pilots blazed away at the enemy with rifles and pistols. They stuffed their pockets with hand grenades

and trailed cables hung with weights and aimed at entangling them in enemy propellers. Bomber crews simply lobbed their missiles over the side.

Primitive though their equipment may have been, the early pilots were quick to learn, and demanded aeroplanes designed and equipped for the tasks in hand. In characteristically British fashion, industry rose to the occasion and produced a succession of highly respectable machines. But the British belief in the invulnerability of their islands persisted, until the Germans forced them to think again.

In January 1915, Zeppelin airships raided the Midlands. Raids on London and other cities followed. Then, in 1917, formations of large, twin-engined Gotha bombers flying by day took over, incensed the people and jolted Prime Minister Lloyd George into asking Jan Christian Smuts, the South African soldier and statesman, to look over the country's air defences. Smuts reported to Parliament in September that they were alarmingly backward. He recommended putting all air Services under a single Ministry. The army and the navy protested, but they were overruled, and on 1 April 1918, the Royal Flying Corps and the Royal Naval Air Service were merged to become the Royal Air Force.

Given their heads, the War Office and the Admiralty would have put a stop to

Coming into RAF service from the spring of 1937, at the same time as German volunteers were trying out the first bomber version of the Heinkel He 111 in the war in Spain, the Armstrong Whitworth Whitley became the RAF's standard night bomber. Intended from the beginning for night work in an RAF tradition going back to the First World War, the Whitley, even when new, was considered too slow to be operated by day. However, the Rolls-Royce Merlin engine, as pictured, in Whitley IVs and Vs transformed the performance of the old warhorse which would give a good account of itself in many roles including anti-submarine, operational paratroop drops, paratroop training and emergency transport.

all ideas of an independent air force when the war ended. But the RAF's Commander-in-Chief, Sir Hugh Trenchard, had other ideas which he put into a Memorandum for Winston Churchill, the Minister for War and Air, who presented it to Parliament on 11 December 1919. In that memorandum he described how a powerful air force could be moulded into a deterrent against future wars. Fifteen years later it provided the basis for the RAF's great expansion, and also became the model for most air forces throughout the world.

Trenchard was, of course, perversely criticised for not resisting Sir Eric Geddes in 1921 in his campaign to reduce the size of the armed forces. But, he reasoned, a war-weary complement of men idling time away was not the right mixture needed for the foundations of an élite service. His vision was of a new breed of Serviceman: a scientist weapon-minder whose indoctrination had to be protected from traditional military influences.

Putting his plans into practice was not so easy. Apart from having to defend his infant Service from the determined and frequently underhand attempts by the army and the navy, in particular, to destroy the RAF, Trenchard was thwarted by Prime Minister Lloyd George's Ten-Year Rule. This was based on the assumption that, with Germany debarred by the Treaty of Versailles from building military aircraft and submarines, it was safe to reduce the RAF to a nucleus for ten years. This became absurd because until 1932 governments continued the policy effectively extending the ten years indefinitely.

What Trenchard did in the circumstances was to create little more than a skeleton force which would easily lend itself to expansion from a reserve established by the creation of the Auxiliary Air Force, the University Air Squadrons and

Shortly after it first flew in 1936, the big-winged Fairey Battle prototype was hailed in the British press as the RAF's new 300mph high-speed bomber capable of flying from its aerodrome near London to the south coast in just over ten minutes. Reality was to be more prosaic. Already the Air Staff knew that the Battle would be of little value in a European war. However, as the next stepping stone in the RAF's light bomber line, following the Hawker Hart and Hind into service, the Battle, stemming from a specification originating in first form in 1931, was a logical stepping stone. Equipped with split-flaps, retractable undercarriage and metal variable-pitch propeller in an all-metal airframe of stressed-skin construction and powered by the first production standard Rolls-Royce Merlin engine to enter service, the Battle became the basic building block of the Expansion Air Force.

the Short Service Commission.

Like the Italian General Giulio Douhet, Trenchard believed in the bomber as a defensive weapon. 'In a democratic country like ours,' he said, '... war cannot be continued unless the bulk of the people support it. If the people are subject to sufficient bombing they will compel the Government to sue for peace.'

In technical terms, he thought taking the war to the enemy was a more effective and cheaper alternative than waiting for enemy bombers to arrive at unpredictable heights, times and locations with their deadly loads, mainly because the only way of catching them was to keep a prohibitively expensive number of fighters and pilots on standing patrols. And even if that could have been done there was no guarantee of complete protection.

He had little faith in fighters, and in 1925 told a Cambridge audience that '... the aeroplane is the most offensive weapon that has ever been invented. It is a shockingly bad weapon of defence.... It is on the bomber offensive that we must rely for defence. It is on the destruction of enemy industries and, above all, on the lowering of morale of enemy nationals

caused by bombing that ultimate victory rests... '.

In the meantime, the Locarno Treaty and the Paris Air Agreement of 1925 had lifted restrictions on Germany building civil aircraft and she was taking full advantage. Teams led by Claudius Dornier, Ernst Heinkel, Hugo Junkers and Willy Messerschmitt were hard and effectively at work.

Throughout the 1920s the Treasury kept the RAF, and the other fighting Services, in a state of penury. Many squadrons had to make do with aeroplanes left over from the war including the de Havilland D.H.9A and its derivative the Westland Wapiti. Indeed, many of them continued in service until well into the 1930s, which meant the aircraft industry was unable to develop properly. Companies managed to survive by turning to other markets at home and abroad. Even so, enough money trickled into the Royal Aircraft Establishment, Farnborough, to keep its staff of under 150 – reduced from 5,000 during the war – going, and into sponsoring the Schneider Trophy winners of 1927 and 1929.

When, however, it was announced that because of the country's economic crisis the 1931 contest could not be officially supported, there was an outcry. Lady Houston, the widow of a shipping magnate, shamed the government into revising its shortsighted decision by offering to contribute £100,000 towards the expenses. Trenchard's attitude to all this was surprising. He was indifferent, even prejudiced against the High-Speed Flight, according to the late Air Cdre Rodwell Banks in his autobiography, *I Kept No Diary*.

Lady Houston's gesture was a curiously British and not unusual expression of patriotism that would be repeated by other wealthy patriots from time to time. Hers paid big dividends: out of that year's winning Vickers Supermarine

S.6B seaplane evolved the Spitfire fighter and the Rolls-Royce Merlin engine.

By all the rules of the game, the RAF ought to have been dismembered out of existence by the older fighting Services after the war. That it was not was due, it can be reasonably supposed, to a fear analytical psychologist Carl Jung might have traced back to a crisis in Trenchard's life when he was 39.

In 1912, Maj Hugh Trenchard of the Royal Scots Fusiliers had a blotted copybook. He was prone to quarrel with superiors and snipe at his juniors. His colonel in Northern Ireland had declared Londonderry too small for them both. His application for posts in Egypt, Macedonia, South Africa, Australia, New Zealand and even Harrods to escape from Londonderry came to nothing, mainly because of his age. Despondent, because there appeared to be no prospects of further, conventional promotion, he applied for leave of absence to take a flying course.

The age limit for pilot training in the RFC was forty. When he enrolled he had a mere four weeks in which to earn his aviator's certificate. Instructed by T O M Sopwith, he qualified, but without distinction, for he had no natural flare for flying. Posted to Central Flying School as adjutant, he soon recognised that the embryonic RFC was the vehicle in which he could exorcise his deep fear of failing to reach high office. As head of the Royal Air Force some years later, there was only one way to complete that exorcism and stay at the top: he had to protect the young Service from dismemberment by the army and the navy and create for it a valid, legal and independent existence of its own.

Thus, when Marshal of the Royal Air Force Sir Hugh Trenchard stepped down at the end of 1929, he bequeathed to the nation an air force that was the embodiment of his own strengths and weaknesses. On the positive side, it was an air force endowed with sufficient independence, individuality, organisation, discipline and *esprit de corps* to provide for quick and efficient expansion in times of need. On the negative side, its aeroplanes were obsolete and below strength, reflecting Trenchard's greater interest in the doctrine of air power rather than in the design and quality of the tools needed to implement it. He appeared to have no affinity with aeroplanes, nor to appreciate their finer technical points; nor did flying excite him.

The emphasis on doctrine coupled with the inference of years of Treasury restraint appears to have curbed those responsible for procurement from believing in what dynamic and imaginative men like Mitchell, Camm, Fairey, de Havilland and others could do for the RAF. It was an attitude that persisted even into the 1930s. For example, Air Marshal Sir Robert Brook Popham, one of the old school, thought that arming fighters with eight guns was going too far. Moreover, he disapproved of closed cockpits.

In Germany, meanwhile, the gliding movement had gathered more than 50,000 adherents; long before Hitler came to power in January 1933, selected members were being streamed through Germany's secret military training facility at Lipetsk in the Soviet Union, and the aircraft industry was experimenting with military types.

A private intelligence network was keeping Winston Churchill secretly abreast of developments in Germany, and although he was regarded as something of a crank for going on about German air rearmament, his warnings were heeded by Parliament when it authorised an air expansion scheme in July 1934. For Churchill, however, it did not go far enough. He complained that the first plan was designed more to impress the Germans than to equip the RAF for early action. Everything ordered was intended for the shop window behind which were no reserves.

In February 1935, the Luftwaffe went public. In March, Hitler announced bluntly that his air force had reached parity with Britain's and that he would go on building until it had reached parity with France's Armée de l'Air which had

The twin-engined Bristol Blenheim, classed as a medium bomber, stemmed from a private initiative between its makers and the proprietor of the *Daily Mail*, Viscount Rothermere, who had, incidentally, been the RAF's first Secretary of State for Air in 1918. To the Air Ministry, the aircraft came into the 'not invented here' category, receiving initially only small orders, enabling Bristol to build up an export business in Blenheims. With the introduction of the Shadow Factories scheme, the Blenheim became the second type, after the Battle, chosen for large-scale production, continuing to be made until long after it was obsolete. However, over one thousand Blenheims had been delivered to the RAF by the autumn of 1939, compared with fewer than fifty of the Luftwaffe's later Junkers Ju 88, with which the earlier Blenheim was often to be unfavourably compared.

1,500 front-line aircraft. Reliable informants in Germany – including Gen Erhard Milch, the Secretary of State for Air – refuted Hitler's claims. British Intelligence reckoned it would take Germany until 1937 to reach parity with France, and a further two years to be ready for war.

By unveiling the Luftwaffe and overstating its strength, Hitler did manage to rouse the British Government. Soon after a further expansion of the RAF was authorised for completion by March 1937. This called for forty-nine squadrons (ie 588 aircraft), more than the earlier programme, and brought the strength of the RAF up to 122 squadrons, containing 1,512 aircraft: twenty heavy bomber, eighteen medium bomber, two torpedo bomber and thirty-five fighter squadrons, with eighteen reconnaissance and other units.

Despite the improvement, it achieved little more than to dress the shop window. There was still no provision for adequate reserves. Only one-third of the squadrons of the home defence force was to be fighter squadrons. The ratio reflected the Trenchard-indoctrinated Air Staff's view that in the long run only offensive power could give the superiority for safety.

A bomber force would be of scant value if the fighters and the rest of the air defences proved too weak to repulse a succession of enemy attacks. Even if the thirty-five fighter squadrons were equipped with the best aircraft available, they would not have time to reach combat height before the attacking force crossed the coast and reached the capital, London. Also, the fighter pilots would not know where to intercept the attackers.

The acoustic detectors built on the south coast were useless, and no one knew whether the experiments in radiolocation which were then beginning at Orfordness on the east coast, would be

successful. In the circumstances, it was not surprising that the big bomber enthusiasts held sway.

Disarmament and the failure of civil aviation to develop led the British aircraft industry to wither. It survived for fifteen years on piecemeal Air Ministry contracts for repair and overhaul and the occasional order for new civil or military aeroplanes.

Although designers were inventive and ingenious, they were compelled to stick to the conservative requirements of the Air Ministry. Thus, when Britain began to rearm, a huge task confronted her industries. New aircraft designs called for different production techniques and new materials. Britain was then producing a mere 15,000 tons of aluminium per annum, whereas Germany and Austria were producing 200,000 tons between them. Plant, machinery and the output of

the industries making vital accessories were also inadequate. The problems were immense.

Lord Swinton was the Secretary of State for Air at the time. His audacity, imagination, understanding and drive about what the expansion programme needed did not always meet with the approval of successive Prime Ministers, and until Germany annexed Austria in February 1938, the Treasury would not release enough money to give the RAF the aircraft it needed.

When he took over the Air Ministry, Swinton found that the procedure was to build a prototype and test it. Once it had been fully tested, a number of aircraft were ordered. The result was that comparatively simple biplanes took five years from prototype to squadron service.

It was obvious to Swinton that changes would have to be made if the 1939

The strange geodetic Vickers Wellington represented the spearhead of Bomber Command's capability in the run-up to war, and was much used in showing the flag, as here over Paris on Bastille Day, 14 July 1939. By the start of the war, less than two months later, Bomber Command deployed six operational squadrons of Wellingtons. Of Hampdens and Blenheims there were also six operational squadrons each and five of Whitleys. In total there were twenty-three bomber squadrons operational from England with a further ten squadrons of Battles and two of Blenheims detached to France. Wellingtons operating in formation by day proved unable to defend themselves against fighter attack, and being without self-sealing petrol tanks suffered disastrously. However, the Wellington went on to become one of the most widely used twin-engined aircraft in the history of war. Before the war, Bomber Command had intended to fly mainly by day, but as 1940 dawned, home-based aircraft were switched to night operations, with the exception of Blenheim squadrons.

deadline for the completion of the 1936 programme (124 home-based squadrons, ie 1,376 front-line aircraft and 225 per cent reserves) was to be met. By eliminating the earlier slow procedure and ordering off the drawing board and incorporating modifications on the production line, he reduced the time lag considerably.

To ensure against failure, several manufacturers were given the chance to meet the specifications with their own designs. Which one was chosen depended on the Air Council. Mistakes were inevitable, the most striking being the choice of the Fairey Battle rather than the Hawker Henley as a light bomber.

The Fairey Battle day bomber carried a pilot, observer and a wireless operator/gunner. It was almost as big as the twin-engined Blenheim, with no more engine power than the Hurricane fighter. Of simple and clean design, it carried a bomb load of 1,000lb over a range of 1,050 miles. It failed in war because the Air Staff did not see the anomaly of ordering multi-gun fighters together with a day bomber with only one fixed gun firing forward and a single Vickers gun firing aft.

Designed as a fast, monoplane replacement for the Hawker Hart biplane series, the Henley's wings were interchangeable with the Hurricane's, which was an ingenious way of assisting production, maintenance and repair. The Air Ministry's approach to the light bomber in 1936 eliminated the Henley from quantity production. Yet, what a difference it might have made during the battles of France and Britain had it been chosen instead of the Battle.

The Handley Page Hampden had similar origins to the Wellington, both stemmng conceptually from a specification of 1932, though the designers' solutions were to be very different. The Hampden's small fuselage cross-sectional area, its fighter-like cockpit and manoeuvrability and comparatively high speed, made it appear to be a most suitable day bomber. However, on its first major encounter with the enemy, five were shot down — one complete formation out of eleven aircraft despatched. The Hampden, an ingenious aeroplane both aerodynamically and in its production engineering, was completely unable to defend itself, though later it was to become a successful mine-layer and even a torpedo bomber.

Two of the Air Staff's better decisions stand out, however. The first, in 1934, was to fit eight guns instead of two to the Hurricane and Spitfire; the second, in 1936, was to go for the heavy bombers — Stirling, Halifax and Manchester/Lancaster — that eventually carried the offensive into Germany. To ensure that production was unhindered by the aircraft industry's fifteen lean years, Lord Swinton's Shadow Factory scheme also stands out. This involved recruiting the automobile industry, particularly Austin and Rootes, to make engines and airframes in new purpose-built factories. In parallel, a vast scheme of subcontracting to the parent manufacturers was also instituted which would eventually involve 15,000 businesses ranging from small garages and engineering workshops to the largest companies in the land.

Looking back to that time more than half a century ago it is entirely reasonable to wonder why things were done as they were. Why, for instance, were many hundreds of obsolescent biplanes, Hawker and Gloster variants, produced after the first 1934 expansion scheme began, for surely the Air Staff did not wish to fight the next war with aircraft recognisably of the vintage of the previous one? The answer to this lies principally in the fact that the seventeen or so firms in the aircraft industry, plus the engine makers Rolls-Royce, Bristol, Armstrong Siddeley, Napier, de Havilland and Blackburn, needed to have their work forces not only kept busy but expanded enormously, to tackle fresh tasks — the new aircraft for the Expansion Air Force. In chronological order of production progress and delivery these were the Armstrong Whitworth Whitley, the Bristol Blenheim and the Fairey Battle, all of which were coming into squadron service by the summer of 1937. Next came the Hawker Hurricane in the winter of 1937/38, the Handley Page Hampden, the Vickers Wellington and the Vickers Supermarine Spitfire in 1938. The Vickers types posed particular production headaches, though in the end they were built in greater numbers than any of the others and continued in service throughout the war.

On paper the results of all these exertions of the expansion years produced the numbers needed to convince an enemy that he was facing the fiercest of tigers. What is so astounding is that when war broke out, Bomber Command was actually no more than a paper tiger. □

Wellington Is and Hampden Is of Bomber Command visit the Armée de l'Air at Salon en Provence to the north of Marseilles while on exercises in June 1939.

September 1939: The Advanced Air Striking Force Goes to France

William Simpson

This article, by a pilot who served with the Royal Air Force's force of Fairey Battle light bombers, based in France, records what happened in the period from the beginning of September 1939 to the following spring. In the field the combination of unsuitable aircraft, untried tactics, bad weather, improvised bases and few aids to aircraft operation made the real world anything but the 'phoney war' which it was later to be dubbed by the armchair brigade.

Fifty years ago it was a lovely summer afternoon on Saturday 2 September 1939 as some 160 Fairey Battle single-engined light bombers of the ten Battle squadrons of the Advanced Air Striking Force (AASF) of the RAF, each flying in tight squadron formations, converged above Shoreham-on-Sea, Sussex, and set course across the Channel to land at five airfields prepared for us by the Armée de l'Air in northern France. We flew to our war stations in anticipation of the declaration of war against Germany by Prime Minister Chamberlain, expected and duly broadcast the following morning. As a Flight Lieutenant and Flight Commander of B Flight No.12 Squadron, long known as the Dirty Dozen, my Battle was somewhere in there with a seasoned crew of Flt Sgt Odell as observer and Cpl Tomlinson as wireless operator and air gunner (WOP/AG) with myself as pilot.

We had flown together many times before, particularly in the last year since the Munich crisis, as indeed had many Battle aircrews, some 480 of us airborne together that day. These were exercises, however; never before had we been on our way to war. Even the air route was familiar, this time making landfall at Le Tréport and passing the airfield at Poix, its name marked in white stones; on again over the flat countryside to land at our destination, Berry-au-Bac, roughly north of Rheims, the cathedral city dominating the fruitful Champagne country. Previously, on air exercises, we had fought off interception by our own and friendly French fighters with nothing more lethal than camera guns; now our guns were loaded and there was always the possibility, remote though it seemed, that unfriendly Luftwaffe fighters could materialise among us. Indeed, although we had prepared for war before both on

The Fairey Battle. Peeling off prewar is B Flight leader, No.12 Squadron, author of this article. (*Via the author*)

paper and in the air, there was a strange unreality about that sunbathed flight where dark foreboding should have reigned.

Since all aircrew of the RAF were then, as now, volunteers who had clamoured to win a place in the air, if only in most cases for love of flying, we were all locked into this new situation with no way out short of a dishonourable desertion more frightful than the fear of duty-bound action – a very real fear, for intelligence about the size and efficiency of the Luftwaffe was good and we knew that no matter how well we had been trained the odds against carrying out our orders successfully and surviving were far from favourable.

I recall that our morale was good on the whole and we had some confidence in our collective skills and in our own fighters for cover while we got on with bombing tasks. Nevertheless, we would have been happier if the Fairey Battle – a stout and agreeable aircraft to fly when not too heavily loaded – had been replaced by Bristol Blenheims before the war began.

Our part of the overall formation was 76 Wing AASF, made up of Nos.12 and 142 Squadrons. We took off from our

Vying for unsuitability in serious war. The Amiot 142 'fighter-bomber-reconnaissance' aircraft of the Armée de l'Air and the RAF's Fairey Battle light bomber. (*Imperial War Museum*)

base at RAF Bicester, Oxfordshire, in sections of four aircraft each. We then formed up into flight, squadron and wing formation. As we watched for black crosses on the wings of Messerschmitt 109s (*we* certainly did not call them Bf 109s) in the later stages of the flight over France our elated feelings arose from a release of pent-up emotions – the excitement of the unknown, fear, regret at leaving loved ones. At least something was happening now. When we landed at Berry-au-Bac, section by section in formation, we knew that elsewhere in France the other Battle squadrons were also going in to their prepared airfields.

At Rheims-Champagne were the headquarters of 72 Wing and No.226 Squadron, with No.105 Squadron of the same wing at Villeneuve le Vertus; at Bethenville the HQ of 71 Wing and No.40 Squadron, with No.15 Squadron at Coudé sur Vraux; at Challerange the HQ of 74 Wing and No.103 Squadron, with No.150 Squadron at Écury; at Auberive the HQ of 75 Wing and No.218 Squadron, with No.88 Squadron at Mourmelon.

Berry-au-Bac, bordered by a military road on one side and woods on another, was in flat, featureless land dotted with red and white villages. Much fought over in the First World War it was now restored to fields of beet and wheat. The River Aisne and a canal were nearby. On the airfield were three aircraft fuelling points – tanks sunk into the ground – and we refuelled by hand pumps; a laborious job, especially as two flawed pump handles broke. It was almost dark by the time our refuelled Battles could be dispersed, covered and picketed down.

We were some 100 aircrew – nearly all

A forlorn tactic. Fairey Battles practise flying a defensive box. (*Via the author*)

officers and NCOs of the two squadrons. The ground personnel of the squadrons and of Wing HQ, some 300 more officers and airmen, had still to arrive by boat, train, road transport (MT) and both large and small requisitioned Imperial Airways and British Airways aircraft. And so to bed, but first into two decrepit buses to find our quarters in the village. Little had been prepared. It was early in the morning after an afternoon landing before the NCOs were on the floor of the school rolled in their own blankets, airmen likewise on straw in barns and stables, and the officers in beds in crowded billets. The friendliness of the few Armée de l'Air officers and men was undoubted, but the level of military readiness for highly technical squadrons of aircraft was laughable.

Why France? Why Me?

After Munich in September 1938, plans for co-operation by the RAF with the navy and for independent air assault against German industry were brought to a state where they could be undertaken. Co-operation between the RAF and the army, however, was more difficult. No one in September 1938 was sure where the army was going to fight. Two British Divisions would probably go to France, but there was no exact plan for their use nor had any promises been made to France. The RAF had a plan to send out an AASF to France but although arrangements for it had been made and its reception, location, maintenance and defence had been discussed with the French, no agreement had been sought about its use. Nevertheless, as its name implied it was to go to France not to help the British or French armies but to get its short-range single-engined bombers within reach of particular targets. These targets had long since been marked on Air Ministry target maps of German industry.

Long before Munich I had been persuaded that since war was inevitable within a few years there was much to be said for joining the armed Services early rather than being called up later. The depression years and a hard won job in

They also serve who only stand and wait. The winter of 1939/40 in France. (*Imperial War Museum*)

an advertising agency in stuffy London could not compete in interest and adventure with a life in the air then offered by the RAF with its Short Service Commissions, learning to fly and being paid for it. Thus when I was twenty I applied, was accepted and was learning to fly in Egypt at No.4 FTS, RAF Abu Suweir by mid-1935. I qualified for my wings on Avro Tutors and Avro 504Ns, the Armstrong Whitworth Atlas and the beautiful Hawker Audax; applied for flying-boats, preferably in the Far East, and was posted immediately to No.12(B) Squadron then temporarily in Aden during the Italian invasion of Abyssinia. There I flew Hawker Hart IIB single-engined bombers on 'hairy' cross-country flights, both singly and in formation, over the villages in the mountains of South Arabia and sometimes over the Red Sea. This taught me a lot about the need to take care as well as the risks and I acquired some skill at landing on small patches of sandy ground not marked on any map and often on top of or among the mountains where mistakes did not bear contemplation. So when No.12 Squadron returned to Andover, its usual base, I felt quite operational. That was in 1936, the RAF Expansion was well on and I was rapidly converted to the delightful Hawker Hind and became a flight commander while still a pilot officer. From then on, although life in the squadrons of the RAF had plenty of youthful hilarity and the flying was both difficult enough to be thrilling as well as unendingly agreeable, there was a lot to learn in an ever changing, ever growing preparation for whatever was to come. In the squadron we grew and flew together and re-equipped with Fairey Battles in March 1938. I had the pleasure not only of collecting our first Battles but also of converting the other pilots from the back (instructor's) seat of our two-seat training Battle.

The Fairey Battle has been much ridiculed as a fighting machine by aviation writers for half a century, and deservedly so. It was quite unsuitable for war in 1939 and 1940: slow, big, presenting a large target to German flak and to the current Messerschmitt 109s of the Luftwaffe. In its earlier form it had no armour plating to protect its crew (although this was added later for the AASF in France), and no self-sealing fuel tanks - though this was not unusual for British bombers of the period. But it was not unique in being outdated in 1939/40. So were the other contemporary single-engined so-called 'fast' monoplane bombers; fast when devised but rendered obsolete and ineffective by the smaller breed of faster monoplane fighters usually equipped with comparable engine power. The Germans, with the brilliant and beautiful Heinkel He 70 of the early 1930s, had finally abandoned the formula as a result of experience during the Spanish Civil War.

The tragedy was that in Britain the formula continued because, as part of the deterrent to war which the successive RAF Expansion Schemes were intended to be, Fairey Battles streamed from factories (including the Austin Shadow factory) in increasing numbers from 1937, reaching a total of 2,185 by September 1940.

Look at one of the many preserved lithe Spitfires with single Rolls-Royce Merlins and think of the big Battle with the same power and its wing span of 54ft – within 2ft of the twin-engined Bristol Blenheim. Flown light, the Battle was highly manoeuvrable, but in war carrying its crew of three and four 250lb bombs in wing cells and 212 gallons of fuel for its single Merlin it was struggling to find much over 200mph cruising compared with its official trials figure of 241mph top speed at 13,000ft (range 1,050 miles with 1,000lb bomb load). What with this modest performance and loaded with one fixed Browning gun in the starboard wing, one free Vickers K gun on a special mounting at the back of the cockpit and the addition of one more backwards and downwards firing gun under the belly for the observer, it presented problems. It was immensely strong, however, as we were soon to have proved.

Air Support in France

Now that we were in France, the Battles of the AASF were a detached forward group of wings, originally No.1 Group of Bomber Command, still an element of that UK-based greater command with its many heavier and larger bombers still building up to what was very much later to become a formidable force. Also in northern France and attached to the British Expeditionary Force (BEF), the so-called Air Component of the BEF was building up to another group of wings which, with its thirteen squadrons of

Prewar parade practice. B Flight No.12 Squadron in echelon stepped up. (*Via the author*)

Better than Battles, a Bristol Blenheim IV of the Air Component prepares to unstick. (*Imperial War Museum*)

Lysanders, Blenheims and Hurricanes, was slightly larger than the AASF by the spring of 1940. Of these there were five army co-operation squadrons of Lysanders (Nos.2, 4, 13, 16 and 26) for tactical reconnaissance and photographic survey at the BEF front line; four Blenheim squadrons (initially two Nos.53 and 59) for strategic reconnaissance up to the Rhine; and four Hurricane squadrons (Nos.1, 73, 85 and 87) for protection of the army and other fighter duties. There was also a further Lysander Squadron (81) for communication and further home-based fighter squadrons reserved for reinforcement. Overall commander of the British Air Forces in France (BAFF) was Air Marshal A S 'Ugly' Barratt, with Air Vice-Marshal P H L Playfair (AASF) and Air Vice-Marshal C H B Blount (Air Component) reporting to him. Logistically, it had been a major effort to get their two forces into the field and ready for battle. For instance, it meant men and equipment for several operational HQs – that of the AASF (later nicknamed 'Panicky Panther' by Germany's Lord Haw Haw after its codename) at Rheims, that of the Air Component at Arras and one for each of the constituent wings. We had between us untold signals units, stores depots, casualty clearing stations, aircraft servicing and salvage units, a filter centre, RDF (radar) stations, mobile offices, ambulances, petrol bowsers, etc. Even the bulk of the office work was done on wheels, and the mobility was wonderful to behold. That part of planning and movement, give or take a few hitches, had been well done. When it came to operational planning it was another story altogether.

In the event of a German attack on land, through the Low Countries, the job of the BEF, the Air Component and the AASF had now been agreed by the British and French and was known as Plan D (D for the River Dyle). While France stood at the Maginot Line, mobile French and British divisions would meet an expected attack by the Germans coming through Belgium. Where the left of the Maginot Line ended near Montmedy, the steep wooded valley of the Meuse was expected to be defensible long enough to allow French and British forces to move north from the Franco-Belgian frontier to hold a line on the River Dyle, from Antwerp in the north to near Sedan in the south.

Settling in

We tucked our aircraft tails into the trees and covered them with leafy branches. Sometimes German reconnaissance aircraft hummed high overhead. Luckily they did not attack for we had only three machine gun posts. We became bronzed and fit in the September sunshine and ready for lunch in the open – tinned bully beef, ship's biscuit, a soldier's ration of wine and hot stew from field kitchens like small black steam engines. We had one French ambulance but no fire tender for many days. Then at last the many vehicles of our own MT column arrived – petrol bowsers, a Chance light for night flying, buses, office, store, water and cooking trailers, radio trucks, etc. We were taking shape as a wing of two operational squadrons, not as smartly turned out as at peacetime Bicester, but at least fit to fight.

The scale of organisation and equipment deployed for war abroad today was also great in 1939. Thanks to Peter Elliott, librarian of the RAF Museum, RAF records are available showing the division of responsibility between the army and the AASF in France. They describe the provision of water, electricity, gas and other essentials; the docking of ships, landing of stores, delivery to rail heads and thereafter to bases, storage or fighting units; and how the army helped the RAF. They shipped, docked, unloaded. There was a base area HQ at Nantes, an RAF Port Detachment, an Aircraft Depot, and so on.

With the arrival of our ground crews and equipment the pilots, observers and WOP/AGs were relieved of our outdoor pioneering work. Our aircraft were in the woods and we had quite sophisticated timber braced dugouts – log roofs, tin sheeting and sometimes even windows, fireplaces and chimneys, covered with soil and flush with the ground. Chimneys smoked cosily as amateur cooks fried up below. Ground crews replaced the branches, soon to lose their leaves, with camouflage netting. Huts were built. We had tractors and even office trailers. I was particularly lucky with a tent. I even had charge of a Miles Magister two-seat elementary trainer, the commander of the wing's runabout which he seldom used, and could fly it, ostensibly to look for emergency landing grounds. Before the end of the first month the Nos.12 and 142 Squadron officers shared the comfort of a château at Guignicourt nearby and progressively all ranks had improved billets; much ingenuity was shown in making them relatively comfortable.

We were indeed settling in and our mobility was off its wheels. Our immobility was not planned, although you might have thought we were there forever, our defences being indeed inadequate: two Hurricane squadrons detached to the AASF at their own request, few anti-aircraft guns, stretched lines of communication with some stores and aircraft parks, vulnerable ports at home and in France, roads and railways much exposed and stores depots in France the same. Had we, as in the First World War, come prepared to stay? Now immobilised, could we become mobile?

Daylight Operations

While the good weather of September lasted we managed to get in a good deal of flying, including reconnaissance operations to check on the German troop dispositions which we photographed from various heights. It was on one of these flights that reality was brought tragically home to us about the vulnerability of our Battles: a formation of five from No.150 Squadron on a high-level flight around 20,000ft and spread out to take a mosaic of air photographs (each aircraft covering a strip of ground overlapping its neighbour) over the German positions was attacked by Messerschmitt 109s. Four crews and aircraft were lost. Only the leader escaped to crash-land back at base. This marked the end of high-level daylight incursions into enemy areas for all of us by decree of the commander of the AASF. That was on the last day of September. I had myself led a low-level reconnaissance of three aircraft crossing the Siegfried Line at between 100 and 300ft with observers and air gunners recording what they could see on cameras and in notes. We came back with many photographs of guns, encampments and fortified positions which created a sensation back at our wing. We also had two bullet holes in the wing of one Battle, although mostly the German troops waved at us. I had exceeded the orders for this operation, going in so far and so low. It must have been a toss up as to whether I would face court-martial or be congratulated. Nothing was said officially, however, in criticism or praise. The temptation to go too far was great; we were young and frustrated. Other rather more decorous reconnaissances at high level were also flown by our and other squadrons, until the ban was imposed. One at 23,000ft over Saarbrücken ran into heavy flak. An observer passed out from lack of oxygen and his pilot was forced to return to base and three other Battles were damaged.

In October the rains came and there was little flying and time was devoted largely to working out tactics for avoiding or defending ourselves against German fighters. We had mock battles with French Morane and Curtiss fighters and with the Hurricanes of Nos.1 and 73 Squadrons. We became very friendly with No.1 Squadron but little of benefit emerged from our trials. These included rough use of rudder to skid sideways when attacked; going into propeller fine pitch and lowering flaps and undercarriage in the hope of forcing fighters to fly past; and forming a line astern tight corkscrew to reach the ground where we would hedgehop at low speed. Other tactics such as flying in box formation, designed to give our rear gunners a combined opportunity to concentrate fire, were soon abandoned. For all our vulnerability, however, the Battle gunners had some remarkable successes later on, and when No.150 Squadron was attacked they hit two Messerschmitt 109s and probably destroyed one of them.

Before the end of October the Battle squadrons had practised formation dive-bombing tactics against our own anti-aircraft guns, and had dropped live bombs at low level on a nearby bombing range. In the meantime we had tasted the delights of nearby Rheims and there was talk of some limited home leave by Christmas if the balloon remained on the ground. There was also an alert in late

October but the raid never came. Some of our Battles came up for 180 hour inspections and had to be flown to Rheims-Champagne where there were special inspection crews and facilities. Finally, by the end of November No.12 Squadron moved to a new airfield at Amifontaine nearby, leaving No.142 at Berry-au-Bac which had proved too small for two Battle squadrons. Amifontaine was a pleasant green airfield bordered by woods and well drained, particularly as it included a sloping area. This made it unsuitable for night flying, which had to be done from Berry-au-Bac, but there was good cover from bordering woods, some Nissan huts and tents. Then the snows began and the winter set in as one of the coldest for years, but at Amifontaine our Battles had their noses tucked into shelter tents, each one stove-heated. There were also tents where aircrew and ground crew could shelter together.

We added Balaclava helmets and other warm clothing to our uniforms. This did not much help the riggers, fitters, electricians and armourers who had to check all the equipment both inside and outside the aircraft each day. I was lucky as I was sent down to Perpignan near the Spanish border in January to open up a small airfield at St Laurent for the use of first No.12 Squadron and then the other Battle squadrons as a base for practising our air gunnery over the sea – a welcome break indeed from the harsh winter in the north.

In due course the squadron's Battles joined me and undertook their exercises. Back at Amifontaine there was bad weather and little flying in February. This was followed by a thaw that reduced our airfield to a quagmire. Many aircraft became unserviceable due to damaged tailwheel mountings as they slithered

Living rough. A Westland Lysander, an ingenious aircraft in search of an effective role, when forming part of the Air Component. (*Imperial War Museum*)

and bounced through the mud, and at one time only six of our Battles were fit to fly.

Joys of Night Flying

In March we started regular night flying from Berry-au-Bac. There were rumours that we were preparing to drop special mines in the Rhine to destroy shipping. Instead, we discovered, it was to be leaflets on German towns. First we had to brush up on our night flying with cross-country flights over France. These were testing in a number of ways. The country was blacked out. There was a complicated arrangement of small light beacons each coded with a Morse letter that changed daily. Our navigation aids were our eyes, maps, bombsight for drift, and wireless telegraphy to the ground *in extremis*. The three crew members could talk to each other if switched on to intercom. There were no luxuries such as friendly voices from ground controllers. It was extremely cold. The lights were never switched on at base; a glimmering few were turned on only when we arrived overhead.

During the day the crew flight tested the aircraft and checked all equipment, such as it was. We then flew to Berry-au-Bac. At dusk the control officer for the night arrived and arranged a path of small glim lights (aptly named) in an L shape, with the long arm of the L into wind and slightly longer than the distance needed by a Battle for take-off. At the downwind end of the L a small mobile searchlight, or Chance light, was parked so that its beam was cast along the line of take-off and landing. The pilot taxied out, stopped, winked out in Morse his aircraft letter with his top signalling light, and was cleared to take off with a green, or wait with a red, from the duty controller of the night, a fellow pilot. Inside the pilot's cockpit, lighting was limited to a small orange glow, just sufficient to read a map. In flight he checked his course and noted each little

light beacon as it came up hopefully flashing the expected Morse letter. Fortunately, the Battle could be trimmed to fly steadily; there was no autopilot. Behind and beneath the pilot the observer lay on his stomach, also checking the course of the flight (visually through his open hatch in the floor) and took drift measurements using the standard bombsight of the period. The observer kept the log and requested course changes as needed. He also made out regular position report forms and passed them to the WOP/AG, manning his radio and gun in the rear part of the 'summer house', who tapped out the information on his Morse key to base. All three members of the crew kept a sharp look out and made mental and written notes on the weather, barrage balloons and French anti-aircraft fire. You could be up there for some three hours and near to the limit of your fuel.

The pilot was the captain, but it was a crew of equal importance, each depending on the other two: on the pilot for flying, the observer for navigation, the WOP/AG for communications and protection.

I flew one of the first two leaflet raids made by No.12 Squadron on the night of 23 March 1940 in my usual Battle L5190 (V for Victor) with, as usual, Flt Sgt Odell and Cpl Tomlinson. We took off on the first suitable moonlit night to fly over Germany and drop leaflets from about 10,000ft. In the operations hut we studied the large-scale map of Germany with its flags and coloured pins showing flak concentrations, enemy fighter aerodromes and balloon barrages. We noted the letter of the night to be flashed to ground if we were challenged to prove our identity to the French AA, and those

Squadron scramble by No.87 Squadron Hawker Hurricanes of the Air Component. Some aircraft still have Watts fixed-pitch airscrews. (*Imperial War Museum*)

Before the balloon went up. No.12 Squadron pilots of B Flight, their names now uncertain. Flg Off Don Garland (back row centre) will win a posthumous VC, while Flt Lt Bill Simpson (front row right) will come through. (*The author*)

flashed by light beacons in France on our way. We were given emergency rations and other material useful in enemy territory. The full moon shone out of a clear sky and the countryside was bathed in soft light. We took off and climbed to 10,000ft on course, passing through scattered cumulus until we reached the area where France, Luxembourg and Germany meet. As we crossed into Germany the clouds cleared to reveal dark forests spreading away to the left, broad stretches of undulating downland and river-filled valleys, small towns and villages all standing out clearly in the moonlight. There was no sign of life and no lights.

As we looked down on the Rhine searchlights suddenly lit up to criss-cross the sky in ranging beams. Groups of little red balls began to pop up like champagne bubbles, then came 'flaming onions' like

orange tadpoles wriggling in the air. We were jinking, turning, climbing, dropping, trying to avoid the lights and heavy flak now bursting round us, while concentrating on reaching the point where we had to throw out our bundles of leaflets to spread downwind onto Coblenz. This done we had to fly low for our second task and at 2,500ft the light flak began, red and green tracers streaking up all around us. As we circled Coblenz at 2,000ft we looked for signs of blast furnaces and electric transformer plants at work. We saw nothing so flew on down the Rhine to Bingen, the river shining bright and broad, and still tracer leaped up at us from the banks. It was eerie. Roads and railways seemed empty and still, groups of barges tied together and other small craft at their moorings. Had it not been for the flak it would have been difficult to believe that any life existed on and around the giant river.

Eventually we broke off and as quickly as possible were on our course for France, had passed Luxembourg and concentrated on the little French winking beacons indicating our way home. Near Metz, however, I was suddenly aware of a dark suspended shape casting its shadow across us – a barrage balloon, one of many we now had to weave between. We may have missed a vital beacon or it may have been unserviceable for at our estimated time of arrival there was no sign of Berry-au-Bac; over to Cpl Tomlinson's Morse key and wireless for a bearing. We were too far south so on a back bearing we steered for home. Soon the Chance light and the flarepath were switched on and we landed. It was team work – Odell had navigated us to the target and Tomlinson brought us safely home. My part as pilot had been made easy. I was to think of this great team spirit in the weeks to come when my two faithful companions proved themselves again and again when things were tough.

Spring 1940

With the spring and first signs of summer we had settled into what seemed like a set routine. Elsewhere the RAF had long been busy, attacking shipping at sea and in anchorages (but not tied up alongside quays), dropping leaflets by night, training, re-equipping, and taking part in bizarre adventures such as the Norwegian campaign following the German invasion of that country on 8 April 1940. Meanwhile, in France the Battle squadrons were ready for whatever was to come. Just as the weather had been sunny and pleasant when we arrived the previous September, so it was the following spring. The Battle squadrons were raring to go. To what and when is the subject for a further article. □

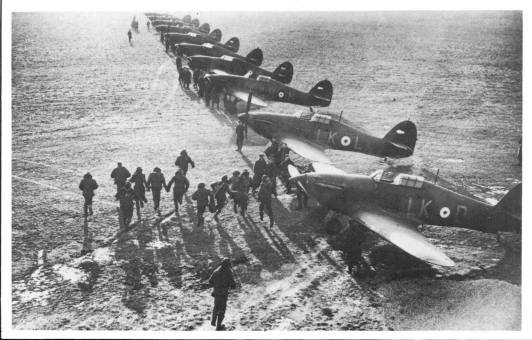

September 1939: Westwards to Whitchurch

Hugh J Yea

To make the best use in war of the, by later standards, meagre number of civil aircraft available, the British Director General of Civil Aviation began to make plans in October 1938. This account traces the behind-the-scenes planning and activity that then went on right up to the start of the war early in the following September.

During the Munich crisis, Britain's 69-year-old Prime Minister Neville Chamberlain made his first-ever flight abroad on 15 September 1938, when he left Heston for Munich and travelled onwards by train to Berchtesgaden to meet Hitler. He and his entourage had used two British Airways Limited Lockheed 10A Electras G-AEPR and G-AFCS for the round-trip, and a week later, when he flew to Cologne to meet Hitler at Bad Godesberg, he used the BA Lockheed 14 Super Electra G-AFGN, and Lockheed 10A Electra G-AEPO was used as the back-up aircraft. The third and final visit to Munich, on 28 September, again to meet Hitler, Mussolini and Daladier, was made by two of BA's Lockheed 14s,

G-AFGN and G-AFGO, both of which featured prominently in the photograph

'Here be Dragons.' Fortunately, in view of what was to happen later, the foresight of the Bristol City Fathers made possible the opening of the municipal airport at Whitchurch, only 3 miles from the city centre, in 1930. Five years later when this picture was taken, the airport played an important role in air communications within the United Kingdom, being particularly well placed in relation to journeys involving water crossings – either short-haul across the channel to Cardiff or on the longer sea crossing to the Channel Islands via Bournemouth or from 1936 between Bristol and Dublin. The journey from Whitchurch to Pengam Moors aerodrome, just 2 miles east of the centre of Cardiff, took 10 minutes by de Havilland D.H.84 Dragon operated by Western Airways at a fare of 9s 6d (47½p) single. (Reece Winstone)

taken on Chamberlain's return to Heston on 30 September, when he emerged from the aircraft waving the notorious joint declaration that he and Hitler had signed that morning, and which he claimed was symbolic of the fact that Britain and Germany would 'never go to war with one another again'.

The extensive publicity accorded to these fast, gleaming, all-metal, twin-engined Lockheed airliners used during this 'shuttle diplomacy' gave rise to criticism voiced in Parliament and elsewhere to the effect that it was a blow to British prestige that the Prime Minister should have been obliged to resort to using American-manufactured aircraft. (It is interesting to note that Chamber-

lain's flights in pursuit of peace were made in the progenitors of an aircraft – the Lockheed Hudson – that was destined to have a memorable career in the war and make a substantial contribution to the end of Hitler's ambitions.)

The crisis had the effect of focusing attention on many areas in need of urgent government attention and action, among which was a dawning recognition of the importance of civil aviation resources to the government, and how ill-prepared Britain was in this respect.

Twelve Months to War

It was therefore no surprise when in October 1938 the Director General of Civil Aviation (DGCA), Sir Francis Shelmerdine, produced an *Appreciation of the Employment of Civil Aviation in War*, constituting a comprehensive review of the state of British Civil Aviation for consideration by the Committee of Imperial Defence. It incorporated a thorough survey of the fleets and staff of Imperial Airways, British Airways and nineteen independent carriers, listing the payload and passenger capacity of every aircraft and a list of operating crews, ground engineers, key administration staff, etc.

The DGCA's recommendations were, first, complete militarisation of all civil aircraft fleets, crews and staff; second, State ownership of all aircraft operated on a civil basis; and third, the operation of a National Air Communications (NAC) organisation using existing air transport companies under State charter. The DGCA favoured this last option.

The regathering crisis engendered swift action, and permission to proceed with planning for an NAC organisation was submitted by January 1939 to the Inter-Departmental Committee of International Air Communications by an Air

From midnight on 31 August 1939, civil aircraft were prohibited from flying 'over that portion of England and Scotland (and territorial waters) east of a line drawn from a point near Poole...' In other words the entire shaded area on the map was out of bounds.

Ministry memorandum, *The Role of the Department in Civil Aviation in the Precautionary Stage and in War*. This concluded, 'It is no longer possible, nor does it appear to be in the national interest, to look upon Civil Aviation as a subsidiary of the Royal Air Force, which in the event of war would automatically close down and

Before the move. 26 August 1939, Croydon. Handley Page H.P.45s *Horatius* with *Heracles* and *Hanno*. Also discernible on the original are four Armstrong Whitworth A.W.27 Ensigns including *Eddystone* and *Ettrick*. (Lettice Curtis)

have its resources dissipated'. The DGCA's favoured recommendation was accepted and his department was authorised to go ahead with planning the NAC system.

By spring 1939 Air Ministry officials had started calling upon the independent carriers, all of which since 1 November 1938 had required a licence to operate issued by the Air Transport Licensing Authority, to outline plans for their co-ordination in the event of a national emergency. They were required to sign a 'dormant contract' and subsequently given sealed orders and code words that would be relayed to them to make ready their aircraft and crews to leave for destinations to be revealed from the sealed orders once they were airborne.

As far as Imperial Airways and British Airways were concerned, they were already subject to being requisitioned by the government under agreements to which they were party. Confidential tripartite discussions between the Air Ministry and senior officials of both companies were held to draw up an immensely complicated set of instructions known as the War Book, detailing policies and plans to control the transition of civil aviation from peace to war. Among the necessary priorities established were, first, the provision of transport for the RAF; second, the carriage of important loads of both passengers and freight; and third, the operation of a surcharged Empire Air Mail Scheme to replace the 'all-up' scheme already in force.

The dispersal of Britain's Merchant Air Fleet was to be based on the Western Plan as civil aircraft were to be prohibited from flying over eastern England and part of the east of Scotland, which was to be under the control of Fighter Command. A zig-zag dividing line was drawn in a northerly direction from the south coast at a point just east of Poole by way of Salisbury, Kingsclere, Oxford, Pershore, west of Birmingham to Cannock, Leek and Skipton to a point in the Pennines 5 miles east of Aysgarth (North Riding, Yorkshire). Thence northwest to a point near the Scottish border some 10 miles south of Hawick, then northeast to a point east of Motherwell, north to Stirling, thence northeast via Perth to Stonehaven on the east coast of Scotland. Civil fleets were to be dispersed to nine aerodromes west of the line in England, Wales, Scotland, the Isle of Man and the Channel Islands.

The headquarters of the new NAC, Imperial Airways and British Airways, would be in Bristol, it having been agreed that in place of Croydon and Heston, the airlines' wartime bases for landplanes would be Whitchurch and Exeter, and for flying-boats, Poole, Falmouth or Pembroke Dock would be used instead of Southampton. Shoreham and several other aerodromes might be called into use as circumstances dictated. Numerous assumptions were necessary, including what the availability of aircraft was likely to be, for in the event of war it had been made clear that all civil aircraft production would be halted, and orders already placed in the USA cancelled, and that civil aviation was to be completely subordinated to the military. The RAF was to be given first call on the airlines'

resources regardless of any long-term commercial considerations, but despite their expressed concern at the absence of any clear-cut indication of the preservation of any commercial operations, Imperial Airways and British Airways were obliged to base their manpower planning upon what the NAC envisaged that they would be required to do.

The planning of future staff requirements was complicated further by the fact that as early as August 1938 a scheme had been worked out for the forthcoming amalgamation of Imperial Airways and British Airways to form the British Overseas Airways Corporation (BOAC), the Bill for which had been introduced to Parliament in June 1938. By the end of 1938 it had been decided that BOAC was to be a public corporation, and the organisation for the new BOAC had been drawn up during the early part of 1939 and the placing of top level staffs for the two companies had been agreed. However, despite the inevitable uncertainties as to the full effects of the new merger, the two companies liaised closely and by June 1939 most staff in Imperial Airways and British Airways had been informed whether their services were to be retained in the event of a national emergency.

Early that month it was announced that the grades of staff to be retained were captains, first officers, radio operators, flight clerks, flight stewards, victualling clerks, station superintendents, station officers and trainees, engineers (other than those engaged in clerical work), coxswains, labourers on maintenance, labourers and seamen at Hythe and Southampton, and motor transport drivers. By 23 June administration and

clerical staff whom the companies wished to retain had received individual letters to this effect, and those staff not receiving such notification were to be regarded 'as being on leave without pay until a period of six months after the cessation of hostilities, or such earlier date that they may resume employment with the company'. (With the start of war, they were notified that they were free to accept alternative employment immediately should they wish.)

Many of the staff whose services were to be retained had navy, army or air force commitments and were informed that official assurance had been given that (excluding the Auxiliary Air Force) they would be 'screened' from call-up, at least at the outset of the emergency. This gave rise to problems as many reservists did not wish to be deferred from going into uniform, and some, although perhaps already in airline uniform, did not relish possible loss of service seniority should they delay or postpone their call to the colours. Still others considered that their expertise would be of much greater value to what became known as the war effort if they transferred to other fields of endeavour, the sort of philosophy that may well have influenced some of British Airways' Lockheed 14 pilots who since February 1939 had been giving flying instruction on the type to the RAF Avro

Gone west. Newly arrived at Whitchurch, the Armstrong Whitworth A.W.27 *Egeria* is in process of losing her bright metal skin in exchange for a coat of wartime camouflage. The British Airways Junkers-Ju 52/3m *Jason*, an all-freighter version, has already received the treatment and has a Union Flag painted on her rudder. (*Capt R T Halliwell via the Science Museum*)

Anson pilots who were to convert to Lockheed Hudsons and who had been attending British Airways Training School at Heston. Just as valuable as experience in the operation of aircraft was the knowledge and experience of flying the routes to Europe that the RAF pilots gained by acting as first officers with British Airways; in the same manner members of the Luftwaffe had long been obtaining route familiarisation and night-flying experience by flying as crew with Deutsche Lufthansa on its Continental routes and those to Britain.

Threshold of War

By early August 1939, most Imperial Airways and British Airways staff received another National Emergency instruction from their establishment officers updating some of the information they had been given two months earlier, but now detailing exactly what they were required to do if war was declared. This now included information about evacuation to, and billeting at, the company's unspecified new bases, and making reference to the eventual possibility of securing accommodation for their families.

On 4 August 1939 the BOAC Bill received royal assent and thus became law, and although BOAC was not established by the Secretary of State for Air until 24 November 1939, and BOAC did not take over Imperial Airways and British Airways until the appointed day of 1 April 1940, the two companies continued to act in concert. Consequently, officials of both were involved in making preparations for the reception of their organisations and fleets from London, Heston and Croydon by numerous visits to Bristol and Whitchurch, and forays to Exeter. Because of their involvement with preparations for the move, quite a large number of staff had become aware that the Grand Spa Hotel in Bristol and Bristol's Municipal Airport at Whitchurch were to become the destinations for most of the several hundred evacuees. Although still not an open secret, it led to many staff contacting relatives and friends in and around the Bristol area from midsummer 1939 for help and advice in seeking accommodation, and many made weekend visits for a personal reconnoitre of the vicinity.

They were by no means alone in making these endeavours, for hundreds of civil servants were likewise attempting to jump the gun when informed that their departments planned to 'go west' as apart from Air Ministry moves to Bristol, there was to be a major evacuation of the Admiralty to Bath, the BBC Variety Department to Bristol and numerous private companies seeking premises and accommodation. (Subsequently, official estimates indicated that about two million people privately evacuated themselves to Devon, Wales, Scotland and other more secluded areas, and that as early as February 1939 a million billets had already been privately booked. From the end of June and the first week of September 1939 some 3½ to 3¾ million people moved from areas thought to be vulnerable to those considered safe.)

The choice of Whitchurch for a major wartime base had been made by Imperial Airways, British Airways and the Air Ministry as having the best location, facilities and potential for improvement under the prevailing circumstances. The airfield had been granted an Air Ministry Operating Licence on 7 February 1930, and it had become Britain's third municipal airport when inaugurated on 31 May 1930. Located 3 miles south by east of Bristol it covered some 300 acres and offered then a main east-west take-off run of some 3,000ft and north-south run of 2,880ft, together with other runs on the quadrantal points of 2,700 and 2,850ft. The single municipal hangar was joined by another early in 1931 when Airwork Limited (which had opened Heston aerodrome in 1929) set up a Bristol Repair Depot, and commercial carrier operations had begun in July 1932 when the British Air Navigation Company flew the first scheduled service to Cardiff with a Fokker F.VII. Several carriers then started using Whitchurch – Norman Edgar (later to become Western Airways), Crilly Airways, Railway Air Services, Channel Air Ferries, Irish Sea Airways, Great Western & Southern Air Lines, etc – linking Bristol with many other provincial cities. By 1933 a new hangar was constructed and a large tarmac apron laid and in 1934 Airwork extended its facilities with additional workshops, and because of the increasing number of regular air services using Bristol, the Airport Committee installed flood and runway lighting. The Air Ministry agreed to install a wireless transmitting station and direction-finding beacon, a teleprinter link with Cardiff was added, and Whitchurch became one of the new Met forecasting stations. In 1935 the Bristol City Council took responsibility for the development of the airport. Soon after the Munich crisis, the Air Ministry made arrangements for additional RAFVR facilities by setting up No.33 Elementary & Reserve Flying Training School, and the personnel and twelve aircraft were installed by December 1938. By January 1939 additional hangars and buildings to house them had been erected, their aircraft strength increased to twenty-six, and a VHF radio beacon was added to the existing facilities.

On 23 August the Soviet-German Non-Aggression Pact was signed, and it was apparent that war was inevitable in a matter of days, so the following day Parliament enacted the Emergency Powers (Defence) Bill that, in short, empowered the government to act as it saw fit without reference to Parliament. Reserves of army, navy and air force were called up, Air Raid Precautions (ARP) put on stand-by, and from 24 August the Royal Observer Corps was on watch day and night. On 25 August a Treaty of Alliance was signed between Britain and Poland, and by 29 August the first of some hundred-plus new regulations under the above Bill was issued.

On that same date, Whitchurch became established as A Base when advance parties from Imperial Airways and British Airways prepared for what now seemed inevitable: acceptance of their air fleets for Air Ministry control.

With effect from midnight on Thursday 31 August 1939 the Air Navigation (Emergency Restrictions) Order promulgated by the Air Ministry came into force, and their NAC organisation, activated that morning, had communicated to all independent operators the first of the three code words that they had been given some nine months earlier, warning them to 'get ready to stand by'. By the evening of 1 September, the day that Hitler's armies invaded Poland, the fleets of all the independent air carriers were *en route* to their wartime bases in the west. By that time, Germany had bombed Polish cities and it was alleged that 1,500 people, including women and children, had been killed in the air raids, and the British and French Governments threatened that they would declare war unless Germany withdrew all its forces from Poland by 1100hrs on 3 September. In Britain, complete mobilisation of the navy, army and air force was ordered, blackout throughout Britain from sunset to sunrise became mandatory, and the official evacuation of thousands of children, mothers and disabled began from London and other major cities designated as vulnerable.

Like the fleets of the independent carriers, those of Imperial Airways and British Airways had come under the control of the NAC from midnight on 31 August and the airlines had received instructions to implement the War Book plans. (As it happened it was not until 5 September that the Air Ministry confirmed by letter that the government had requisitioned their aircraft from that time and date, by which time the NAC headquarters had been set up under the control of the Deputy Director General of

A Whitchurch miscellany. Left to right in the background are a British Airways Junkers-Ju 52/3m, a D.H. Albatross, two A.W. Ensigns and a Lockheed 14 Super Electra. Another British Airways Lockheed 14 occupies the centre foreground with a Lockheed 10A Electra behind. (*Capt R T Halliwell via the Science Museum*)

Civil Aviation (W P Hildred) at Stoke Bishop, Bristol.)

Operations of British and foreign airlines between the United Kingdom and the Continent were suspended and all British civil landplanes abroad on service returned to the United Kingdom. The Air Navigation (Restriction in Time of War) Order 1939 specified that the whole of Britain east of the dividing line, already mentioned, would become a restricted area for civil aircraft, as would certain other specified areas. These were the neighbourhood of Invergordon, Castleton (Weymouth), Plymouth and Scapa Flow. All landplanes entering the United Kingdom from abroad were allowed to do so only at Shoreham Airport (where an enclave existed for the purpose), Belfast Harbour Airport, Whitchurch Airport, Speke Airport (Liverpool) or Perth. Flying-boats and seaplanes entering the country could do so only at Poole Harbour Airport or Pembroke Dock Airport, and all aircraft arriving from overseas were to do so by specified approach lanes. Aircraft were directed to fly no higher than 3,000ft and no lower than 1,000ft but were obliged to remain visible from the ground, and no flying was permitted between sunset and sunrise. Civil flying anywhere over the United Kingdom required a special permit, and private and club flying were banned.

However, special provision had been made for two maintenance bases for Imperial Airways located in the eastern prohibited area: Bramcote, near Nottingham, for landplanes and Hythe, near Southampton, which would continue initially to provide overhaul facilities for the twenty-four Short C class flying-boats operating Atlantic and Empire services, and two other marine aircraft.

In addition to Imperial Airways' twenty-five flying-boats and one seaplane, NAC assumed immediate control of their forty landplanes, sixteen of which were based overseas. These were three Armstrong Whitworth A.W.XV Atalantas based at Karachi; nine D.H.86 Diana class (three of which were based at Bangkok and six at Khartoum) and four Handley Page H.P.42s based at Cairo. The twenty-four based in the United Kingdom for dispersal to wartime bases were one Avro Ten, eleven A.W.27 Ensigns, seven D.H.91 Frobisher class, three H.P.45s* and two Short L.17s. From British Airways NAC took control of sixteen landplanes: one Fokker F.XII, three Junkers-Ju 52/3ms, five Lockheed 10A Electras and seven Lockheed 14 Super Electras.

Before NAC took over the fleets, however, several of Imperial Airways' and British Airways' aircraft had been pressed into military duties, the first being the former company's H.P.42W *Horatius* which flew the first consignment of RAF supplies to Paris (Le Bourget) in late August, and on 31 August a D.H.91 Albatross departed from Britain for a fast round-trip to Karachi carrying senior army personnel. She returned after the

*At the time the Handley Pages were known as H.P.42E, or Eastern, and H.P.42W, or Western. Subsequently, it was discovered that the Handley Page designations were H.P.42 and 45 respectively.

declaration of war to find the Albatross fleet dispersed to B Base at Bramcote.

Committed to War

On 1 September No.1 Group Bomber Command was ordered to implement the pre-arranged plan to detach an air formation of ten squadrons of Fairey Battle short-range bombers to France (the Advanced Air Striking Force – AASF). Separately, more squadrons of Hawker Hurricanes and Westland Lysanders were to constitute the Air Component of the British Expeditionary Force. As war approached and preparations for the build-up of British forces in France proceeded, the RAF's No.24 (Communications) Squadron, Hendon, found itself unable to cope with the demands being made upon its motley assortment of aircraft (despite its motto, 'Prepared for all things') and on 1 September a British Airways Lockheed 14 was requisitioned to fly RAF signals personnel to Le Bourget, thus becoming the first to comply with the mandatory control stops at Shoreham in both directions. Other Lockheed aircraft were either pressed into service, or ordered to stand by at Heston to be available at short notice.

Similarly, when Imperial Airways' fleets were evacuated from Croydon on 1 September, two Ensigns remained behind as, early the following morning, Capt Perry was required to fly to France in advance of the AASF to carry thirty RAF personnel and equipment to an open field near Rheims in preparation for the arrival of some of their 160 Fairey Battles. Perry made a second return trip to France that same afternoon. On 2 September a second Ensign positioned from Croydon to RAF Benson to depart early on 3 September with equipment and stores, and was returning from Amiens when war was declared.

However, most of the Imperial Airways' aircraft vacated Croydon on Friday 1 September, the first to leave being the two Short Scylla L.17 landplanes *Scylla* and *Syrinx* that had been withdrawn from storage at Southampton (Eastleigh) Airport and ferried to Croydon a few days previously. They led the exodus by taking equipment and advance parties to Whitchurch, which had by then been taken over by the Air Ministry when the E & RFTS had been disbanded and its aircraft dispersed, as had those of the independent operators. Having discharged their loads, *Scylla* and *Syrinx* were dispersed to Exeter, to become quickly involved with the transport of military stores and RAF ground staff to AASF bases in France.

As the majority of airport and engineering staff would not be in position at Whitchurch until Monday 4 September, the arrival there of the aircraft fleets needed to be staggered, so on 1 September the D.H.91 Albatrosses were first flown to Bramcote, to be repositioned to Whitchurch over the next few days.

Similarly, seven of the A.W.27 Ensign fleet were flown up to Baginton at Coventry, their arrival there having been so secret that the aerodrome manager was taken by surprise and he was horrified to have his camouflage scheme completely ruined by the dispersal of seven large shining silver airliners. His relief was abundantly apparent when he learned that they were stopping over only briefly before proceeding 'elsewhere'.

Meanwhile, two Ensigns, three H.P.42Ws and the Avro Ten followed *Scylla* and *Syrinx* from Croydon to Whitchurch where British Airways' Gatwick-based foreign contingent of one Fokker F.XII and three Junkers-Ju 52/3ms were soon to join them. Several of British Airways' Lockheeds had been involved in No.24 Squadron's activities and remained temporarily at Heston when the rest departed on 1 September, soon to be dispersed around the perimeter at Whitchurch.

On Saturday 2 September, as over fifty RAF fighter aircraft streamed into Croydon, Imperial Airways staff were intent upon the mammoth task of loading into lorries and two goods trains the hundreds of tons of equipment and stores, much of which had been packed, crated and coded during the preceding months. By this time they knew that most of it was destined for Bristol and Whitchurch, for a British Airways announcement on 1 September on their evacuation from London and Heston had been reported in the Saturday newspapers.

All that day a constant stream of vehicles shuttled between Croydon Airport, the Imperial Airways offices in Stafford Road and Waddon station, carrying the boxes and crates and loading them on to the trains in the sidings, while dozens of Carter Patterson vans were used to take material required to be in position at the wartime bases before the bulk of some 600 staff arrived.

Many staff required to assist with the reception arrangements at the Grand Spa Hotel at Clifton, Bristol, and to augment the advance parties at Whitchurch, were despatched on Saturday by Green Line coaches, and others left in their own transport. The majority, however, were instructed to report to Waddon station at 0700hrs on Sunday 3 September where they boarded a special passenger train for Temple Meads, Bristol, where fleets of vehicles awaited to distribute them to various hotels and billets around the city and environs. By that time war had been declared.

The civil aircraft and staff were now in position to provide maximum support to the war effort. They were now under the direct control of NAC and space on all aircraft was at the disposal of the government. The Air Ministry instituted a system of priority of loads, administered by the DGCA's staff at Bristol.

It had enlisted not only the hundreds of civil aircraft and staff, but had at its disposal the accumulated professional experience and expertise of airline management, operation and aircraft maintenance of what was soon to be known as 'the Merchant Air Service'.

These significant resources were to prove of inestimable value in the war years ahead. □

British Airways Limited Fleet (representative data) September 1939

Aircraft	All-up weight	Span	Length	Typical no. of passengers	Cruising Speed	No. of engines	Powerplant type
Lockheed 10A Electra	10,500lb	55ft	38ft 7in	10	180mph	2	400hp Pratt & Whitney Wasp Junior
Lockheed 14 Super Electra	17,500lb	65ft 6in	44ft 4in	14	212mph	2	760hp Wright Cyclone GR-1820-F62
Fokker F.XII	19,836lb	88ft 6in	60ft 8in	16	130mph	3	500hp Pratt & Whitney Wasp T1D1
Junkers-Ju 52/3m	23,150lb	95ft 10in	62ft	15	152mph	3	770hp Pratt & Whitney Hornet (BMW 132H)

Aircraft

Lockheed 10A Electra	Lockheed 14 Super Electra	Fokker F.XII
G-AEPN	G-AFGP	G-AEOS
G-AEPO	G-AFGR	
G-AEPR	G-AFKD	
G-AFCS	G-AFKE	**Junkers-Ju 52/3m**
G-AFEB	G-AFMO	G-AERU *Juno*
	G-AFMR	G-AERX *Jupiter*
	G-AFYU	G-AFAP *Jason**

* Freighter

Imperial Airways Limited Fleet (representative data) September 1939

Aircraft	Class	All-up weight	Span	Length	Typical no. of passengers	Cruising speed	No. of engines	Powerplant type
Avro 618 Ten		10,225lb	71ft 3in	47ft 6in	8	95mph	3	215hp A S Lynx IVC
De Havilland D.H.86	Diana	10,250lb	64ft 6in	46ft 1in	10-12	141mph	4	200hp D.H. Gipsy Six
Short S.20 *Mercury*		15,500lb*	73ft	51ft	—	180mph	4	370hp Napier Rapier VI
Armstrong Whitworth A.W.XV	Atalanta	21,000lb	90ft	71ft 6in	9-11	120mph	4	340hp A S Serval III
Handley Page H.P.42	Hannibal	28,000lb	130ft	89ft 9in	24	95mph	4	490hp Bristol Jupiter XIF
Handley Page H.P.45	Heracles	29,500lb	130ft	89ft 9in	38	95mph	4	555hp Bristol Jupiter XFBM
De Havilland D.H.91	Frobisher	29,500lb	105ft	71ft 6in	21	193mph	4	525hp D.H. Gipsy Twelve
Short L.17	Scylla	33,500lb	113ft	83ft 10in	39	90mph	4	660hp Bristol Pegasus XC
Short S.21 *Maia*		38,000lb**	114ft	84ft 11in	—	164mph	4	920hp Bristol Pegasus XC
Short S.23 Empire Boat	C class	40,500lb	114ft	88ft	16-24	164mph	4	920hp Bristol Pegasus XC
Short S.30 Empire Boat	C class	48,000lb***	114ft	88ft	16-24	164mph	4	890hp Bristol Pegasus XII
Armstrong Whitworth A.W.27	Ensign	49,000lb	123ft	114ft	40	170mph	4	850hp AS Tiger IX

* AUW 20,800lb for normal Composite launching

** AUW limited to 27,700lb for Composite launching

*** AUW 53,000lb when flight refuelled

Aircraft

Avro 618 Ten
G-AASP *Achilles*

De Havilland D.H.86
G-ACPL *Delphinus*
G-ACWC *Delia*
G-ACWD *Dorado*
G-ADFF *Dione*
G-ADUE *Dardanus*
G-ADUF *Dido*
G-ADUG *Danae*
G-ADUI *Denebola*
G-AEAP *Demeter*

Handley Page H.P.42
G-AAGX *Hannibal*
G-AAUC *Horsa*
G-AAUE *Hadrian*
G-AAXF *Helena*

Handley Page H.P.45
G-AAXC *Heracles*
G-AAXD *Horatius*

G-AAUD *Hanno*

Armstrong Whitworth A.W.XV
G-ABTI *Atalanta*
G-ABTJ *Artemis*
G-ABTL *Astraea*

Armstrong Whitworth A.W.27
G-ADSR *Ensign*
G-ADSS *Egeria*
G-ADST *Elsinore*
G-ADSU *Euterpe*
G-ADSV *Explorer*
G-ADSW *Eddystone*
G-ADSX *Ettrick*
G-ADSY *Empyrean*
G-ADSZ *Elysian*
G-ADTA *Euryalus*
G-ADTB *Echo*

Short L.17
G-ACJJ *Scylla*
G-ACJK *Syrinx*

De Havilland D.H.91
G-AEVV *Faraday*
G-AEVW *Franklin*
G-AFDI *Frobisher*
G-AFDJ *Falcon*
G-AFDK *Fortuna*
G-AFDL *Fingal*
G-AFDM *Fiona*

Short S.20
G-ADHJ *Mercury*

Short S.21
G-ADHK *Maia*

Short S.30
G-AFCT *Champion*
G-AFCU *Cabot*
G-AFCV *Caribou*
G-AFCX *Clyde*
G-AFCY *Aotearoa*
G-AFCZ *Australia/Clare*
G-AFDA *Awarua*

Short S.23
G-ADHL *Canopus*
G-ADHM *Caledonia*
G-ADUT *Centaurus*
G-ADUV *Cambria*
G-ADUW *Castor*
G-ADUX *Cassiopeia*
G-ADVB *Corsair*
G-AETV *Coriolanus*
G-AETX *Ceres*
G-AETY *Clio*
G-AETZ *Circe*
G-AEUA *Calypso*
G-AEUB *Camilla*
G-AEUC *Corinna*
G-AEUD *Cordelia*
G-AEUE *Cameronian*
G-AEUF *Corinthian*

The Birth of Air Transport

John Stroud

The Review respects and acknowledges all of aviation's pioneers. The matter of who did exactly what and when, in air transport, and of course where, is set down here, in detail, by John Stroud. Before rushing for pens or word processors, readers are invited to note the qualifications made.

Orville Wright made the first controlled and sustained flight with a power-driven heavier-than-air craft at Kitty Hawk, North Carolina, on 17 December 1903. This is , true, but only because the statement includes the qualifications 'controlled' and 'sustained' and 'with a power-driven heavier-than-air craft'.

The establishment of firsts in air transport presents us with the same need for qualifications. In this case six are required to establish the beginning of regular, daily, scheduled, international passenger-carrying commercial air services: passenger, civil, daily, regular, international and sustained. (Even the term daily needs qualification because it was frequently used to denote Monday to Saturday inclusive.)

The first scheduled passenger air service, opened on 1 January 1914 between St Petersburg and Tampa by St Petersburg-Tampa Airboat Line, met four of the qualifications but was neither international nor sustained. The Austro–Hungarian air mail service between Vienna and Kiev, begun on 20 March 1918, was not operated daily, did not carry passengers and was, at least in part, a military operation – but it was the first regular scheduled international air mail service. On 15 May 1918, the United States Post Office began New York–Washington and Washington–New York mail services. These met four of the qualifications but were not international and did not carry passengers. When Deutsche Luft-Reederei began its Berlin–Weimar service on 5 February 1919, it met all the qualifications except international. Lignes Aériennes Farman began Paris–Brussels services on 22 March 1919, but these were weekly and so failed to meet the qualification daily.

Finally, on 25 August 1919, the British airline Aircraft Transport & Travel met all six qualifications when it inaugurated a regular, sustained, civil, daily (Monday to Saturday) passenger service between London and Paris.

Many of these pioneering ventures will be discussed more fully but first it is necessary to look further back than the second decade of the twentieth century.

We shall never know who first thought of air travel, although the Greek legend of Daedalus and Icarus suggested a flight plan for a journey between two islands. The flight ended in disaster and in any case should probably come under the heading of private flying.

However, in 1842-43 William Samuel

The Benoist XIV leaving St Petersburg, Florida, on the first-ever scheduled passenger air service, on 1 January 1914.

The Zeppelin LZ 120 *Bodensee* operated scheduled services between Friedrichshafen and Berlin during 1919. (*Luftschiffbau Zeppelin*)

Henson and John Stringfellow in Britain were working on the design of their steam-powered twin-screw 150ft span *Aerial* or Aerial Steam Carriage. They also tried to promote the Aerial Transit Co to finance the building and operation of aircraft – the first airline? D E Columbine acted as what would now be termed publicity manager and achieved widespread distribution of illustrations depicting the Aerial Steam Carriage flying in various parts of the world. It was shown over London, leaving the English coast at Dover, over the Pyramids and even as far away as India.

The first company actually established for the aerial carriage of passengers was Delag (Deutsche Luftschiffahrts AG), founded at Frankfurt-am-Main on 16 November 1909. Delag ordered a passenger Zeppelin rigid airship, LZ 7 *Deutschland* – the world's first passenger aircraft – which made its first flight on 19 June 1910. The company set up a number of airship stations throughout Germany and its first cruise was scheduled to leave Düsseldorf on 28 June. The *Deutschland* set out on a three-hour cruise with twenty-three journalists as passengers but after a flight of about nine hours the Zeppelin, as well as the commander's reputation, was wrecked but no-one was hurt. They were hungry, however, having had only a champagne and caviar breakfast.

A replacement Zeppelin was acquired and before operations were ended by the start of the First World War, Delag had operated seven Zeppelins which made 1,588 flights covering 172,535km and carrying 33,722 passengers and crew. No passenger was ever injured, although four of the airships were lost through various causes.

Although the history of airship transport extends well beyond our subject year of 1919, I shall complete the story

before leaving the field for heavier-than-air craft.

During the period of its prewar operations, Delag did not operate any scheduled services despite having published maps showing a German route network. After the war, however, the company ordered two new Zeppelins, LZ 120 *Bodensee* and LZ 121 *Nordstern*. The twenty-passenger *Bodensee* began operating regular scheduled services over the Friedrichshafen–Berlin route on 24 August 1919, usually flying the route in opposite directions on alternate days. The last flight was on 5 December after which the service ceased for the winter. By that time, *Bodensee* had made seventy-eight flights over the route and carried 2,253 revenue passengers.

It was planned that in 1920 the *Bodensee*, lengthened and with twenty-four seats, would resume working the Friedrichshafen–Berlin route and that the newer *Nordstern* would operate between Berlin and Stockholm, but the Inter-Allied Commission of Control demanded that the two airships be handed over to the Allies.

However, this was not the end of the airship as a passenger transport. Britain wasted a lot of money on an airship scheme for Empire communications, and the aeroplane's limited range and payload for a time left doubts as to whether the long-range air transport vehicle would be the aeroplane or the airship. Except in Germany the question was settled when the British rigid airship R101 crashed in France at the start of its voyage to India in October 1930.

Germany built the LZ 127 *Graf Zeppelin* for training and research into transocean navigation and it became the first transocean airliner when it began operating South Atlantic services on 20 March 1932. The bigger LZ 129 *Hindenburg* was the first Zeppelin designed to carry passengers on transocean air services and it opened the North Atlantic route to New York on 6 May 1936. Its destruction a year later at the start of the 1937 season finally removed the airship from the air transport scene.

Because the St Petersburg-Tampa Airboat Line service was the first scheduled airline operation, it is desirable to record a few more facts about it. The route across Tampa Bay was 21 miles long and drastically cut the surface journey time between these two Florida points. The first service left St Petersburg at 1000 hrs on the first day of 1914. It was operated by a small two-seat 75hp Benoist Type XIV biplane flying-boat piloted by Anthony Jannus, and for the next three months operated twice a day. A two-

Officers of No. 1 (Communication) Squadron RAF at Hendon in December 1918. Maj Cyril Patteson, the CO, is in the centre wearing a leather jacket, and Lieut H 'Jerry' Shaw is first from the left. The aeroplane is an Airco D.H.4.

passenger Benoist was also used on the route and regularity was good. When the service was withdrawn at the end of March, some 11,000 miles had been flown and more than 1,200 passengers carried. Considering the primitive equipment used, only eleven years after the Wright brothers' first powered flights,

F5764, an Airco D.H.4A, was used by the RAF Communication squadrons and passed to Handley Page Transport as G-EAWH.

the Florida operation was a great achievement.

Nevertheless, it must be said that the aeroplanes of the pre-First World War period were anything but suitable for sustained, to say nothing of economic, air services. It was the war which forced development of the aeroplane and aero engine, produced large numbers of experienced pilots and aviation mechanics, and led to the provision of a large number of aerodromes.

Strangely, the United States did not play a major role in the early days of air transport. The US Air Mail Service began on 15 May 1918, operated by military aeroplanes with army pilots, and did not enjoy a flawless inauguration. The first flight from New York to Washington, with a change of aircraft at Philadelphia, made the journey in 3hr 20min but the pilot of the northbound flight set off in the wrong direction. That August, the Post Office took over the services and on 15 May 1919 opened the Chicago–Cleveland sector of what became the complete transcontinental mail service on 8 September 1920.

There were a few short-lived passenger services in the USA but the real birth of air transport took place in Europe, with France, Germany and the United Kingdom as the main participants.

It was in Britain and Germany that the

Lieut H 'Jerry' Shaw who flew the London–Paris charter on 15 July 1919. He is seen here with passengers disembarking from an Aircraft Transport & Travel Airco D.H.16 at Hounslow. (*Flight*)

first two airlines were formed. As early as 5 October 1916, with great optimism, George Holt Thomas founded Aircraft Transport & Travel with £50,000 capital and with the object of 'establishing and maintaining lines ... and to enter into contracts for the carriage of mails, passengers, goods and cattle ...'. On 15 November 1918, four days after the Armistice, Holt Thomas announced that arrangements were being made for the operation of London–Paris services by Aircraft Transport & Travel.

Holt Thomas had earlier founded the Aircraft Manufacturing Co (Airco), employing Geoffrey de Havilland as desig-

Part of Aircraft Transport & Travel's fleet at Hounslow in 1919. Left to right: Airco D.H.4A G-EAJD, Airco D.H.16 G-EACT which flew the first scheduled London–Paris service, and Airco D.H.4A G-EAJC which E H 'Bill' Lawford flew to Paris as AT&T's first operation on 25 August 1919.

ner and test pilot. This eventually led to the founding of the de Havilland aircraft and engine companies.

Air transport has always depended to a large extent on governments, sometimes with disastrous effects, and in May 1917 the Air Board had appointed the Civil Aerial Transport Committee to report the steps necessary to develop aviation for civil and commercial purposes. On 7

February 1918, the Committee made its report and advocated 'State action for the development of aerial transport services', stressing the need for Empire services and the necessary route surveys.

At the end of October 1919 the Advisory Committee on Civil Aviation recommended the establishment of certain main trunk airlines connecting Canada, Newfoundland, South Africa,

Pages from the Aircraft Transport & Travel log for the first week of scheduled cross-Channel services, showing the operations in both directions. (*RAF Museum Archives*)

August 1919.   Paris - London.

Aug 25ᵗʰ. Airco 9. (Lt McMullin) left le Bourget at 12.30 with 1 passenger and parcels - Landed Hounslow 2.40

Aug 26ᵗʰ. Airco 16 (Maj Pattison) left le Bourget at 12.30 with 3 passengers   Landed Hounslow. 3.39.

Aug. 27ᵗʰ. Airco 4a (Lt Lawford) left le Bourget at 12.30 with 2 passengers   Landed Hounslow 3.20.

Aug 28ᵗʰ. Airco 4a (Lt Shaw) left le Bourget at 12.30 with 2 passengers and parcels - Landed Hounslow. 2.20.

Aug. 29ᵗʰ. Airco 9 (Lt McMullin) left le Bourget at 12.30 with 2 passengers and parcels. Landed Kenley 3pm.   Passengers + parcels by road to London

Aug. 30   Airco 4a (Capt Riley) left le Bourget. 12.30 with 2 passengers and parcels. Landed Hounslow. 3.10

   Airco 9. (Capt Baylis left le Bourget 12.30 with 1 passenger. Landed Hounslow 3.20

   Airco 9. (Lt McMullin) landed Hounslow from Kenley 10.15.

August 1919   London - Paris.

Aug. 23ʳᵈ. Airco. 9. (Lt McMullin) left Hounslow at 1.15 with spares for Paris Service Landed le Bourget 3.40

Aug. 25ᵗʰ. Airco 4a. (Lt Lawford) left Hounslow at 9.5 with 1 passenger and Goods Landed le Bourget 11.40

Aug. 25ᵗʰ. Airco 16 (Maj Pattison) left Hounslow at 12.40 with 4 passengers Landed le Bourget 3.5

Aug 26ᵗʰ. Airco 4a. (Lt Shaw) left Hounslow at 1.5. with 2 passengers and parcels. Landed le Bourget

Aug 27. Airco 9. (Lt McMullin) left Hounslow at 1.43 with 1 passenger and parcels. Landed le Bourget

Aug 29ᵗʰ. Airco 4a (Capt Riley) left Hounslow at 8.58 with 2 passengers and parcels. Landed le Bourget

   Airco 9. (Capt Baylis) left Hounslow at 12.36 with 2 passengers and parcels Landed le Bourget

   Airco 16 (Lt Lawford) left Hounslow at 12.26 with 4 passengers Landed le Bourget

Aug. 30ᵗʰ. Airco 4a. (Capt. Gathergood) left Hounslow at 12.30 with 2 passengers - Landed le Bourget

India, Australia and New Zealand to the United Kingdom by air and stated that 'the proper place for initial action' was the route to India and ultimately thence to Australia, 'to be followed by a service to South Africa and the development of these Imperial routes should be by private enterprise backed by State assistance'.

Holt Thomas was thinking along similar lines but considered that the place to start was across the Channel between London and Paris. The just over 200 miles separating London and Paris was considered about the right distance for practical non-stop operation of the aeroplanes of the period, there was good traffic potential, and such a route would provide a good demonstration of regular airline operation in an area not noted for good weather. Most important, it involved a sea crossing. Although Holt Thomas decided to begin with London–Paris services, even at that time he was

An *Illustrated London News* drawing made during the first scheduled London–Paris flight. The pilot (Maj Patteson) is passing a message to the front-seat passenger. The army officer seen in the embarkation photograph is on the right.

looking to extensions that would bring Madrid and Vienna within eight hours of London and Rome within ten hours. He wrote, 'when our airway extends link by link as far as Constantinople we shall bring that city within sixteen hours of London instead of fifty-six – the latter time ... being for prewar transit'.

Although Aircraft Transport & Travel was undoubtedly the pioneer British airline, it had an advantage not enjoyed by DLR in Germany – military operating experience over cross-Channel routes.

On 13 December 1918 a Communication Flight of the Royal Air Force stationed at Hendon was expanded to form No. 1 (Communication) Squadron under the command of Maj Cyril Patteson (please note his name). At the same time the 86th (Communication) Wing was formed. This wing comprised No. 1 (Communication) Squadron and a detachment at Buc, near Paris. The squadron was equipped with single-engined Airco D.H.4s and, later, with a few twin-engined Handley Page O/400s, and these aeroplanes were used to carry members of HM Government between London and Paris for the Peace Conference, regular services beginning on 10 January 1919.

On 23 March 1919 No. 1 Squadron was joined by No. 2 and the operation continued until September by which time 749 flights had been made with 91 per cent regularity, and 934 passengers, 1,008 bags of mail and 46 despatches had been carried. Ninety per cent of the two squadrons' flights had been over the London–Paris route. There was one accident, however, which caused a passenger fatality.

In February 1919 an air parcel service was begun between Folkestone (Hawkinge aerodrome) and Ghent. This was organised by Aircraft Transport & Travel at the request of the Belgian Government and operated by RAF D.H.9s flown by Service pilots but with AT&T's name appearing on stickers on the fuselages.

Then on 1 March 1919 No. 120 Squadron of the RAF began an air mail service linking Hawkinge with Maisoncelle where Nos. 18 and 110 Squadrons took over to continue the service to Cologne. This service was to provide communication with the British Army of Occupation, and the D.H.4s, 9s, 9As and 10s continued this work until August when 1,842 flights had been made with 96 per cent regularity and 90 tons of mail. On 15 August AT&T took over the operation which ceased in June 1920.

The Ban is Lifted

On 1 May 1919 civil flying in the British Isles was allowed for the first time since 1914. Numerous flights were made that

Passengers embarking at Hounslow for the first scheduled London–Paris flight, on 25 August 1919.

day but the most important were those of Handley Page, the manufacturer of large bombers. Using O/400s, Handley Page carried passengers and newspapers to various cities and continued these flights until the twelfth of the month. These were to lead to the founding of the second British airline, Handley Page Transport.

The first regular daily service in Britain was begun on 24 May by Avro Civil Aviation Service. It operated from Alexandra Park Aerodrome, Manchester, to the South Shore, Blackpool, via Birkdale Sands at Southport. Single-engined Avro 504K, 504M and 536 biplanes were used and the fares were £5 5s single and £9 return. The service operated without subsidy and lasted eighteen weeks during which 222 flights were scheduled and 194 completed.

During 1919 a number of short-lived services were operated within Britain, although some of the claims to have operated regular services cannot be confirmed. No international flights were allowed until 13 July when for a week civil flying between London and Paris was allowed in connection with the Peace celebrations. The first such flight over

The Airco D.H.16 K-130. Seen here with Airco markings and the national colours on the rudder, it was repainted before flying the first scheduled London–Paris service. On 25 August 1919 it carried the national identification letter G on its rudder but retained the temporary registration K-130. It became G-EACT.

the route was made on 15 July when Lieut H 'Jerry' Shaw, formerly of No. 1 (Communication) Squadron, flew a charter for AT&T's Aviation Hire Department. He used the D.H.9 K-109 and carried one passenger. On 14 June Handley Page Transport had been incorporated and, on the day of Jerry Shaw's flight, made a flight from Cricklewood, North London, to Paris.

The most important date in the history

Hounslow aerodrome before it became the first London airport. The largest hangar in the centre became the customs shed.

Above: Handley Page Transport's Handley Page O/7 G-EAAF at Buc on 25 August 1919. The registration is painted on a fabric panel which has been attached to the fuselage over the ornamental style of Handley Page's name.

Below: Handley Page Transport's Handley Page O/400 G-EAAE photographed with short engine nacelles and internal fuel tanks.

of international air transport was Monday 25 August 1919, for that was the day on which Aircraft Transport & Travel began scheduled cross-Channel services between London (Hounslow) and Paris (Le Bourget).

The AT&T services were scheduled to leave Hounslow and Le Bourget six days a week at 1230hrs with a two and a half hour block to block timing. The inaugural service, with four passengers including the editor of the *Illustrated London News* and an army officer, left Hounslow at 1240hrs and arrived at Le Bourget at 1505hrs. The aeroplane used was the Airco D.H.16 G-EACT but it still bore its temporary registration K-130. The pilot was Maj Cyril Patteson, previously mentioned as the CO of No. 1 (Communication) Squadron.

The *Illustrated London News* published a page of pictures of this occasion and there is film of the D.H.16 departing, but controversy has raged over this event for seventy years. It has been claimed that Lieut E H 'Bill' Lawford made the first flight and in fact he did make the first flight that day, but not to the advertised schedule. He left Hounslow at 0905hrs

and arrived at Le Bourget at 1140hrs. He was flying the Airco D.H.4A G-EAJC and carrying G M Stevenson-Reece of the *Evening Standard* plus a load of newspapers, leather, grouse and jars of Devonshire cream. We know that Lawford was given the weather as 'squaggy with bolsoms in the Channel' but we do not know why the flight was made.

The first scheduled flight from Paris was made by Lieut J McMullin in a D.H.9 carrying Stevenson-Reece and Lawford as passengers. It arrived at Hounslow at 1440hrs, so Lawford and Stevenson-Reece had made the out and back flight with an elapsed time of 5hr 35min. The single fare was £21 although it was soon reduced to £15. Strangely, the Air Ministry's statistics for cross-Channel services started only on 26 August.

It is interesting to record that on 26 August the departure from Hounslow was 35 minutes late and that on the twenty-seventh 1hr 13min late, but that on the twenty-ninth AT&T made three flights, at 0858, 1226 and 1236hrs with a total of eight passengers and some parcels, and that three types of aircraft were used, D.H.4A, D.H.9 and D.H.16.

On 30 August the departure was on time at 1230hrs. There is no mention in the AT&T flight log of a Paris flight on the twenty-eighth but in the first week of operation there were eighteen passengers.

There were flights from Paris every day that first week and all left on time, the Saturday service being duplicated. Block times varied between 1hr 50min and 3hr 9min. However, the Friday flight landed at Henley from where the two passengers and the parcels were taken to London by road. Seven pilots were used during the first week's operations.

On 10 November AT&T began carrying mail on the London–Paris route, with a surcharge of 2s 6d (12½p) an ounce, of which I believe the airline received only 6d (2½p). That day an air mail pennant was attached to the rudder of the D.H.4A G-EAHF at Hounslow and Lieut J McMullin took off with the first official mail load, but because of weather he only got as far as Epsom before having to return; however, the service got through the next day.

To complete the essential story of AT&T it should be recorded that in April 1920 it introduced the first of the eight-passenger D.H.18s and on 17 May 1920 began a thrice-weekly London–Amsterdam service in conjunction with KLM. On 17 December that year, however, it ceased operation. In 1922 AT&T was succeeded by the Daimler Airway.

Handley Page Transport had wanted to start its Paris service on 25 August 1919 but was not ready to begin regular operation until 2 September when Lt-Col W Sholto-Douglas (later Marshal of the Royal Air Force Lord Douglas of Kirtleside and ultimately chairman of British

View forward through the cabin of a Handley Page O/7. Note the diagonal structural members, the clock and the calendar. The cabin even had curtains.

Departure scene at Cricklewood in the early days of Handley Page Transport. The aeroplanes are Handley Page O/400s and the step-ladder for passenger embarkation is almost certainly one of the six which Handley Page bought for a guinea (£1.05). The writer swears that he saw one of these step-ladders at Cricklewood when HP was building Victor jet bombers.

European Airways) made the inaugural flight with the H.P.O/7 G-EAAF with seven passengers. However, Handley Page Transport did make one-off flights to Paris on 25 August, Maj E L Foot flying G-EAAE with seven journalists and Sholto-Douglas flying G-EAAF with his wife and Miss Gertrude Bacon as the only passengers on the return flight on the twenty-sixth – reportedly the first women to cross the Channel by air.

On 23 September Handley Page Transport opened a thrice-weekly London–Brussels passenger and goods service and on 11 October introduced 3s (15p) lunch baskets.

On 27 September 1919 the British railway system was brought to a standstill by a strike which led to a number of short-lived air transport operations. On the twenty-eighth Supermarine Aviation Works began a passenger service between Southampton and Le Havre with single-engined Supermarine Channel-type flying-boats and mail was carried from the second day. This service was maintained until the end of the railway strike on 5 October. There have been numerous reports that Supermarine also operated Southampton–Cowes and Southampton–Bournemouth services with these flying-boats, but I have found no hard evidence that this was true; they seem to have been positioning flights for joyriding visits. However, Supermarine Sea Eagles were used on Southampton–Guernsey services from September 1923.

From 29 September, because of the strike, AT&T and Handley Page Transport carried mail on their respective London–Paris and London–Brussels services, while within the United Kingdom the RAF flew mail services linking London, Bristol, Birmingham, Manchester, Newcastle and Glasgow.

A Leeds–Hounslow mail service was begun on 30 September by North Sea Aerial Navigation using Blackburn Kangaroos. These also had accommodation for eight passengers, plus the 400lb of mail or cargo. Fares of £15 15s single and £30 return were charged. (It was during this strike that Vickers Ltd used a non-rigid airship to carry company mail between Barrow and Sheffield.)

A major event in British air transport history took place on 13 October 1919 when the shipowners S Instone & Co began a private air service linking Cardiff, Hounslow and Paris for the carriage of staff and documents. The first flight was made by the D.H.4 G-EAMU flown by F L Barnard.

In February 1920 Instone began operating the Hounslow–Paris sector as a public service. This led to the development of a major British airline which, on 12 December 1921, became The Instone Air Line. Instone set high standards (probably the first airline to introduce staff uniforms) and on 31 March 1924 became one of the major constituents of the new national airline Imperial Airways. The other constituent airlines were Handley Page Transport, Daimler Airway (successor to AT&T) and British Marine Air Navigation, the company which operated the flying-boat service to Guernsey from 1923.

Results

According to Air Ministry figures British airlines made 149 flights across the Channel in the period 26 August to 30 September 1919, with a total of 359 passengers. Between 1 October 1919 and 31 March 1920 there were 605 flights with 796 passengers. In August and September 1919 the value of air-freighted exports to France was £2,158 plus £844 for re-exports while imports were worth £4,501 of which £76 bore duty. From the beginning of cross-Channel mail carriage in November 1919 to the end of March 1920 it was estimated that there were 5,360 letters to Paris and 4,739 inbound.

Reliability and punctuality figures are not available for 1919 but those for the

The caption on the negative of this photograph read, 'Loading Mails for abroad in a Handley Page aeroplane at Hounslow Aerodrome'. This may have been taken during the railway strike in 1919 or it may have been a posed publicity photograph. The aeroplane is an unconverted Handley Page O/400 which retains its forward gun mounting.

S Instone & Co's Airco D.H.4 G-EAMU at Hounslow, almost certainly on 13 October 1919 when the company began a private service to Paris. The company's house flag appears on the fuselage between the cockpits. The aeroplane in the background is a Breguet 14T of the French airline CMA.

1920 Aircraft Transport & Travel 'was just able to break even with the fare at £5 for the London–Paris service, if every seat was taken on every flight'. At that time there was no government subsidy.

Founding of IATA

Although not an airline operation, there was one very important event in 1919 which must not be overlooked. On 25 August, the day on which AT&T began its London–Paris services, the organisational meeting of the International Air Traffic Association was called in The Hague. The invitations to attend the meeting were sent out by George Holt Thomas, chairman of AT&T, and the meeting was chaired by the airline's managing director, Sir Sefton Brancker.

One of Deutsche Luft-Reederei's large fleet of L.V.G. C VIs with which the airline began operation in February 1919. (*Lufthansa*)

period 1 January to 31 March 1920 show that 251 flights were begun over the London–Paris route (presumably the total for both directions) and of these a hundred reached their destination on the same day in under three hours, fifty-one took three to four hours and sixteen took over four hours. Forty were completed on the second day or not at all, and the time taken by thirty-five flights was not recorded. Allowing that half the flights for which the times are not known actually made the journey in under four hours, the percentage of flights achieving this time was 70.7. The other nine flights were not accounted for in the Air Ministry figures.

Unfortunately, no financial figures appear to have survived but Robin Higham in his *Britain's Imperial Air Routes 1918 to 1939* stated that in the late summer of

A Breguet 14T Salon of Cie des Messageries Aériennes at Hounslow in the autumn of 1919.
(*Flight*)

The other airlines represented were Det Danske Luftfartselskab, Deutsche Luft-Reederei, Det Norske Luftfartrederi, Svenska Lufttrafik and Nederlandsche en Koloniale Luchtverkeer Maatschappij (then being formed and to become famous as KLM). The agreement forming the first free association of international airlines (not all then actually operating) was signed on 28 August. This was the beginning of IATA which today, as the International Air Transport Association, has some 185 members which in total carry well over 600 million passengers a year.

Germany Makes an Early Start

Pride of place for starting sustained regular passenger, mail and goods services must go to Deutsche Luft-Reederei (DLR) which on 5 February 1919 opened a daily service linking Berlin (Johannisthal aerodrome) and Weimar, the temporary seat of the new National Assembly, via Leipzig.

DLR, like Aircraft Transport & Travel, was founded during the First World War, on 13 December 1917. The founders were Allgemeinen Elektricitäts-Gesellschaft (AEG) and Luftschiffbau Zeppelin in co-operation with the Hamburg-Amerika Line which looked after the bookings. The new company's first task was to study the potential development of air transport and it received Air Traffic Permit No. 1 on 8 January 1919. A fleet of aeroplanes, together with pilots and other staff, was assembled and less than a month after receiving its operating permit the airline was in business.

One of the most interesting aspects of DLR was its fleet. By far the majority of its aeroplanes were two-seat single-engined biplanes which had been used for reconnaissance during the war or intended for this work but not delivered

One of the Haefeli DH 3s used on the Swiss Army's mail and passenger services during 1919. (*Swissair-Photo AG*)

when the war ended. According to the present Lufthansa, DLR had seventy-one of these single-engined types and fifteen twin-engined 'high capacity' aircraft which could carry two crew, mail, freight and up to six passengers. In fact, DLR was to operate well in excess of this total of eighty-six aeroplanes.

More than fifty of the aeroplanes were L.V.G. C VIs and these were used for mail, freight and passenger services. The open rear cockpit could take two passengers and it seems that the airline provided them with warm clothing, helmet and goggles. A.E.G. J IIs were also used in numbers and some of these were fitted with enclosed cabins. The twin-engined types were Friedrichshafen G IIIa and A.E.G. G V bombers. These, too, had open cockpits although at least one of the Friedrichshafens was converted to have an enclosed cabin for six passengers.

DLR also had some small seaplanes for coastal services and for a time in 1919 operated some five-engined Staaken R.XIVa biplanes on special missions to the Ukraine.

Following the opening of the Berlin–Leipzig–Weimar service, nearly three months before civil flying was allowed in Britain, DLR began a Berlin–Hamburg service on 1 March, and in mid-March began a Berlin–Brunswick–Hanover–Gelsenkirchen service to link the capital

with the Ruhr. Gelsenkirchen was used as a main distribution point, with motorcycles radiating over a wide area to act as feeders to and from the air service. In July the first holiday resort services were opened, from Hamburg to Westerland on the island of Sylt and Berlin to Swinemünde on the Baltic. DLR's operations were undertaken with great difficulty, largely because of the scarcity and high price of fuel, and there were periods when services had to be suspended.

Although DLR was the first and by far the biggest of German airlines, a number of other companies began services during 1919. DLR was merged with Lloyd Luftdienst to form Deutscher Aero Lloyd and in 1926 this company was to become a constituent of Deutsche Luft Hansa, a predecessor of the magnificent Lufthansa of today. The words Luft Hansa, sometimes with a hyphen, were joined to become one word from 1 January 1934.

France's Contribution

In this review of the birth of air transport, I have placed France third among the three European pioneers of air transport because Britain began the first daily international passenger services while Germany operated the first daily domestic passenger services. This choice does not reflect my appreciation of French aviation pioneering. France has made an enormous contribution to the development of world aviation, and its opening up of the air route from France to South America was an achievement beyond praise. But according to the terms laid down at the start of this article, France came third.

On 8 February 1919 Lieut Lucien Bossoutrot flew a Farman Goliath twin-engined transport biplane from Toussus le Noble, one of the Paris aerodromes, to Kenley, south of London, and this has

often been claimed as the start of Paris–London air services. It was not a service but a one-off flight and the eleven passengers were all military gentlemen. Two days later a converted Caudron C.23 night-bomber was flown from Paris to Brussels with five journalists aboard but that flight, also, was not the start of a regular service.

Even earlier, at Christmas 1918, Latécoère had begun its great pioneering venture when the first proving flight left Toulouse for Barcelona. This was the first step in developing the route to South America, and by June 1925 the service was in full operation as far as Dakar.

Early in 1919 a group of French aeroplane and aero engine manufacturers formed Cie des Messageries Aériennes (CMA), initially for the carriage of goods between Paris and Lille. This service opened on 18 April 1919 and two months later the operation was extended to Brussels.

On 16 September CMA began a Paris (Le Bourget)–London (Hounslow) service, using single-engined Breguet 14T biplanes, with a Mons Massot making the first flight. The first French service from London to Paris was on 19 September and soon the route was extended to Cricklewood and worked in pool with Handley Page Transport.

For a time most French air services were worked by large fleets of single-engined biplanes, mostly open-cockpit military aeroplanes, but some, like the Breguet 14T, had rudimentary passenger cabins. At about the time the war ended, however, Farman had built its Goliath twin-engined bomber which was successfully converted to a passenger carrier with two separate cabins, and it is almost certain that when Lignes Aériennes Farman (the Farman Line) opened its weekly Paris–Brussels service on 22 March 1919 it employed Farman Goliaths. A Farman advertisement which appeared in the *Overseas Daily Mail* earlier that month stated that Goliaths would be used between Toussus le Noble and Brussels. The scheduled time was 2hr 45min and the single journey fare 365 francs.

Under the heading 'Les grandes dates des Lignes Farman' and based on documents in the Musée Air France, in *Icare* No. 82 it is stated that the first Paris–London and Paris–Brussels flights by Goliath were on 8 and 12 February 1919 respectively. On both occasions Lucien Bossoutrot was the pilot, having joined Farman on 1 January 1919. The Brussels flight, with twenty-four passengers, seems, like the flight to Kenley, to have been a one-off before the start of regular operation on 22 March.

A strange aspect of the Kenley flight, which had only military passengers (civil flying was not then allowed), is that there exists a cheque for £15 15s, dated 18 January 1919, made out to M Henri Farman and signed by Frank H Butler. This cheque is said to be for the first Paris–London flight, and at some unknown time was mounted on a card bearing the words 'pour le Premier Voyage Paris Londres' and 'du Premier Voyageur dans le premier Aérobus-Transport'. The card also bore a decorative version of the Farman badge. It nevertheless seems extremely doubtful that Frank Butler was on that flight.

A later French airline was Cie des Grands Express Aériens which began a Paris–London service with Farman Goliaths on 29 March 1920. In 1923 Grands Express and CMA merged to form Air Union which in turn, in 1933, became a constituent of the newly formed Air France, now one of the world's great airlines.

The southwest corner of Handley Page's Cricklewood aerodrome. The photograph was taken from one of the Handley Page O/400 family.

A Swiss Venture

Not to be overlooked as a pioneering venture of 1919 is the Swiss experimental military mail service which was opened between Zürich and Berne on 8 January 1919 and extended to Lausanne on 1 February. This operation was badly affected by weather. By the end of January only fourteen return flights had been made, and the extension to Lausanne was only flown on sixteen days out of the first twenty-eight. It became a public mail service on 5 May, weather having prevented this on 1 May, and from June until the closure of the service at the end of October 246 passengers had travelled in the open cockpits of the two-seat Haefeli DH 3 single-engined biplanes.

The Aeroplanes

Most of the aeroplanes used on the first air services in 1919 were single-engined military biplanes, some with modifications to provide enclosed accommodation for two to four passengers. Most were of wooden construction with fabric covering although a few types, such as the Breguet 14, had metal structures with fabric covering. All had water-cooled engines with the associated plumbing which contributed to the large number of forced landings. Because of the low landing speeds few of these forced landings caused serious damage to the aeroplanes or injury to their occupants.

The small number of twin-engined

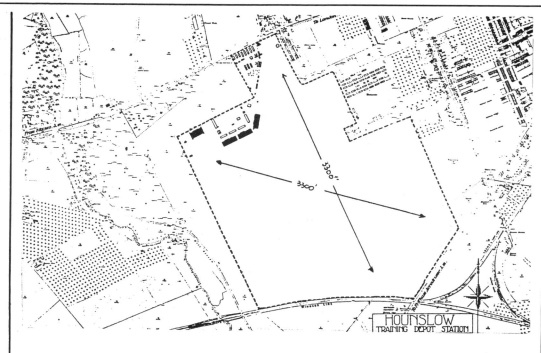

Hounslow aerodrome with the hangars shown as black rectangles. The entrance, served by tram, was at the narrow section near the top of the drawing. (*Courtesy A J L Hickox*)

aeroplanes used were again converted bombers or bomber designs, the most successful of which was undoubtedly the Farman Goliath. The table on page 176 shows typical data for a representative range of types in airline service in 1919.

The Airports

All the early air services operated from aerodromes which were simply grass fields with hangars for the aeroplanes and often huts for passenger handling.

London's first terminal airport was situated on Hounslow Heath almost under what is now the approach to Heathrow's 27L runway. The aerodrome had been prepared during the war and some of the wartime hangars were used at the birth of civil air transport, one serving as the customs shed. The take-off and landing distances available were 3,300ft NNW–SSE and WNW–ESE. A light beacon was installed for civil flying, and in October 1919 the Air Ministry announced that radio communication was available on the Hounslow–Le Bourget route from 0645 to 1900hrs GMT. On 29 March 1920 Croydon aerodrome became the London Customs Airport and Hounslow was closed.

Handley Page Transport and the French CMA used the Handley Page works aerodrome at Cricklewood. This was a rather restricted area and to get the longest take-off run into the prevailing wind aircraft had to start uphill, with the remainder of the aerodrome out of sight. The factory buildings were at the south-west corner immediately adjacent to the London and North Western Railway, its carriage sheds, and numerous telephone wires. Initially, aircraft using Cricklewood had to land at Hounslow for customs clearance but on 18 February 1920 Cricklewood was appointed an Air Port with Customs and a Civil Aviation Transport Officer. Handley Page Transport did not transfer all its services to Croydon until 27 May 1921.

Le Bourget, the Paris airport, was established as a military aerodrome in 1917. Situated 12km north-northeast of the city, it was a roughly triangular area, with civil facilities along the southwest–northeast aligned Route Nationale No. 2. It remained a Paris airport until March 1977 since when it has handled only third-level airlines, general aviation and military communication flights.

The Brussels aerodrome used by the

Cricklewood aerodrome grandly titled Cricklewood Airport. Its position was given as 5134N 0012W. (*Courtesy A J L Hickox*)

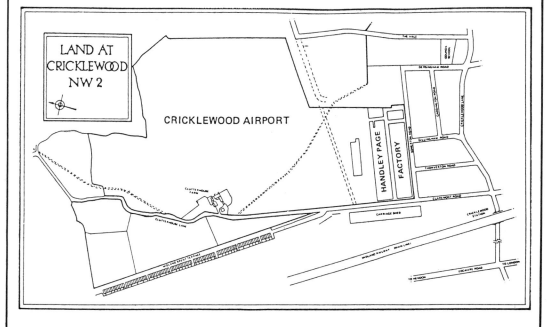

pioneer airlines was Evère, also known as Haren. This was a wartime aerodrome and continued to serve as the Belgian capital's airport until the Second World War. Even after the war Evère remained Sabena's main base for some time and the airline's aircraft used to taxi from the Evère maintenance base to the German-built Melsbroek which was close by. Melsbroek became the present Brussels National Airport.

The Junkers-F 13

Although it did not see airline service in 1919, mention must be made of the Junkers-F 13 which first flew on 29 June 1919. This was a low-wing cantilever monoplane of all-metal construction with open cockpits for two crew and a well furnished enclosed cabin for four passengers. It originally had a 160/170hp Mercedes D.IIIa water-cooled engine but most early production aeroplanes had the 185hp BMW IIIa.

In excess of 320 Junkers-F 13s were built and they were used by large numbers of airlines in Europe, Asia, the USSR and the Americas, some serving

The first Junkers-F 13. Seen here as D-1 *Nachtigall* (Nightingale) at Berlin-Tempelhof, it was originally D-183 *Herta*. (*Lufthansa*)

with the US Post Office. The famous Junkers-Ju 52/3m of the 1930s employed the same structure and corrugated metal skin as the F 13. The little F 13 was of great importance, being the first all-metal transport aeroplane to go into production. There were sixty or seventy variants and a few survived until after the Second World War. □

	D.H.4A	D.H.16	Breguet 14T	L.V.G. C VI	Handley Page O/7	Farman Goliath
Engines	360hp Rolls-Royce Eagle VIII	360hp Rolls-Royce Eagle VIII	300hp Renault 12Fe	200hp Benz Bz IV	Two 360hp Rolls-Royce Eagle VIII	Two 260hp Salmson CM.9
Span	42ft 5⅝in	46ft 5⅞in	47ft 1½in	42ft 7¾in	100ft	86ft 10in
Length	30ft 6in	32ft	29ft 6in	24ft 7in	62ft 10in	47ft
Wing area	434sq ft	489.75sq ft	538.2sq ft	372.4sq ft	1,648sq ft	1,733sq ft
Empty weight	2,600lb	2,963lb	2,664lb	2,127lb	8,326lb	5,511lb
Loaded weight	3,720lb	4,378lb	4,166lb	3,086lb	12,050lb	10,515lb
Cruising speed	100mph	100mph	77mph	100mph	70–80mph	75mph
Range	300 miles	450 miles	285 miles	350 miles	500 miles	248 miles
Passengers	2	4	2–4	2	14*	12

Types in airline service, 1919

* Two passengers were carried in the open nose cockpit.
There were considerable variations in the weights of these aircraft.

The Sopwith Century that Shook the World

Dr Norman Barfield

Sir Thomas Sopwith, himself an early pilot, unerringly chose the right men to do difficult jobs, the creation of weapons for air defence. His witnesses are the Sopwith and Hawker fighters of two world wars and the Harriers of today, a concept still unique almost twenty-nine years after it was first flown.

The last link with Britain's, and indeed the world's, original pioneer aviation signatories was severed when Sir Thomas Sopwith – legendary founder and leader of the great Sopwith/Hawker/ Hawker Siddeley aeronautical and engineering dynasty from the early years of this century through massive and decisive contributions in two world wars to the brilliant Harrier of today – died on 27 January this year, nine days after his 101st birthday and a uniquely fulfilled life in both business and leisure pursuits.

Whereas, when monarchs or statesmen die, nations and often the whole world mourn, the passing of this exceptional centenarian – long forgotten by many, even in the aviation business where he wrought so much influence – has engendered a new and deep reflection on just how expansive was his life story and its achievements and, in particular, what an incalculable debt the British nation owes to him.

Great national, political and military figures cause and fight wars, but it was Sopwith's quietly decisive industrial and business prowess which was an essential ingredient in winning them, highlighted by the fact that he led the companies which designed and built outstanding and high-scoring fighter aircraft in both world wars, the Sopwith Camel and Hawker Hurricane respectively.

This is the more astonishing when it is remembered that Tom Sopwith was already approaching his sixteenth birthday when the American Wright brothers made the first sustained and controlled flights with a power-driven heavier-than-air craft on 17 December 1903; earned one of the earliest pilot's licences in the world in November 1910, having taught himself to fly; did business with the Wrights and numbered among his aviation contemporaries the Hon Charles Rolls (of Rolls-Royce), Blériot, Grahame-White, Cody, Paulhan and

T O M Sopwith in the 1940s. Awarded the CBE in 1918, he was knighted in 1953.

Hamel; went on to found and lead the great British aeronautical and engineering combine Hawker Siddeley; sponsored the revolutionary V/STOL Harrier; and witnessed Concorde, men walking on the moon and the space shuttle – in fact, the entire spectrum of what we term aerospace.

All this occurred in a crowded and colourful lifetime in which he was, at various times, sportsman, racing motorist and yachtsman, as well as engineer, industrialist and, above all, airman, and all with high proficiency.

Cast in a unique mould, Tom Sopwith entered a world where what was to become one of his great passions, the motor car, had only just been invented and powered aeroplane flight, which was to become the central theme of his

uniquely varied personal and professional life, was still fifteen years away.

Born on 18 January 1888, Thomas Octave Murdoch Sopwith came from a wealthy Northumberland family, being given his unusual second name because he was the first son following seven daughters (five of whom would also pass ninety).

Tracing his mechanical aptitude and engineering interests from his father and grandfather, who ran successful lead mining companies, Tom Sopwith's enthusiasm for aviation began in 1906 at the age of eighteen with a half share in a balloon built by the Short brothers, who were then official balloon-makers to the Aero Club of Great Britain (later the Royal Aero Club). He thus joined a small circle of motoring and ballooning personalities, notably including Frank Hedges Butler (who in 1901 had been the first to suggest the formation of an Aero Club in Britain), Charles Rolls (then selling Panhard and Minerva cars), J T C Moore-Brabazon (who held the Royal Aero Club's first pilot's certificate) and Griffith Brewer. However, his initial application for an aeronaut's licence was refused because of his youth.

The Flying Bug

This was but an appetiser and Tom Sopwith's urge to fly a powered aeroplane came in the summer of 1910 via his already well-developed seafaring interest. At that time, he shared the ownership of the 166-ton schooner *Neva* with Maj Bill Eyre, the engineer in charge of the auxiliary engine being Fred Sigrist, a triumvirate partnership which was to be sustained through many and varied experiences in boating and aviation for the next thirty years.

Arriving at Dover harbour from Dieppe, they were attracted by a Blériot monoplane which an American named Johnnie Moissant (who had made the first flight from Paris to London and was the first to cross the Channel with a passenger) was flying nearby, barely a year after Blériot himself had flown the Channel for the first time.

As a result, Sopwith declared, 'That started something inside me. I was incurably bitten by the aviation bug, then and there. And I never recovered'. Deciding that he, too, must fly, he soon made his way to Brooklands, near Weybridge, Surrey, the birthplace and early haven of British aviation. There, Mrs Maurice Hewlett, wife of a well-known author, was selling flights at £5 a time in a Henry Farman biplane flown by the Frenchman Gustav Blondeau, and provided Sopwith with two sedate circuits of the already famous 2¾-mile motor-racing track.

Inspired and determined to become an aeroplane pilot, Sopwith immediately bought his own machine, a Howard Wright (no relation to the Americans) 40hp Avis monoplane, designed by W O Manning. It having only one seat, he had to teach himself. So rudimentary was his mount that getting airborne consisted simply of rolling along the ground until becoming separated from it. At this first

The Sopwith *Tabloid* on the Thames at Glovers Island, Richmond, on the day of its maiden flight on floats. The date is 8 April 1914 and the pilot Howard Pixton. Within six days this combination, supervised personally by T O M Sopwith, had won the second ever Schneider Trophy contest at Monaco, over a 280km course, at an average speed of 85.5mph, bringing international glory to British aviation in a sphere of activity previously dominated by France.

attempt he stalled and crashed from 40ft (his first of many to come) but walked away from the wreckage completely unscathed: 'because it all happened so slowly and gently', he said.

Undeterred, he then bought a 60hp biplane and two months later passed his flying test to be awarded Royal Aero Club Certificate Number 31 on Tuesday 22 November 1910.

On the Saturday of that same week he established a new British aerial duration record of 3hr 12min and a distance record of 107 miles, flying around Brooklands. The day ended with a ride for his sister, May, was who to be such a great support to him in the years ahead. An impressive week by any standards.

Success in Two Worlds

Before the end of 1910, Sopwith had also won the Baron de Forest Prize of £4,000 for flying the longest distance from England to continental Europe by a British pilot in a British aeroplane. This enabled him to show his prowess as a sporting pilot during the next two years on both sides of the Atlantic, whence he broke many more records and won more prizes, both in aeroplanes and high-speed motor-boats.

While in the United States he survived three more crashes (including 'spinning in') and, for good measure, won the Harmsworth Trophy for motor-boats, for Britain, in New York in 1912, driving Mackay Edgar's *Maple Leaf IV* built by Samuel Saunders with his patented seaplane hull construction system. His very

successful participation in numerous flying competitions was not without some antagonism, from the Wright brothers in particular. However, he diplomatically overcame this by buying a Burgess-Wright biplane which, together with his substantial prize monies and his original British biplane, he returned to England to establish a Flying School at Brooklands in 1912: 'teaching people to do something I know very little about myself', he said.

Yet among his many pupils was a middle-aged major in the Scots Fusiliers, Hugh 'Boom' Trenchard, who went on to become 'Father of the Royal Air Force'. Another was one of his ground crew, a young Australian named Harry George Hawker, son of a blacksmith and already a motoring expert, who asked Sopwith if he would teach him to fly for £50 (the money which Hawker had saved for his return fare home) instead of the usual fee of £75. Sopwith agreed and said that 'Hawker soon became a very good pilot'.

From all of this, Tom Sopwith began to develop an interest in aeroplane design and construction, as well as piloting, and later that same year built his first aeroplane, with Fred Sigrist. It was a three-seat tractor biplane with a 70hp Gnome engine. Not only did he succeed in building and flying it but, even more important, managed to sell it to the Royal Navy at Eastchurch for £900 (and thereby initiated a link which has been sustained through to the modern-day Sea Harrier that made a decisive contribution in the Falklands campaign).

This success led him to set up the Sopwith Aviation Company in June 1912: 'for the design and manufacture of land aeroplanes, seaplanes and flying-boats', the first factory being a former skating rink in Canbury Park Road, Kingston-upon-Thames, Surrey. Replacing his flying school activities, the new company quickly flourished under Tom Sopwith's leadership and with the support of his fellow directors Fred Sigrist, the engineer, and Harry Hawker, the test pilot, each of whom had put up £5,000 share money. Their most successful early type was the *Tabloid* which, fitted with floats, won the Schneider Trophy Race at Monaco in 1914.

It was also during that year that Tom Sopwith married the Hon Beatrix Mary Hore-Ruthven.

Building a new factory at Canbury Park Road one month later at the start of the First World War, during the next four years the company, together with numerous sub-contractors, brought mass production to a fine art and built more than 16,000 Sopwith aeroplanes – Pups, Camels, 1½ Strutters, Dolphins and Snipes – with around 4,000 being built in France. The most prolific was the immortal Camel of which nearly 6,000 were built and by the end of the conflict more than half of the Allied air strength consisted of Sopwith types.

Sopwith himself has always insisted that the design of these early types – the full series of which was popularly known as the Sopwith Zoo – was a joint effort. But there is no doubt that through his own flying experiences he made a distinct personal mark on all of them. (He gave up serious flying when the war started: 'before the air-speed indicator had been invented', as he put it.)

Renaissance

Paradoxically, despite this massive achievement, after the cessation of hostilities, Sopwith was to spend the rest of his industrial business life leading a company under the Hawker name rather than his own. This was because, after managing to carry on for a while by recycling war surplus aeroplanes, by the middle of 1920, with no more orders and claims for Excess War Profits Duty from the Treasury, he prudently wound up his company, still able to pay his creditors the full 20 shillings in the pound.

Undismayed, later that year he established a new company, ostensibly to make motor-cycles, by renting back a corner of his former premises, and called it the H.G.Hawker Engineering Company Ltd – 'so that there could be no confusion with the Sopwith Company in liquidation, and named after Harry Hawker who had done such great work throughout the war as our test pilot', Sopwith said. Sadly, Hawker himself did not live to see the first Hawker aeroplane, being killed in July 1921 in an air accident, but not before he had dramatically but unsuccessfully competed in what was in effect the *Daily Mail* trans-atlantic air race in 1919 (won by Alcock and Brown) in the specially adapted Sopwith *Atlantic*.

The single-seat Sopwith Triplane, with its single Vickers gun, went into action on the Western Front in the spring of 1917, some six months before the much better known Fokker Dr I triplane, a favourite mount for Manfred von Richthofen.

Sopwith was always a good picker of men (witness his celebrated lineage of test pilots – Hawker, Raynham, Bulman, Sayer, Lucas, Wade, Duke, Bedford and Simpson) but never more so than when in 1923 he hired a young draughtsman named Sydney Camm who had served his apprenticeship with Martinsydes at Brooklands. Camm was made chief designer in 1924 and, as Sopwith's technical supremo for the next forty years, he led the Hawker technical team in distinctive style, embracing an order book of almost 30,000 aircraft of fifty-two different types, from the diminutive Cygnet light competition biplane through to the conception of the radically different and significant Harrier.

The following year, as well as directing his business from Kingston, Sopwith became chairman of the Society of British Aircraft Constructors (SBAC), which appointment he held for the next three years until 1927. By 1930, more records and successful aircraft had followed, most notably the Hart biplane day-bomber which went on to become the standard squadron equipment of the Royal Air Force with around 3,000 built in eight variants.

Unhappily, Sopwith suffered the loss of his first wife that same year, but two years later married Phyllis Brodie Leslie and Thomas Edward Brodie Sopwith was born in November 1932.

The Great Expansion

The outstanding success of the Hart resulted in Sopwith needing to expand significantly and to bring about major corporate changes in the 1930s and he acquired additional capital to establish a public company, Hawker Aircraft Limited, which replaced the previous title on 18 May 1933. The Gloster Aircraft Company was also purchased outright in 1934. Then on 11 July 1935 Sopwith took the momentous decision to form a Trust to acquire the shares of the Hawker Siddeley Aircraft Company by amalgamating half the assets of Hawker Aircraft Limited with the total assets of the Armstrong Siddeley Development Company Limited. Within what had now become the world's largest and most formidable aviation Group were thus incorporated Armstrong Siddeley Motors, Armstrong Whitworth Aircraft, A V Roe and Air Service Training (in addition to the Gloster and Hawker companies).

Alongside all this, Sopwith was making bold headlines with his nautical exploits in his two challenges for the Americas Cup, in 1934 and 1937, representing the Royal Yacht Squadron against the New York Yacht Club with his J class sloops *Endeavour* and *Endeavour II*. He narrowly failed, but such was the fervour with which his valiant efforts were followed that, in sharp contrast to his quietly efficient business style, he became a national hero.

Meanwhile, on the world scene the menacing rise of Hitler's Nazi regime foreshadowed another major war. However, because money was not being expended on re-equipping the Royal Air Force to the standard already possessed by the Luftwaffe, Sopwith was perceptively planning, as a private venture, the production of a thousand Hurricane fighters without any guarantee of an Air Ministry order, a brave and far-sighted decision which almost certainly proved positive in winning the Battle of Britain in 1940.

National Salvation

First flown at Brooklands in November 1935, the Hurricane was designed with the classic simplicity that was the hallmark of all Camm's work. No production orders were forthcoming, however, and in 1936 Sopwith led the Hawker Siddeley Board into the nationally vital decision to sanction the necessary expenditure to lay down tooling and order materials as a private venture. Although the official order did eventually materialise, there is no doubt that Sopwith's action gave the RAF pilots hundreds more fighting machines when they needed them most.

Hawker Harts in production in the 1930s at Brooklands.

A further initiative resulted in the first Canadian-built Hurricane, by the Canadian Car and Foundry Co of Montreal, flying on 10 January 1940, just one year and six days from the issue of the specification, and eleven months after delivery of the first tooling from Britain.

At the beginning of the war on 3 September 1939 the Hawker factories at Kingston, together with extra factory space at Langley, a dozen miles away near Slough, Buckinghamshire (which had been acquired in 1938 to augment the Kingston and Brooklands output) and between them employing just over 4,000 people, were engaged in Hurricane production, 500 having been delivered to the RAF. Gloster-built Hurricanes also started to enter squadron service before the end of 1939 and in the first ten months of the war the British factories had together delivered a further 1,350, more than all other fighters combined. By the time the Battle of Britain began on 8 August 1940, RAF Fighter Command had equipped fifty-two interceptor squadrons, of which thirty-two had Hurricanes.

The final Hurricane (suitably inscribed *The Last of the Many!*) was purchased by the company in mid-1944 as an ever-present reminder of, and tribute to, the more than 15,000 Hurricanes which flew and fought in more parts of the world than any other aircraft. This has since been maintained in immaculate condition by the Battle of Britain Memo-

rial Flight (notably flying into Brooklands in November 1985 as part of the celebrations of the fiftieth anniversary of the first flight there of the Hurricane prototype).

Another primary wartime aeroplane also under the ultimate control of Sopwith within Hawker Siddeley was the Avro Lancaster, conceived by the great, but greatly underrated, Roy Chadwick, and of which more than 7,000 were built, bearing the brunt of the British heavy bomber attack.

Hawker Siddeley had thus played a monumental role in the war effort, supplying over 40,000 aero engines as well as these key interceptor fighters and heavy bombers. In addition to the main factories at Kingston, Langley, Hamble, Gloucester, Coventry, Manchester and Yeadon, Sopwith's Group had dispersal sites in at least forty locations ranging from Sedbergh in the north to Paignton in the south.

After Hawker had vacated Brooklands early in the war, production facilities were concentrated at Canbury Park Road and at the now-enlarged factory and airfield at Langley. When the twenty-year lease on the more extensive factory premises at Richmond Road, Kingston, held by Leyland Motors since 1928, expired Sopwith lost no time in enabling Hawker Aircraft to take over the site, although the administrative offices, design department and machine shops remained at Canbury Park Road for a further ten years.

Postwar Success

In 1948, the Hawker Siddeley Aircraft Company was renamed the Hawker Siddeley Group which, through the Hawker Siddeley Development Company, controlled twenty-five wholly owned subsidiaries by 1951, seven of which were principally involved in airframe or aero engine manufacture.

The inadequacy of the flight test facilities at Langley, especially with the rapid growth of commercial air traffic at nearby London Heathrow Airport, then led Sopwith to acquire the tenancy of the ex-wartime airfield at Dunsfold, near Guildford, and about 20 miles from Kingston, in 1950.

With the experimental Gloster E.28/39 pioneering the jet age for Britain in 1941 and that company producing the RAF's first operational jet fighter, the Meteor, which saw service in the closing stages of the war, the Hawker lineage continued with the rocket-firing Typhoon and Tempest and the Sea Fury as its ultimate piston-engined types before moving on to jets with the Sea Hawk which began service in 1953.

Taking advantage of the great versatility of production capacity that the Hawker Siddeley Group had afforded from the outset more than a decade earlier, Sopwith and his team continued to arrange widespread dispersal of production throughout the Group. This he did while allowing all the constituent companies to continue to operate separately under their original names, to compete with each other for first Air Ministry and later Ministry of Supply contracts and build their own prototypes. So it was that in the 1950s Armstrong Whitworth at Coventry was designated to handle the majority of Sea Hawk and Hunter jet fighter work. Valuable export earnings also resulted from licence agreements for the Hunter to be produced in Belgium

A youthful T O M Sopwith with the original Sopwith Three-seater, a member of the Sopwith *Tabloid*, Schneider, and Baby family, drawing attention to the Schneider Trophy win of 1914.

and The Netherlands, making it Hawker's most successful postwar type with more than 2,000 produced.

In 1953, while chairman of the Hawker Siddeley Group, Sopwith received his knighthood in the Queen's Coronation Honours for his services to aviation. That year the Hunter also captured a new World Airspeed Record of 727.6mph, a remarkable advance compared with the less than 100mph capability of the Sopwith types of forty years earlier.

Sir Thomas was successively chairman

and president of the Hawker Siddeley Group from its foundation in 1935, and through its further enlargement at the time of the major government-sponsored rationalisation in 1960 to include Blackburn, Folland and de Havilland, until he retired in January 1963 at the age of seventy-five, and from the Board in 1978, whence be became founder president for the remainder of his life (another twenty-six years).

Able to conduct his industrial leadership and sporting activities in equally brilliant but unobtrusive style, he invariably pondered and decided much of his business policy away from his office, which technique he combined with a near genius for picking outstanding managers – Sigrist, Eyre, Jones, Hill, Spriggs, Dobson, Lidbury and Hall – and to delegate.

It is also an abiding tribute to him that the huge consortium of aviation talent, resources and products which he assembled remains vigorously active and flourishing in characteristic style as a major component of British Aerospace, while those in other areas of mechanical and electrical engineering are similarly active with the continuing Hawker Sid-

deley Group of companies. Other centenarians have obviously witnessed great and profound changes in their lifetimes but no other man in history has actually presided over such massive technical and industrial advances as did Sopwith.

Sopwith the Man

Lady Sopwith, his great companion and supporter, died in 1978, after which Sir Thomas continued to live quietly in his beautiful Georgian country home, Compton Manor at Kings Somborne, near Winchester, Hampshire, with his businessman and sportsman son Tommy and his family only a few miles away.

Despite his many achievements Tom Sopwith was a very private and self-effacing person (his entry in *Who's Who* consisting of only seven lines) and he always directed credit to those directly involved rather than taking it himself. Perhaps the best way to sum up his personal qualities is to quote Sir George

Hawker's own Hurricane, *The .Last of the Many!*, flies over Brooklands in 1985, symbolising the Sopwith contribution to national survival in 1940 and the first flight of the prototype just five years before that.

Edwards, himself a pioneer and mentor at Sopwith's original base at Brooklands, in his characteristic address and tribute at the Memorial Service for Sir Thomas, appropriately held at St Clement Danes, the Church of the RAF, in London on 12 April 1989. Sir George said:

> Essentially Tom Sopwith was an Englishman of the type that we always produce when we need them most. He was a man possessed of great physical and moral courage, but still with the courtesy and modesty of an English gentleman. He had a strong sense of humour but no great tolerance of fools. He was not an ambitious man. He did what he thought needed doing, got on with it and invariably got it right. He picked the best men he could find and then let *them* get on with the job, with no interference. His ever-calm attitude showed in one of his rare personal pronouncements when he said, 'If worry will help, then worry, but not if it doesn't'.

'He was also a countryman at heart, who took great delight in his family life, and the Sopwith family is altogether unusual', Sir George asserted. The signature of the first Thomas (his grandfather) appears on page 71 of the 300-year-old

Following the end of the First World War strenuous efforts were made to create a new lineage of Sopwith aircraft for civil use. The first, the Dove, pictured in single-seater form, more usually a two-seater, was an adaptation from the Sopwith Pup. Also produced before the firm went into voluntary liquidation were the Sopwith Atlantic, Wallaby, Antelope, Gnu, Grasshopper and the Schneider built for the 1919 contest. (*W A Sherwood*)

held, together and led with such sustained acumen, and the unique lineage of Sopwith, Hawker and Hawker Siddeley and now British Aerospace aircraft that he inspired (more than 45,000 of them of 112 different types) throughout the Sopwith century that shook the world. □

book of the *Obligation of the Fellows* of the Royal Society to which he was elected in 1845. The *Guinness Book of Records* includes Sir Thomas himself as holding the record for an aeroplane quick start, from wing-tip to take-off in nine seconds, which he won all those years ago in the United States and remarkably still stands today. It also contains records held by young Tom for speed-boat racing and his wife Gina for ski-ing, Sir George pointed out.

Epitaph

Asked at the time of his 100th birthday on 18 January 1988, celebrated in appropriate style at Brooklands Museum by a unique gathering of distinguished names from British aviation and the Armed Services, and at his home with his family, to what he attributed his success and longevity, Sir Thomas replied, 'Pure luck'.

But manifoldly more lucky are the many who continue to benefit from the huge and hugely valuable legacy that he has bequeathed to the British and, indeed, the world aviation scene. Without him and his extraordinarily pervasive influence, this and much of our general way of life would be much the poorer. Above all, he will be remembered with respect, affection and admiration for the formidable teams which he brought, and

Almost a must at the best air shows, a British Aerospace Sea Harrier FRS 1 bows, made possible by a blend of unique technologies, developed by teams working under Hawker's Sir Sydney Camm and Sir Stanley Hooker of Rolls-Royce, both now dead but to whom the Harrier is their finest memorial. It is salutory to recall that the Hawker P.1127, the Harrier's direct forebear, made its first untethered hovering flight on 19 November 1960.

All pictures by courtesy of British Aerospace, Kingston.

Regional Airport: Brussels National

Guy Roberty

Brussels National, primarily an important European international airport and intercontinental hub, has now assumed the status of hub airport at the capital of the European Economic Community (EEC). From 1993, this important grouping of nations bids to become the most important domestic trading block in the world. It is in this context that the *Review* places Brussels National the key regional airport of Europe.

Most Belgian citizens will be surprised if not hurt to find Belgium's main airport described as a regional airport! Most, quite rightly, look upon Brussels National Airport, BN in short, as an international airport of world class and an important European and intercontinental hub. Despite the obvious limitations of its name, Brussels National, which reflects more on Belgium's linguistic division, the airport of Brussels has a true international vocation. Since the small dimensions of the country do not allow the viability of a domestic network, all commercial flights in and out of Brussels National are international. At the same time, the small size of the country combined with the excellent network of motorways and railways extend BN's hinterland beyond the borders of Belgium and includes large potential in the

Brussels National today.

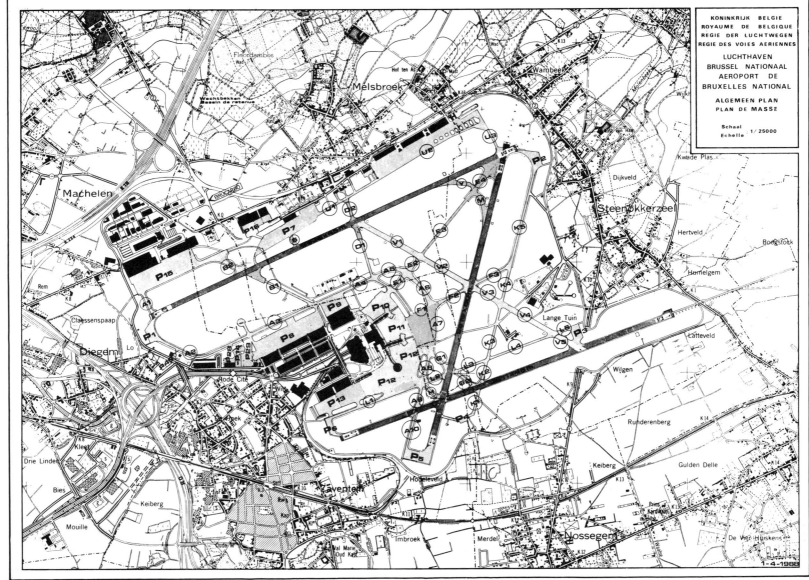

Brussels National   5054N 0429E

ICAO location indicator: EBBR

IATA location indicator: BRU

Elevation: 55m

Relation to city centre: 16km NE
Potential market served: 20 million
Runways: 07L/25R 3638 x 45m
  07R/25L 3211 x 45m
  02/20   2819 x 50m
Operational: 24hrs with some limitations at
  night due to noise abatement
Aids: ICAO Category III Approach and
  Runway lighting
  ILS Category III on Runway 25L
  VOR/DME, NDB, VDF, SSR, RSR, SRE, TAR
Owned and operated by: Belgian State, via
  Régie des Voies Aériennes/Regie der
  Luchtwegen
Passengers: 7,087,227 (1988)
Principal users: Sabena and about 60 other
  airlines
Movements: 161,076 (1988)
Scheduled services destinations: All
  European capitals and many provincial
  towns linked by 47 European airlines,
  including all major carriers, and commuter/
  feeder airlines
  North America is served by Sabena, six US
  and one Canadian carrier
  Africa, more than 20 destinations in Africa
  by Sabena and five African carriers
  Middle and Far East by nine carriers
  Central and South America: two carriers
Holiday charter traffic: Eight specialised
  charter airlines link Brussels to the
  European sun and more distant
  destinations. Twenty-two tour operators
  have offices/desks at the airport
How to get there: train service (every 20min)
  from Central or North stations
  Car or taxi: 20min ride to centre of city
  Regular coaches to Antwerp, Ghent and
  Liège
Hotel: Sheraton opposite terminal building.
  More hotels close to the motorway to city.
  Courtesy transport to/from hotel
Customs: Full coverage

neighbouring countries of France, West Germany, Luxembourg, The Netherlands and even the United Kingdom.

Brussels National Airport, conveniently close (20 minutes by car or railway) to the capital city of Belgium, of the EEC and NATO, has experienced an extraordinary development due to the growth of Brussels in world affairs and to the liberal approach of the Belgian Government to traffic rights. Major expansion plans were launched in 1983 but were only finalised last year and foresee three

major phases until the year 2010 in order to cope with the consequences of European integration and deregulation. It is of great significance to note that while the airport will remain under State ownership and overall control, the expansion of the passenger terminal and its commercial facilities will be financed and operated by private enterprise.

An Airport for All

Brussels Airport serves various sectors of aviation. Its main activity is, of course, international air transport, with at present no fewer than fifty-nine airlines using BN's facilities.

Cargo follows close behind and Brucargo, an entirely separate freight complex, is retained as the European distribution hub by several major international cargo airlines. More recently, the Express mail and parcels delivery services have taken over older buildings and the leaders of that specialised industry have selected BN as their European and African distribution centre. New buildings will soon confirm the permanency of that activity which has also motivated the Belgian postal services to organise at BN a major international sorting centre in order to compete with the private sector.

The Belgian Air Force is also present at Brussels National where it occupies three large hangars at the northeast corner. A large camp just outside the confines of the airport houses the 15th Wing (Transport) which operates twelve Lockheed C-130 Hercules, two Boeing 727-129s and about a dozen other smaller aircraft. The 15th Wing also handles many foreign military aircraft on NATO business.

The overhaul and maintenance industry is present in force at Brussels National represented by Sabena Technics, the engineering subsidiary of Sabena, the main user of BN. That company not only looks after the Sabena fleet but is also a member of the Atlas consortium and has many foreign customers. More recently, a newcomer, Teamco, has built at BN a large and modern plant which, besides looking after the needs of TEA's fleet (TEA is a large Belgian charter airline), has an important contract to maintain, modify and overhaul the numerous helicopters of the US Forces in Europe.

General aviation is welcomed at BN and enjoys a separated terminal operated by Abelag, a well-established Belgian company offering the complete range of general aviation services. BN is also the European headquarters of Cessna with its own facilities.

All those activities expressed themselves in the 1988 provisional figures (1987 official figures between brackets)

of 161,076 (142,319) aircraft movements, 7 million passengers (6·4 million), 249,808 (227,728) tons of cargo and 42,904 (28,293) tons of mail.

To illustrate the progression since 1947, the year of the partial move from Evère/Haren, the old airport of Brussels, now disused, to Melsbroek as BN was then known, we have selected from the airport statistics the figures relating to 1947 and 1967 and repeat those for 1987:

	1947	1967	1987
Aircraft movements	8,778	71,605	142,319
Passengers	73,995	1,925,640	6,468,341
Cargo (tons)	1,268	60,065	227,728
Mail (tons)	286	4,960	28,293

Of course, these figures, important as they are for the Belgian authorities and users of the airport, would not impress those dealing with the realities of London, Paris or Frankfurt Airports and maybe this justifies our inclusion of Brussels amongst the regional airports of Europe. However, if one comes back to recent evolution, it appears that Brussels National is expanding at a higher rate than most European airports and, while it obviously will never match the airports of the big capital cities in Europe, it is promised a much bigger role than previously envisaged. Certainly being the capital of Europe will help, although some compare the limited effects of somewhat similar circumstances (with correction for scale) on Canberra, Ottawa, Brasília and even Washington DC. But the opening of most of western Europe as a vast, highly populated, domestic continent surely guarantees much increased activity at Brussels National. Everybody in the airline industry is aware of this and so are the Belgian authorities.

Towards a New Tomorrow

While Brussels Airport has been upgraded throughout its 50-year history from the hard runways of the Luftwaffe during the Second World War, the switching of the main terminal from Melsbroek on the north side to Zaventhem in the southwest corner for the 1958 International Fair and the additions of the 1970s and '80s, it is now heavily congested and its future needed a complete reappraisal. Under the tenure of M De Croo, a progressive Minister of Communications under the last government, a comprehensive study of the future needs of BN was launched in 1983. It is appropriate at this time to explain that Brussels National is owned by the Belgian State and is the responsibility of the Régie des Voies Aériennes/Regie der Luchtwegen, the civil service organisation which looks after the eight State airports in Belgium and Air Traffic Control for the Ministry of Communica-

tions. Incidentally, another separate organisation, Administration de l'Aéronautique also under the same ministry is responsible for aircraft airworthiness, registration and ground and aircrew licensing. The same department with the ministry of external affairs looks after aviation law and granting traffic rights.

The Régie des Voies Aériennes, RVA in short, has an ongoing review programme of airport development, mainly of course for Brussels National as the commercial transport activities in other Belgian airports are limited, with the possible exception of Deurne (Antwerp) and Ostend. Attempts to develop Bierset (Liège) and Gosselies (Charleroi) for scheduled and charter traffic have not met with great success.

In 1983, it was perceived that the future prospects of BN needed more than updating and that the socio-economic impact of air transport on Belgium in all its aspects warranted a fresh, independent in-depth study. There was no doubt that evolution towards a more integrated European Economic Community, if not European Political Community, and the advancing deregulation of air transport within Europe would have a tremendous impact not only on Brussels Airport but also on Air Traffic Control (ATC) and the Belgian State controlled airline Sabena. On the other hand, the consideration of limited funds and available subsidies, the government of the time as well as the present one being committed to the drastic reduction of its deficit, combined with the increased costs of ATC, security and environmental constrictions, did propose a very delicate equation.

The problem of Sabena and of its semi-privatisation (the Belgian State remains the majority shareholder) was quickly if ruthlessly solved and does not really belong here although one cannot dissociate Brussels National from Sabena, its most important customer and tenant. Let us say that the Sabena of today is a very different, leaner company and profitable. The new Sabena still has major hurdles to jump and its search for partners may have a significant effect on the development of its base airport.

ATC, so far, cannot be privatised and remains a total government responsibility. It must be stated that Belgium has been a promoter and staunch defender of Eurocontrol and remains firmly committed to it. Happily, those countries which attempted to kill Eurocontrol are now coming to see the advantages of a total reorganisation of European ATC, if only because there is no other solution to the appalling mess nationalism has created for that vital and integral part of international air transport.

While overflying traffic at or above 20,000ft remains the responsibility of the Eurocontrol centre at Maastricht (The Netherlands), all traffic over Belgium up to 20,000ft is controlled by the ACC (Area Control Centre) based at BN. A new powerful AEG/Telefunken radar was ordered in 1983 to replace older equipment and serves the needs of both the Belgian ACC and Eurocontrol in Maastricht. A new computer integrated system is being studied and applied progressively with the help of the Massachusetts Institute of Technology. This system, called CANAC, should bridge the gap until, at last, a European integrated system is developed. There is no denying that ATC remains a major problem everywhere in the world, and this includes Belgium. Great pressure of work was placed upon ATC staff but it seems some respite has been given at the national level, and the medium- and

The terminal building in 1980. Since then a hotel has been built over the car park opposite the terminal.

long-term solutions rest with a European integration.

The third segment of the ministerial review and the one which interests us particularly here is the needs of the airport itself. M De Croo called in the American firm of Burns & McDonnell to study and propose a basic programme of extension for BN.

The conclusions were delivered in 1984 and were projecting three phases covering the next 20 years up to the year 2005. The basic problem concerned terminal facilities and gates rather than runways. The three runways can cope with increased movements and only needed better equipment leading to Category III operations. So the pattern of runways will remain the same with the two parallel runways 07L/25R and 07R/25L, respectively of 3,638m by 45m and 3,211m by 45m. The third, transverse, runway is 02/20 with 2,819m by 50m and being across the prevailing winds offers a precious alternative in case of high crosswinds.

Large expanses of aprons will be added necessitating in some places new routings for taxiways, but the major work is in a considerable increase in terminal buildings and piers/gates capacity.

This plan, excellent in itself, was going to run into a finance barrier. The RVA was and still is showing good profits on its running of Brussels Airport, even if those profits were nearly wiped out by the huge deficits of the other smaller provincial airports. The State budget did cater for about 1.5 to 2 milliard francs investment every year in its National Airport, but it was clear that a major

20-year development plan would mean a quantum leap in investment unacceptable to the strict tight financial policies pursued by the government. While looking for alternatives the prospects of decentralisation of the Belgian State into some kind of federalism of the two main linguistic regions and a separate semi-autonomous Brussels region promised to relieve the national government of the burden of provincial airports. The acceleration of European integration and the probability of a large domestic European network, added to whatever the consequences of deregulation might be, brought about the need to update the 1983 plan, if only because the need for customs and immigration controls would be considerably less while the existence of duty-free shops may see a drastic reappraisal in an integrated (domestic) Europe.

However, an important, if politically controversial, decision taken in 1986 has given the responsibility for all passenger amenities to the private sector.

Parliament changed the statutes of the RVA in order to allow its association with private interests with the aim of developing, financing, building and operating airports and their facilities. It is far from the privatisation of BAA in the United Kingdom, but it is a big step in the same direction.

Enter the BATC

A major Belgian bank had privately made a viability study of such a project and was instructed to proceed with the constitution of a private company in which the RVA would keep a controlling interest. This new company, NV Brussels Airport Terminal Company SA, in short BATC, was officially created in December 1987 with a capital of 2 milliard francs. At the moment the RVA holds 50 per cent and

Belgian banks and financial holdings have the same. However, it is intended in the near future to enlarge the shareholding with the RVA keeping a blocking minority of 30 per cent, banks 20 per cent, financial holdings 20 per cent, airport users and RVA staff 10 per cent, and the public 20 per cent.

The dismay of some major users of the airport is great as some of them are willing and eager to play a major role in the expansion programme. At least they will be allowed some representation on a consultative council of the BATC, but the real power will rest with the BATC Board and its management.

Clearly and obviously, the main and most urgent task of the new organisation is the expansion of the terminal which is now saturated as the traffic has developed much more quickly than foreseen in the 1984 study. The BATC selected the US experts Greiner (in fact the same people who did the 1984 study, now operating under a different banner) to update the master plan taking account of EEC, deregulation and anti-terrorism security, and other aspects having dramatically changed in the meantime.

Greiner delivered the new study in August 1988 and proposed new traffic projections demonstrating that the passenger figures of 1988 will nearly double to 12.5 million by the year 2000 and will reach 16.1 million by 2010.

Several scenarios have been envisaged to establish mainly how many gates will be domestic (EEC) and how many international (non-EEC Europe and intercontinental). Again, three major phases are planned and the first crucial one which will start immediately should be completed in 1993. Basically, the first phase covers the extension of the present terminal towards the north with the

Brucargo, the entirely separate freight complex at Brussels National, shortly before it came into use in 1980.

construction of a new Y-shaped pier. Improvement of the present terminal, more car parks and new access ways are also included. The first phase will cost around 12 milliard francs. Later, stretching until 2010, the two other phases will be realised, the more visible being further extension of the 1993 terminal, towards the north-northwest, this time, with its piers, gates and apron.

The BATC will sub-contract most if not all of the work and is arranging the financing. In return, the BATC will 'own' the terminal for 30 years and will operate it commercially and receive all rents due by the users. BATC will hire the RVA staff in order to run the terminal. All passenger embarkation taxes and air-bridge charges will also go to BATC. In fact, expertise, staff and control will remain with the RVA directly or through the BATC. The input of private management should improve the speed of results and rapidity is now the key word as four crucial years have been lost as the traffic increased way beyond the 1984 projection. At some time between now and 2010, congestion will occur and an alternative temporary solution for handling peak charter traffic will have to be found. Of course, this is the plight of many airports throughout the world and one must remember that it is after all the price of success.

History

Until the Second World War, Brussels' airport was a grass field shared equally by military and civilian aviation. Its name was Evère/Haren from the two councils on whose land it was located. The end of the war found the field too small for modern aircraft operation and much too close to an expanding capital.

In 1938, the Aéronautique Militaire (now the Belgian Air Force) had purchased land near the village of Melsbroek, about 10km from Evère/Haren on the same northeast track but further from Brussels. In 1942, during the occupation, the Luftwaffe expropriated many farmers' land and vastly expanded Melsbroek airfield. The Todt organisation built three hard runways and associated technical hangars and housing. At the liberation in September 1944, Melsbroek fell to the Allies and was operated by the RAF. It reverted to Belgian control in 1945 but the transfer from the prewar airport to Melsbroek was progressive and it was only from Christmas 1948 that Melsbroek became the new airport for Brussels. At that time, and for ten years, the small terminal was in a German-built cluster of buildings looking, for camouflage reasons, like local farms.

Two of the three runways were reinforced and lengthened: 07/25 to 2,400m and 02/20 to 2,100m. The third runway, 12/30, kept its original length of 1,300m. Between 1948 and 1956 three big hangars

Main deck air cargo containers (318 x 244 x 244cm) on wide-body jets require equally efficient transport on the roads. Since 1983, Sabena has used these transporters for carrying freight between Germany and Belgium. Three big containers are housed together with a further four of the slightly smaller underfloor hold containers (318 x 244 x 160cm), on two levels, in the trailer. Alternatively, a total of five of the large size may be carried.

and six workshops were built for Sabena but in 1956 the plans were implemented for a major reconstruction, in the expectation of the heavy increase of air traffic in 1958, the year of the International Fair in Brussels. It was then that the pattern of runways still in use today was decided and built. A brand new runway, parallel to the German 07/25, was laid. Both of those parallel runways were extended to their present length, while the shorter 12/30 was transformed into a taxiway. Much more visible was the construction of a new terminal between the western end of the two parallel runways. This terminal, the core of today's expanded building, was quite exceptional in 1958 and was greatly admired at the time. The writer remembers some disgruntled taxpayer complaining that the RVA had eyes that were too big. In 1955, a railway spur was built underground in the new terminal and is still very much in use today. The new installation of 1958 being built near the village of Zaventhem, Brussels Airport, at least to the local citizenry, changed its popular name from Melsbroek to Zaventhem, although its official name was and still is Brussels National. It is true the RVA had big eyes but that excellent planning saw BN into the jet age and the expansion of the 1960s with only minor adjustments. The explosion of cargo traffic in the same period did, however, necessitate the building of a cargo terminal in 1966, adjacent to the passenger terminal. The 1970s and the wide-bodied aircraft were handled with airbridges added to the 1958 piers and the building of a new pier towards the south leading to

a satellite building with fourteen airbridges. In 1976, the main terminal was extended in depth, tripling the area of the superimposed arrival/departure halls.

Cargo and Express

The cargo terminal of 1966 needed expansion but its proximity to the passenger building dictated a radical departure. With great courage, it was decided to do for cargo what 1958 had done for passengers. A vast modern cargo town was built on the northwest side of the airport and inaugurated as Brucargo in 1980. It is an airport of its own and there is plenty of ground available to support expansion well into the next century. The old cargo building did not remain unused for long; while it was envisaged as a separated terminal for holiday charters, it rapidly became the home of the Express and parcels carriers whose great expansion has led the RVA to rehouse them near Brucargo, their present building being marked for demolition to allow the phase II extension of the passenger terminal.

Federal Express of Memphis did not lose any time and will invest 210 million dollars in its new European terminal west of Brucargo. DHL, TNT and others will no doubt follow and the only brake on their expansion will be the growing complaints about noise levels at night.

Brussels National is close to the city which has also grown in its direction. Take-off towards the west at night can be painful at times. For many years, aircraft movements at night, especially of

Above: Gentler ways on sunny days. Guided tours of the airport were introduced at Melsbroek in 1951.

Below: Away from its principal habitat is this Sabena Sikorsky S-55 seen at Melsbroek in 1953. For over ten years the airline maintained an economically costly pioneering experiment to provide city centre air travel between Brussels and up to a dozen cities in Belgium, The Netherlands, West Germany and France.

A Sabena Douglas DC-4 at Melsbroek in 1952.

intercontinental jets, have been severely limited and monitored. The environment problem was kept to a minimum by those limitations, the high rate of climb of modern jets and the co-operation of ATC in departure routings. It appears that most of the recent complaints are due to medium to small aircraft, the kind used by the Express courier companies for the redistribution of their transatlantic loads. It is thus a difficult balancing act for the present Minister of Communications, M Dehaene, caught between the economic benefits of Brussels being the

No.1 European hub of the various international Express couriers and the increased pressures on the environment. However, the problem is recognised by both the authorities and the companies involved. Noise monitoring has been increased and operators and manufacturers are looking for technical and operational improvements. Nevertheless, while there cannot be any question of curtailing the activities of those who have invested in a European hub at Brussels Airport, it is possible that newcomers will be discouraged from operating at night.

In conclusion, looking back at the development of Brussels National and

accepting that lead times are greater for airports than airlines, it could be said that the RVA has coped well with the development of BN. For the future, the present plan, although somewhat late, is good and it is hoped that future historians will be able to add the investment of 1988 to the success and vision of the passenger terminal of 1958 and Brucargo in 1980.

I would like to thank M J P Emplit, advisor/general administrator of the RVA, who kindly gave me all requested information. The opinions are my own, partly formed by more than thirty years' experience of Brussels National as a passenger. □

Forum Foretaste

To start the ball rolling we showed galley proofs of Dr René Francillon's article about Grumman aircraft in service with the Royal Navy in the Second World War (*Review* No. 2 July 1989) to an RN pilot of that period, Capt E M Brown who, incidentally, shot down two Focke-Wulf Fw 200 Condors when operating from HMS *Audacity*, the very first escort carrier, with Martlets. He has commented:

The Grumman aircraft contribution to the Allied victory in the war at sea was immense. The company produced rugged aircraft totally suited to the demanding environment of carrier operations and thereby reduced critical losses in deck landing accidents. The splendid Wildcat appeared on the naval scene when British and US carriers relied on obsolescent fighters for their defence. The Wildcat revolutionised this situation with its ex-

cellent performance, huge endurance, and mighty firepower. The Avenger introduced a new concept of shipborne torpedo-bomber in the matter of load-carrying ability and defensive firepower. But it was the Hellcat that played Grumman's ace in the pack by making its entry into the Pacific theatre at a time when the ubiquitous Japanese Zero had ruled the tropical skies for four years. The Hellcat ousted the Zero from its position of dominance, thereby decisively altering the course of the Pacific War at sea.

Capt Brown also pointed out that with external drop tanks the 830-mile range, quoted in our tables, for Martlets/Wildcats could be extended by a further 320 miles. The official top speeds (USN and RN), he said, for both the Martlet II at 328mph and the Wildcat IV at 330mph were significantly higher than our figures.

John Fay, an early convert to piloting helicopters, points out in relation to the article on Grumman Aircraft in Wartime Service with the Fleet Air Arm (PAR 2) that the origins of 832 Squadron, in which he served, went back a little earlier than we said. He writes '..The squadron [832] was formed in 1941 and operated Albacores from HMS *Victorious* until the arrival at Norfolk, Virginia, in 1943 where the crews were given conversion courses on Avengers. The squadron re-embarked in *Victorious* and sailed to Hawaii where full working up took place. The transfer to the *Saratoga* did not occur until after arriving in New Caledonia and was for one month to cover the New Georgia landings. The squadron returned to *Victorious* for the journey home to Liverpool where the ship was refitted. The squadron did not embark in HMS *Illustrious* and HMS *Begum* until some months later.'

The Aircrew Association

The Aircrew Association, now in its twelfth year, has a membership of some 13,000 of whom 1,500 are overseas, members of 87 separate branches.

Membership is open to all those who have been awarded a flying badge in any of the British armed services (the Royal Air Force, the Fleet Air Arm or the Army Air Corps), in any of the many aircrew specialisations past or present. Membership is also open to members of the Commonwealth and Allied air forces.

The association aims to foster that particular fellowship and comradeship experienced among those of like mind and experience and to co-operate with and financially assist Service charities.

Without paid staff of any kind and currently with a subscription of just £5 per year, which includes the quarterly magazine *Intercom*, the Aircrew Association is actively seeking more members. Last year it distributed £12,000 to Service charities and this work continues.

The association is run through a small executive committee under the chairmanship of Sqn Ldr Peter Crouch AFM, RAF retd acting on behalf of a main committee of fifteen. Top level support is assured by the approval of the Air Force Board and a list of distinguished vice-presidents under the president Air Marshal Sir John Curtiss.

Enquiries about membership can be made to Fred McMillan, Hon Registrar, 16 Marescroft Road, Britwell Estate, Slough, Berkshire SL2 2LW, England.

Ventura rescued

Paul Squires writing from Yellowknife in Canada's North West Territories tells us that last year he was a member of the team which recovered a Lockheed Ventura GR.V from the bush 50 miles NW of Yellowknife.

Serving with No. 149 Sqn RCAF between 1943 and 1945 the aircraft was then placed in storage for seven years

A Chinook of 440 Sqn Canadian Armed Forces, provides lift-off for the prized Ventura; probably weighing some nine tons. Note the windsock below the aircraft, in use as a stabilising drogue. (*Tom Kiez*)

before being sold for 600 dollars, with less than 200 airframe hours, and converted for aerial survey work.

In August 1953 the Ventura was force landed following engine problems. Of the crew of four, one was hurt and evacuated from the otherwise inaccessible site by helicopter – the first such rescue in the North West Territories. It was entirely appropriate therefore when the Ventura itself was lifted out by Boeing-Vertol Chinook thirty-five years later. The aircraft now lies dismantled and in store at Edmonton, Alberta.

Ventura RCAF serial 2195 will now be restored by the Ventura Memorial Flight Association, to fly again, as a memorial to Canadian Servicemen who served in Canada and Alaska during the Second World War.

The Association wants to hear from anyone who built, serviced or flew Venturas, the ultimate conceptual derivative from the prewar commercial Lockheed 14, via the RAF's Lockheed Hudson and the Model 18 Lodestar. Information please to Tony Jarvis 14210-24A Street, Edmonton, Alberta, Canada T5Y 1L7. ◻

Just after being set down, the remarkably good state of preservation of this particular Ventura is readily apparent. (*Paul Squires*)

BOOK REVIEWS

An Ancient Air
Harald Penrose

Airlife Publishing, Shrewsbury, England.
Hardback, 155 x 225mm, 183pp,
illustrated, £15.95.

This is at once a fascinating and a frustrating book which must have provided some frustrations for its author, too. It is the biography of John Stringfellow of Chard, Somerset, an inventive engineer in the lace-making industry who became involved in a far-sighted project to build a full-sized flying machine in the 1840s, and continued experimenting with steam engines and model aeroplanes almost until his death in December 1883.

The author's frustration was clearly the incompleteness of the surviving correspondence between Stringfellow and other aeronautical pioneers such as Henson, Brearey and Chapman. The last-named gentleman, incidentally, has not hitherto been closely associated with Stringfellow, but this book shows that he contributed significantly to the Aerial Steam Carriage of 1843. There is little doubt that this amazing project was the brainchild of W S Henson and that its steam engine was also largely Henson's conception. Stringfellow, however, did much to improve the engine. Penrose seems to push Henson into the shade and stress Stringfellow's role, probably disproportionately. While there is no doubt that he continued to build model aeroplanes and engines after Henson had left for America, the degree of his success is for the reader to judge. However, the author tends to select evidence in Stringfellow's favour and neglect that against him.

The book suffers from its author's habit of lapsing into the general history of the period. This can be useful for background but technical descriptions of the aeroplanes and engines would have been welcome. Anyone seeking these will need recourse to *Henson and Stringfellow* by M J B Davy (1931) and the 1956 paper by Pritchard and Ballantyne in the Journal of the RAeS, but neither reference is readily to hand for most students. Although Henson's celebrated 1842 Patent has already been published in full, the last occasion was over thirty years ago and it would have made a useful appendix to this book.

The book contains no adequate summary and evaluation of Stringfellow's work, which would have been nice, and other pioneers such as Cayley, Moy, and Harte are often inadequately introduced, the author failing to explain where they fit into the story or what they did. To say that Brearey based his 'undulating-wing' propulsion system on a 'swimming fish' is not as graphically precise as stating that the fish was a skate, which has its own peculiar way of swimming.

All faults aside, however, this is undoubtedly the most complete biography of Stringfellow, and adds much valuable information to that previously published. Three eight-page glossy sections contain a reasonable selection of pictures, and the whole is good value at current prices. **PJ**

Valiant Wings
Norman Franks

William Kimber, Thorsons Publishing Group, Wellingborough, England.
Hardback, 240 x 160mm, 308pp,
illustrated, £14.95.

Three headstones mark the graves of a Fairey Battle crew of No.12 Squadron RAF at Heverlee cemetery in Belgium. Two were awarded the Victoria Cross posthumously but the third received nothing. Their names are Flg Off Don Garland VC, pilot; Sgt 'Dolly' Gray VC, observer; and Leading Aircraftman Roy Reynolds, wireless operator/air gunner.

As they flew and died together so they lie together, a complete RAF crew (regulars in the air force of pre-Second World War) who had relied entirely on each other's skill and courage. Only this RAF crew of all others and in all wars of the air was awarded two VCs – a remarkable stroke of good sense and appreciation for a single brave act of sacrifice. Sadly, it was there that sense stopped and official-dom intervened. For the two VCs awarded posthumously to Garland and Gray could not, apparently, be increased by one for Reynolds. Reynolds' achievements remained unsung, because at that time the VC was the only award that could be given posthumously.

This poignant story sums up the mixture of courage, hopeless endeavour and the clumsiness of political and senior staff thinking that makes the story of the ten Battle squadrons of the Advanced Air Striking Force (AASF) so bitter sweet in courage and folly. These three are buried in Belgium close to the bridge they bombed as the Germans swept through in 1940. Their story is fully remembered in print, not for the first time, in Norman Franks' detailed tribute to a little known war within a war – the Battle of France and the Low Countries, one of the most spectacular disasters since the Charge of the Light Brigade. All credit to the author for bringing it back to light after fifty years by using the verbal recollections of the RAF airmen who survived. Moreover, it is a good read about a lesser known campaign of rather long ago. **WS**

From the Flightdeck 4: Apollo 11 Moon Landing
David J Shayler

Ian Allan Ltd
Paperback, 235 x 171mm, 80pp,
illustrated, including colour, £6.95.

One of a number of books published twenty years after the first landing by man on the moon, this one covers the period from five days before lift-off, on 16 July 1969, until splashdown in the Pacific, just over 195½ hours later, mission accomplished. Much of the text is concerned with the verbal exchanges between Mission Control and the crew Neil Armstrong, Michael Collins and Edwin Aldrin. The banality of much of this dialogue contrasts sharply with the brilliance both technical and human of the achievement. Unlike the building of the pyramids 'a magnificently purposeless symptom of national power and wealth' as one review in a respected heavy newspaper had it – the manned flights to the moon in the period 1969 to 1972 with their return payloads of lunar rock and soil samples may well form the basis, in due time, of further space exploration using resources mined and converted on the moon into useful materials for constructing space stations for the exploration of near space.

At the least it is for consideration that those concerned with helping to open the minds of the next generation of world citizens see that this book, which represents excellent value, be placed on school library shelves as a primer for further discussion. **JHM**

THE PUTNAM AERONAUTICAL REVIEW

ISSUE NUMBER FOUR DECEMBER 1989

Editor: John Motum

Consultant Editor: John Stroud

Assistant Editor: Linda Jones

Design: Swanston Graphics Ltd

Managing Editor: Julian Mannering

Publisher: William R Blackmore

© Conway Maritime Press Ltd

ISBN 0 85177 528 4
ISSN 0955-7822

Published quarterly by:
Conway Maritime Press Ltd
24 Bride Lane, Fleet Street
London EC4Y 8DR

Telephone: 01 583 2412

**Annual subscription for
four issues:**
£16.00 UK, £18.00 overseas,
$33.00 USA and $39.00 Canada.
All post paid.

Typeset by Swanston Graphics Ltd, Derby.
Printed and bound in Great Britain
by Page Bros (Norwich) Ltd.

CONTRIBUTORS

Stanley Brogden, doyen of Australian aviation writers, has a lifetime of experience meeting and reporting the illustrious deeds of Australia's air pioneers. Director of public relations for the RAAF as a Group Captain in 1950-59, he has also been associate editor of the Australian magazine *Aircraft*.

Jim Cownie is a freelance aerospace and technology writer. He served an engineering apprenticeship with Handley Page and gained a London University BSc (Eng); he is also a chartered engineer and a member of the RAeS. He spent three and a half years on the editorial staff of the *Aeroplane* and twenty-three years with Rolls-Royce in PR and marketing publicity posts.

Peter Cunningham, capacity planning manager (airspace and airports) for Dan-Air London, was previously IATA's regional technical director for Europe, having held similar appointments in the Asia/Pacific region and Africa. He has spent a working lifetime in commercial aviation following RAF service as a pilot in 1941-46. He is a master pilot, holds the Queen's commendation for valuable service in the air and is a fellow of the Chartered Institute of Transport.

David W H Godfrey is an aeronautical engineer, academic and writer. From the Miles Technical School, he joined de Havilland before joining the *Aeroplane*. He worked subsequently with Canadair (Montreal) and Lockheed (Burbank). Since 1971, he has been a professor at Ryerson Polytechnical Institute in Toronto and contributes to aviation journals in a number of countries.

David J Smith, air traffic controller and author, returns in this issue to report on a United Kingdom regional airport, Birmingham International.

Anthony Vandyk based in Geneva is international editor of *Air Transport World* of Washington DC. Public relations director of IATA for fourteen years in Montreal and then Geneva, his aviation writing career began with Interavia in 1946, and continued later with American Aviation Publications.

Front cover: A Swissair Douglas DC-2 over the Alps.

Back cover: The Bermuda Islands with the airport in the foreground.

Editorial

Some pictures, like the one on our cover, manage to capture a whole era rather than just one particular moment. From early 1935 the Douglas DC-2s of Swissair became an increasingly familiar sight at European airports, to be partnered by the airline's DC-3s just over two years later. The DC-2, it will be recalled, was the first twin-engined transport designed to maintain height on one engine, albeit not over high Alps or North American Rockies. It may all seem a far cry from today, crossing Switzerland at 550mph, with Europe's highest mountain, Mont Blanc, more than 3 miles below, and few passengers apparently concerned to give the matter a passing thought. That's progress, and perhaps is as it should be.

Those of us with a reflective turn of mind cast back to how it all began. Swissair's first managers in 1931, from the constituent companies out of which the airline was created, knew that for air transport to prosper, from geographically small, landlocked Switzerland, it would need to sell speed. The real competition came from the railways of Europe which were comparatively fast, frequent and unaffected by bad weather.

Accordingly, the decision was taken to shop in the United States for the fastest aircraft available, the single-engined Lockheed Orion, a courageous decision. To operate in safety over mountains at the time required three engines, if flight was to be continued following the failure of one. The Swiss regulating authority backed the judgement of Swissair's founder managers, Walter Mittelholzer and Balthasar Zimmerman, the two Orions being operated safely, initially without radio, for the next four years before being sold. When the Douglas DC-2 came on the market, followed by the DC-3, Swissair was an early purchaser of both. But in 1937 Mittelholzer was killed in a climbing accident and Zimmerman died.

However, the soundness of their equipment decisions was still apparent following the end of the Second World War when Swissair was able to continue its unbroken experience with the DC-2/DC-3 by topping up with war surplus C-47s. From that point it was wise to stay with the known and proven piston-engined powerplant in the best new equipment becoming available at the end of the 1940s, Convair 240s, replacing them with developed Convair 440s later.

The story of Swissair's equipment choices is told briefly in this issue together with an observer's view from the flight deck of some of the later ones.

For the past forty years, Swissair has experienced a continuity of management leadership which is probably unique. Dr Walter Berchtold took over the presidency in 1950, remaining until the early 1970s when Armin Baltensweiler, his deputy, took over the leadership. He is executive chairman today. Robert Staubli, elected president in 1988 until his retirement, joined the airline to fly a DC-2 and indeed he captained two of the four Convair 880 sectors flown by John Stroud, recorded in this issue.

With a fleet of over fifty aircraft ranging in size from the Fokker 100 to the Boeing 747, and over 18,000 employees, Swissair today serves over a hundred cities in seventy-two countries. While there are many bigger airlines, few have more continuous experience of selling air travel's greatest asset, speed.

Innovation continues to be a theme throughout this issue. Without a great deal of flair, imagination and sound foundations Embraer of Brazil would not now enjoy world ranking as an aircraft manufacturer. In Britain, near tragedy at the beginning of the Rolls-Royce RB.211 programme has turned to considerable success which bids fair to continue.

There was a time, until quite recently, in fact, when air journey time depended only on the operating speed of the aeroplane and, marginally, the weather. And the time required at the airport at each end could be predicted. Now, in Europe, and maybe elsewhere, at holiday periods certainly, nothing can be taken for granted. As with the dilemma facing Swissair in 1931, an expanding high-speed European railway network, spearheaded by the French TGV (*train à grande vitesse*), is beginning to look a much better proposition for the 1990s than some parts of the airline network.

On this occasion, the limitations of air travel have nothing to do with the imperfections of the aeroplane and almost everything to do with man's inability to manage the airspace, that environment which used to be known as 'free as air'. Now the ability of airspace to carry traffic is only as good as the bottlenecks allow.

In order better to understand the current problems, we asked Peter Cunningham to explain how we had reached this situation. Personal frustration due to delays is easy; reasoned explanation more difficult.

It is no bad thing that, as *Review* readers, we understand the basics of a specialist and difficult craft, air traffic management, because so much of the initial impetus to unblock the system must come about through a massive input of international political will. And within the EEC each one of us has two elected representatives. With the growth rates in traffic predicted for the next decade, more financial investment, but well spread, will be needed and, just as much, an investment in international goodwill, commonsense and twenty-first century thinking.

There are signs of movement towards improvement; too much is at stake for there not to be. Improvement will need to be continuous, however, and never again sporadic as in the past. There can be no doubt that whichever article in this issue personal preferences may dictate as the most informative or enjoyable, the most important is that by Peter Cunningham. Should we continue to get our air traffic management wrong, much more anguish would result.

Eventually, air travel customers, like any other customer being ill-served, will find their own solutions and vote with their feet. Whatever the rights or wrongs, the airline industry may well have to try even harder and pay more to protect its own investment. This in turn would feed back to the customer. And it might be a small price to pay to be able to travel to a destination of personal choice, efficiently. What is the alternative? To pay an admission fee to visit our transport aircraft in a museum rather than on the departure ramp?

However, there is no need to finish on such a note. Also in this issue Stanley Brogden sets out the story of the very first flight made between Europe and Australia in 1919, to overcome the 'tyranny of distance'. That flight took twenty-eight days to reach Darwin and a further seventy-four days to Melbourne.

On 17 August 1989 a new Qantas Boeing 747-400 (four Rolls-Royce RB.211-524Gs) on delivery from Seattle, the long way round, completed the 11,250 miles from London to Sydney non-stop in 20hr 9min, indicating that the Australians have not lost their touch in long-range operations. Even more to the point, commercial flights have been scheduled by Qantas from both London and Manchester to Melbourne, with one stop at Singapore, in an elapsed time of 22hr 25min, since October.

In Brazil Embraer has announced its intention to build the EMB-145, a regional jet in the overall 50 seater class, for service from 1992, incorporating a high percentage of airframe and systems parts from the EMB-120 Brasilia. Embraer's origins and other current commitments, including joint programmes with Italy and Argentina, are described beginning on page 216. □

Swissair, Innovators for More Than Half a Century

Swiss air transport is celebrating its seventieth anniversary this year and it has an interesting history. Apart from its short-lived military mail and passenger services operated with open cockpit biplanes, most of the pioneer work was done by two companies, Ad Astra-Aero and Balair. Ad Astra mainly employed all-metal Dornier and Junkers aeroplanes, after a brief flirtation with flying-boats on non-scheduled operations, and Balair established itself as a Fokker airline.

The metal Dorniers and Junkers and the wooden-winged Fokkers were the most advanced transport aeroplanes of their period and, after the merger of the two pioneer airlines, the newly formed Swissair continued to lead the way with modern equipment, a policy which has continued throughout its nearly sixty years of operation.

Here, Anthony Vandyk briefly traces the fleet development of Swissair and John Stroud describes his personal experience of the airline's jet equipment and presents operating studies of some of the leading types.

Ad Astra-Aero's Lynx-powered Messerschmitt M 18d passed to Swissair and became HB-IME.

Anthony Vandyk

The merger of Ad Astra and Balair as Swissair on 26 March 1931 resulted in the new company having a fleet of eight three-engined Fokker F.VIIb-3ms, one single-engined F.VIIa, two Dornier Merkurs, a Messerschmitt M 18d and a Comte A.C. 4 Gentleman, providing a total passenger seating capacity of eighty-six. There were ten pilots, seven radio operators and eight flight engineers. Total staff numbered sixty-four. The 4,203km route network was flown only from March to October, operations being suspended in the winter months.

In April 1932 Swissair became the first European company to introduce the Lockheed Orion which had a cruising

One of the two red and white Lockheed Orion 9Bs which Swissair introduced in 1932, the first modern low-wing monoplanes with retractable undercarriages used by a European airline.

speed of 260km/hr, 100km/hr faster than any other aircraft in airline service in Europe. The low-wing single-engined (Wright Cyclone) monoplane had a range of 950km and could carry a payload of 815kg (including the pilot and four passengers). Introduced on the Zürich–Munich–Vienna route on 2 May 1932, Swissair's two Orions were an immediate success, with load factors running over 80 per cent. In 1936 both aircraft found their way to the Spanish Republican Air Force via a French broker. A similar fate awaited one of the two

Swissair's Fokker F.VIIa CH 157 was acquired by Balair in 1927 and is preserved at the Transport Museum in Lucerne.

This Clark GA-43A was acquired by Swissair in 1934 and went to Spain in 1936 for use in the Spanish Civil War.

The short-lived, ill-fated Swissair Curtiss Condor which suffered structural failure in July 1934.

ten-passenger Clark GA-43As operated by Swissair from 1934 and one of the airline's Douglas DC-2s.

In 1934 Swissair introduced the sixteen-passenger Curtiss Condor but after three months in service it suffered structural failure and crashed in Germany. Then began the airline's long relationship with the Douglas Aircraft Company. Five fourteen-passenger DC-2s were bought through Fokker, then Douglas' European agent, and delivered between December 1934 and July 1935. A sixth was acquired from the Austrian Government in April 1936. The DC-2 started Swissair's first service beyond the European continent; a route from Zürich to London via Basle opened on 1 April 1935 and operated in pool with Imperial Airways. With the new aircraft Swissair continued operations through the winter for the first time. Two of the DC-2s remained in service until 1952.

The diesel-powered Junkers-Ju 86 operated night mail services.

In 1936 Swissair acquired its first Junkers-Ju 86. This was a ten-passenger all-metal monoplane powered by two Junkers-Jumo diesel engines. Although the Ju 86 was a passenger aircraft, Swissair used it on the Zürich–Frankfurt-am-Main night mail service, making Swissair the second airline to operate a diesel-engined aeroplane. However, it was not

One of Swissair's original fleet of Douglas DC-2s.

HB-IRM was one of the Douglas C-47Bs acquired by Swissair in 1947. It cost 283,000 Swiss francs.

much of a success and the airline's final Ju 86 operations employed orthodox BMW 132 radial air-cooled engines.

The prototype de Havilland D.H.89 Dragon Rapide was bought second-hand in March 1937 and remained in service until June 1954, as did two D.H.89As purchased in 1948. Another de Havilland aircraft operated by Swissair was a photographic-reconnaissance Mosquito P.R. Mk. IV leased from the Swiss Air Force from January to August 1945.

Before recounting the Swissair DC-3 saga it might be useful to record that at the outbreak of the Second World War the airline had a fleet comprising five DC-3s, three DC-2s, one Dragon Rapide, one Fokker F.VIIa and one Comte A.C.4. After maintaining limited services (from Zürich to Munich and to Berlin and from Locarno in southern Switzerland to Rome and to Barcelona) Swissair was forced to cease flying operations. Staff was cut to ninety-three employees and the main activity was maintaining DC-3s for Lufthansa and various types of aircraft for the Swiss Air Force. On 30 July 1945 Swissair resumed flying operations with the DC-3 as the main component of the fleet.

Swissair bought five DC-3s between

Swissair's first Douglas DC-4. It cost 2,259,000 Swiss francs.

Swissair's first Douglas DC-6B, acquired in 1951.

1937 and 1939. One was sold to ABA Swedish Air Lines in 1940 but the other four continued in service with the Swiss carrier until 1955. Between 1946 and 1947 ten more DC-3s (actually C-47Bs) were acquired and in 1951 a C-47A was leased from Ethiopian Airlines for thirteen months. Scheduled DC-3 operations ended on 31 March 1964.

The DC-3 was the mainstay of the Swissair short-/medium-range fleet in the postwar years. For long-distance operations three DC-4s and one C-54 were purchased and two of them remained in service until 1959 when they were transferred to Swissair's subsidiary, Balair. Although a large number of non-scheduled flights were operated to New York from May 1947 scheduled transatlantic service did not start until

The second Swissair Convair 440 Metropolitan.

Swissair's Sud-Aviation Caravelle III *Uri* was leased from SAS.

1949. A pressurised aircraft was badly needed by then but Swissair's financial problems at that time precluded the outright purchase of new equipment. An arrangement to lease two Lockheed Constellations from KLM was cancelled as the aircraft, already painted in Swissair colours, were being readied for delivery. Instead the Swiss Government bought two Douglas DC-6Bs and leased them to the airline. The two aircraft remained in service from the summer of 1951 until 1962 when they were sold to Denmark's Sterling Airways. Swissair bought four more DC-6Bs in the early 1950s and chartered a fifth from 1958 to 1961. A

Swissair's first Douglas DC-7C was acquired in 1956 and sold to SAS in 1962.

DC-6A freighter was bought in 1958 and sold three years later.

To provide a pressurised replacement for the DC-3 Swissair chose the Convair 240. Four were bought from the manufacturer in 1949 and four more acquired from KLM in 1953 and 1954. The entire 240 fleet was sold to Mohawk Airlines for delivery in 1956 and 1957 and twelve of the developed 440 Metropolitan version ordered as its replacement. In its turn the Convair 440 fleet was withdrawn from service in the late 1960s. One of them remained in Switzerland, converted as a Dart propeller-turbine-powered 680 operated by the charter airline SATA.

Three Fokker F.27s were bought in the 1960s for operation by Balair, a -200 and two -400s. These were sold in the early 1970s. Fokker aircraft were in the origin-

al Swissair fleet and the airline returned as a customer of the Dutch company in the mid-1980s when it ordered a fleet of Fokker 100s.

Now it is time to record the story of Swissair's last piston-engined aircraft, the Douglas DC-7C. Five aircraft were delivered between November 1956 and November 1958 but they were only to remain in service a relatively short time as jets were soon available to take their place; two were sold in 1960, one in 1961 and the last two in 1962.

Swissair's entry into the jet age came later than that of many other airlines although as early as 1956 it had started a 'DC-8 paper operation' to gain experience through simulating North Atlantic flights. It was not until April 1960, however, that the first aircraft was delivered, enabling scheduled jet operations to New York to start on 30 May. Two more DC-8s, all -32s but along with the first converted to -33s in 1961, were received during the year. At the same time as the DC-8s were being delivered

Swissair received four Sud-Aviation Caravelle IIIs on lease from SAS enabling short-/medium-range jet operations to start in the spring of 1960. The four aircraft were subsequently purchased and four more ordered from the manufacturer for delivery in 1962. An additional Caravelle was leased from Air France from February 1964 to March 1966. With the introduction of the Douglas DC-9 in mid-1966 Caravelle operations were scaled down and the last of the French aircraft was sold in 1971.

Delivery of Swissair's third jet type, the Convair 990A, was delayed and in part compensation the manufacturer provided two Convair 880s on lease from August 1961 to the spring of 1962. Delivery of the 990, called Coronado by Swissair, finally got under way in January 1962 and the airline received a total of eight aircraft by the spring of 1964. In 1962 two were leased to SAS, one for two years and one for four years; the latter was subsequently leased to Balair for three years starting in March 1968.

Swissair's Douglas DC-8-62 *Schwyz*.

Convair 990A Coronado.

The second Swissair McDonnell Douglas DC-10-30 *Schaffhausen*.

Swissair Boeing 747-357 *Basel-Stadt*.

Swissair McDonnell Douglas MD-81 over its home territory.

Mainly used on routes to the Far East, West Africa and South America, the 990 was the fastest commercial transport of its time but economically the 100-passenger aircraft was inferior to the DC-8-33 which carried 132 passengers. Operations with the 990 ended in early 1975. One of the aircraft is on display in the Swiss Transport Museum, Lucerne.

The DC-8 fleet underwent various changes during the 1960s: a -53 was added in the autumn of 1963 and in the following year one of the three original -32/33s was converted to a -53; two of the three originals were sold, one in 1967, one in 1968 and the last in 1976 (this one remained in Switzerland as the buyer was Swiss charter operator SATA). The first of the DC-8-62s (two of them -62CFs)

The McDonnell Douglas MD-81 *Lugano*.

was delivered in November 1967 and the last in February 1970. Swissair DC-8 operations ended in April 1984 when the last aircraft was phased out.

A standardised configuration for the DC-10-30 was agreed by Swissair and its KSSU partners (KLM, SAS and UTA). The Swiss carrier received eight of the aircraft between November 1972 and February 1975. A ninth was delivered in 1977 and two more in early 1980. Another two DC-10s were received in 1982. Four of the present ten-aircraft (one was sold in 1988) fleet are the extended-range -30ER version.

Swissair has been operating DC-9/MD-80 series aircraft since 1966 and is likely to keep some of its -81s in service into the next century. Originally, five DC-9-15s were leased from Douglas in 1966 and 1967. The first of Swissair's own -32s was delivered in October 1967 and by the end of 1968 fifteen had been delivered although five were immediately leased to SAS for nine months. Two more were bought from the German charter airline Südflug in 1968 while another six were delivered by Douglas,

two in 1969 and four in 1970. A DC-9-33F freighter was bought in 1969; a year later it was sold to a Canadian company but leased back until 1984. One -32 was sold in November 1976 and another in May 1979. Meanwhile, during 1975–79 Douglas delivered twelve of the stretched -51 version of the DC-9 to Swissair.

In October 1979 Swissair became the launch customer for the MD-81, ordering fifteen of this developed version of the DC-9 for delivery between 1980 and 1982 and optioning another five. The airline currently has a fleet of twenty-two MD-81s with one more on order. The last DC-9s – five -32s and four -51s – were sold in 1988.

The Swissair Fokker 100.

A Crossair Saab 340.

Inflight impression of the Saab 2000. Crossair, with twenty-five on order, is the launch customer.

The Fokker 50, in Crossair colours.

Swissair joined the ranks of Boeing 747 operators in early 1971, using two 747Bs to link Switzerland with New York thirteen times a week. The airline was a launch customer for the stretched upper deck -300 series, receiving two of the aircraft on lease in 1983. The Swissair 747 fleet currently comprises five of the -300 series including two combis.

Swissair was also a launch customer for the Airbus A310, receiving the first of five of the -200 series in March 1983 and three of the 'Intercontinental' -300 series two years later. The airline was the launch customer for the Fokker 100 introduced into service in early 1988. Eight of the Dutch jets are operated.

The next new aircraft type to be introduced by Swissair will be the MD-11 which will replace the DC-10. Delivery of twelve aircraft is scheduled to start next year.

The entire Swissair fleet has Category III autoland capability and also meets international noise regulations. As at the end of 1988 the average age of its aircraft was 5.7 years.

This brief account of the development of Swissair's fleet would be incomplete without mention of the aircraft of its three affiliates, Balair, Crossair and CTA (Compagnie de Transport Aérien). Basle-based Balair has one DC-10, one Airbus A300 and three MD-82s. It also operates an F.27 owned by the Swiss Government. Geneva-based CTA's fleet comprises four MD-83s. Crossair, with offices in Zürich and Basle, had nineteen Saab 340As with ten 340Bs on order for delivery by the end of this year. It has also contracted to buy twenty-five of the fifty-passenger Saab 2000s for delivery starting in 1993. For operation pending the arrival of the Saab 2000s, Crossair has ordered five Fokker 50s to be delivered next year.

Seeing Swissair in Close-up

John Stroud

Although it took place well over fifty years ago, I have never forgotten my first encounter with Swissair. I entered one of the Croydon hangars by a back door and was confronted with a large shiny metal monoplane with the red and white colours of Switzerland on its tail.

It may seem odd that the first view of a Swissair Douglas DC-2 should leave such an impression, but I will explain. The DC-2 was a clean low-wing cantilever monoplane with retractable undercarriage, and being Swiss-owned and -operated it was in immaculate condition. Outside the hangar were the aeroplanes with which I was familiar: Imperial Airways' Handley Page 42s, Short Scyllas and de Havilland 86s, all biplanes, and Lufthansa's Junkers-Ju 52/3ms, monoplanes but with fixed undercarriages and dull corrugated metal skins.

Just as Swissair's DC-2 made its im-

pact on me, so its advanced procurement had an impact on other European airlines and European aircraft manufacturers. When Swissair ordered its two Lockheed Orions, fear of competition led Luft Hansa (as it was then termed) to get Heinkel and Junkers to produce aeroplanes of comparable performance, the He 70 and Ju 160. These aeroplanes in turn paved the way for the He 111 and Ju 86 twin-engined developments and, incidentally, one of the Luftwaffe's most successful bombers.

This enlightened aircraft procurement policy continued through the decades with, in more recent times, Swissair being the launch customer for the Fokker 100 and with Lufthansa (as it is now) for the Airbus 310.

I have long had a very high regard for Swissair as, what I call, a 'technical airline'. I have flown in most of Swissair's

aircraft from the Douglas DC-6B and Convair 240 to the Douglas DC-9s and DC-10s, the Boeing 747 and the Airbus A310, have been privileged to travel on the flight deck on many occasions and received unstinted help from Swissair's crews in making operating studies of most of the airline's jets. I have also seen the airline's maintenance and training and am proud that I 'landed' the Coronado simulator without feeling the mainwheels touch the 'ground'.

Convair 880M

To meet competition on the Far East routes it was essential that Swissair replace its piston-engined Douglas DC-6Bs and the decision was taken to order a fleet of Convair 990As, which the airline named Coronado. Unfortunately, there were delays in production of the Coronado and initially it was below its

Swissair scheduled service SR 500
Convair 880M (Model 22M-3) HB-ICL
19/20 September 1961   Capt Schaerer to Karachi, Staubli to Bangkok

Sector	Zürich – Cairo	Cairo – Karachi	Karachi – Bombay	Bombay – Bangkok
Stage length	1,554nm	1,969nm	501nm	1,703nm
Flight plan	3hr 34min	4hr 21min	1hr 18min	3hr 52min
Block time	3hr 39min	4hr 35min	1hr 29min	4hr 10min
Airborne time	3hr 31min	4hr 25min	1hr 20min	3hr 56min
Take-off weight (brake release)	190,000lb	190,500lb	160,000lb	185,000lb
Runway	34	23	07	27
Runway length	11,300ft	10,100ft	10,500ft	9,780ft
Runway elevation	1,414ft	311ft	81ft	35ft
Temperature at take-off	24 °C	25 °C	32 °C	29 °C
Runway length required	8,694ft (to 35ft)	8,103ft (to 35ft)	6,233ft (to 35ft)	8,000ft (to 35ft)
V1	140kt	139kt	125kt	136kt
VR	145kt	144kt	129kt	142kt
V2	156kt	156kt	143kt	154kt
Time to 100kt	30sec	29sec	24sec	28sec
Max permissible t/o weight	Max AUW	193,000lb	192,500lb	192,500lb
Fuel	72,000lb	76,000lb	48,000lb	77,000lb
Endurance	5hr 22min	6hr 36min	3hr 40min	6hr 5min
Commercial load	21,660lb	18,331lb	16,000lb	11,800lb
Max sector load	24,260lb			
Cruise level	FL 290↗330	FL 340	FL 300	FL 330↗340
Cruise level temperature	−34 °C	−37 °C	−21 °C	−32 °C
Cruise level average wind	−13kt	+5kt	SE 10-15kt	−18kt
Average cruise TAS	488/496kt	486/496kt	505kt	470kt
Average cruise Mach	0.82	0.8/0.82	0.82	0.78
Cruise technique	One step climb	Constant altitude	M:0.82	Constant Mach
Estimated landing weight	142,000lb	140,000lb	140,000lb	140,000lb
Estimated fuel over destination	23,100lb	24,800lb	28,000lb	32,000lb
Actual fuel after landing	24,500lb	23,700lb	30,500lb	31,500lb
Terminal weather	060/10kt 10km clear	Fair	Cloudy with showers	Cloudy, occasional thunderstorms
Alternates	Beirut	Bombay	Karachi	Saïgon
Average cruise consumption	3,000lb/eng/hr	2,900lb/eng/hr	3,100lb/eng/hr	2,600lb/eng/hr

Copyright: John Stroud

performance guarantees. To bridge the gap until the Coronado was ready Swissair leased two Convair 880Ms. These were introduced on the Zürich–Tokyo route on 10 September 1961 and I was invited to fly from Zürich to Bangkok on one of the first services.

In their early period the 880s had a

Swissair Convair 880M HB-ICM at Zürich on arrival from Tokyo in September 1961. The covered passenger steps had been recently introduced at Zürich and Geneva.

Swissair scheduled service SR 110
Boeing 747-257B HB-IGA *Genève* Capt W Stierli
6/7 May 1971

Sector	Geneva – New York
Stage length	3,532nm
Flight plan	7hr 44min
Block time	8hr 29min
Airborne time	7hr 57min
Take-off weight (brake release)	323,400kg
Runway	05
Runway length	12,795ft
Runway elevation	1,411ft
Temperature at take-off	21 °C
Wind at take-off	060/3kt
Runway length required CAR	10,825ft
V1	158kt
VR	164kt
V2	171kt
Max take-off weight for conditions	337,000kg, obstacle limited
Fuel (at brake release)	112,600kg
Endurance	9hr 40min
Commercial load	39,957kg
Cruise level	FL 310⁄330
Cruise level temperature	−46 °C at FL 310
Cruise level wind (average calculated)	−8kt
Average cruise TAS	*
Average Mach	0.84
Cruise technique	0.84 constant
Estimated landing weight	231,900kg
Estimated fuel over destination	24,000kg
Actual fuel at end of landing run	18,500kg
Terminal weather	350/9kt vis 7 miles 4,000ft scattered 9,000ft overcast 62 °F 29.81
Alternate	Boston 191nm
Average cruise consumption	2,750kg/eng/hr
Alternate fuel	7,800kg
Max payload for sector	54,000kg
Number on board	271 passengers, crew 3 + 17

*Spot check TAS 489kt GS 495kt

Route: Geneva – Dijon – Biggin – Shannon – 54N 15W – Cartwright – Port Menier (Anticosti) – Presque Is – Bangor – New York

Swissair scheduled service SR 801
Douglas DC-9-32 HB-IFS *Winterthur* Capt Jan Dekker and F/O Peter Ernst
5 May 1971

Sector	London – Zürich
Stage length	485nm
Flight plan	1hr 12min
Block time	1hr 32min
Airborne time	1hr 12min
Take-off weight (brake release)	44,800kg
Runway	10R
Runway length	12,000ft
Runway elevation	80ft
Temperature at take-off	15 °C
Wind at take-off	070/6kt
Runway length required	7,100ft
V1	138kt
VR	142kt
V2	150kt
Max take-off weight for conditions	49,000kg
Fuel (at brake release)	9,800kg
Endurance	3hr 28min
Commercial load	6,770kg
Cruise level	FL 290
Cruise level temperature	−48 °C
Cruise level wind	Light and variable
Average cruise TAS	468kt
Average Mach	0.78
Cruise technique	Constant speed
Estimated landing weight	40,500kg
Estimated fuel over destination	5,700kg
Actual fuel at end of landing run	5,800kg
Terminal weather	180/5kt vis 5,000m 3/23 7/60 13 °C Dew point 10° 1015mb NS
Alternates	1. Geneva 124nm 2. Basle 71nm
Average cruise consumption	1,420kg/eng/hr
Rotate to 17½° on take-off + climb to 3,000ft	
Rate of climb 1/2,000ft/min	

certain amount of engine trouble and my flight suffered a fairly long delay before leaving Zürich. In those more care-free days when security was unnecessary, I was actually on the apron and helped shut and secure the cowling of the troublesome engine before I climbed aboard via Swissair's impressive covered passenger stairs.

I found that it took me several sectors to get used to the 880's ways, especially on landing and take-off, but I quickly grew to like the aeroplane. At 31,000ft and Mach 0.82 over India at the end of the monsoon I felt that the 880 behaved something like a wild horse, but there was nothing frightening or unpleasant about it.

Douglas DC-9-32

In May 1971 I was invited to make a transatlantic flight in Swissair's first Boeing 747 and took the opportunity of the ferry flights to and from Zürich to study the Douglas DC-9-32 and at last the Convair 990 Coronado.

The smaller versions of the DC-9s were old friends and I found the -32 little different but it was nice on arrival in Zürich to have a go on it in the simulator. I found that it took a bit of nerve to rotate it to 17½ degrees on take-off to simulate compliance with London Heathrow's noise abatement requirements. The DC-9's control responses were such that I thought it would make quite a good fighter.

Boeing 747-257 B

Anybody's early flights in Boeing 747s must stand out as highlights. I had flown in a 747 before embarking on Swissair's

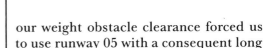
Approaching Bombay in a Convair 880M. The leading-edge slats are in the fully open position.

Genève at Zürich but it was still a very impressive aeroplane to climb aboard. We started with the 26-minute hop to Geneva; the weather was quite warm and at low level there was some turbulence. I had become used to what I call 'pod nod' on Boeing 707s and 720s, where the engine nacelles nod in the vertical plane in turbulence, but in the 747 they nodded horizontally and from my seat I could clearly see the change in the ellipses of the cowlings.

We left Geneva at a weight of 323,400kg with 291 aboard and in a temperature of 21 °C. The runway in use was 23 but at

our weight obstacle clearance forced us to use runway 05 with a consequent long wait for take-off clearance.

There is little to say about the flight but on the return aboard *Zürich*, the second Swissair 747, I had the pleasure of spending the night over the moonlit ocean in the comfort of the upstairs lounge – a rather odd experience to be the sole occupant of the upper cabin with not a sign of the two or three hundred

The three-man flight deck of a Swissair Boeing 747-257B.

One of Swissair's first two McDonnell Douglas DC-10-30s.

Schwyz, Swissair's third Airbus A310-221 landing at Zürich.

Swissair scheduled service SR 804
Convair 990-30A HB-ICE *Canton de Vaud*   Capt Heinrich Maurer
11 May 1971

Sector	Zürich – London
Stage length	455nm
Flight plan	1hr 10min
Block time	1hr 23min
Airborne time	1hr 6min
Take-off weight (brake release)	79,600kg
Runway	34
Runway length	12,139ft
Runway elevation	1,414ft
Temperature at take-off	22 °C
Wind at take-off	030/15kt
Runway length required	5,250ft
V1	134kt
VR	134kt
V2	145kt
Max take-off weight for conditions	114,700kg
Fuel (at brake release)	16,500kg
Endurance	3hr 6min
Commercial load	6,700kg
Cruise level	FL 310
Cruise level temperature	−44 °C
Cruise level wind	080/30kt
Average cruise TAS	490kt
Average Mach	0.83
Cruise technique	M:0.83 constant
Estimated landing weight	72,500kg
Estimated fuel over destination	10,100kg
Actual fuel at end of landing run	9,500kg
Terminal weather	090/5kt vis 10km 2/8 3,500ft
	7/8 5,000ft 21 °C Dew point 13°
	1023mb Trend: no significant change
Alternates	1. Gatwick 83nm
	2. Manchester 156nm
Average cruise consumption	1,300kg/eng/hr

Swissair scheduled service SR 100
Douglas DC-10-30 HB-IHA *St Gallen*   Capts Grob and E Troehler
1 February 1973

Sector	Zürich – New York JFK
Stage length	3,410nm Great Circle
	3,721nm ESAD
Flight plan	7hr 50min
Block time	8hr 4min
Airborne time	7hr 50min
Take-off weight (brake release)	234,040kg
Runway	34
Runway length	12,140ft
Runway elevation	1,416ft average
Runway gradient	+0.1%
Temperature at take-off	+1 °C
Wind at take-off	060/8kt
Take-off flap	9°
V1	170kt
VR	177kt
V2	187kt
Max take-off weight for conditions	235,000kg
Fuel (at brake release)	80,000kg
Endurance	9hr 0min
Basic equipped service weight	120,350kg
Commercial load	33,690kg
Max payload for sector	34,650kg
Cruise level	FL 280↗310↗370↘350
Cruise level temperature	−41/51 at 310 −56 at 370 −58 at 350
Cruise level average wind component	−26kt
Average cruise TAS	485kt
Average cruise Mach	0.84
Cruise technique	Constant Mach
Estimated landing weight	165,040kg
Estimated fuel over destination	11,000kg
Actual fuel at end of landing run	11,000kg
Total burn-off from ramp	70,000kg
VAT	151kt approx
Alternate*	Bradley
Alternate distance	107nm
Alternate fuel	3,300kg
Average cruise consumption	3,000kg/eng/hr

*Flight planned: Zürich–Halifax with alternate Boston. Replanned Halifax–JFK with alternate Bradley

An 'ordinary' DC-9 with the very long MD-80.

people downstairs. On that return flight our flight plan was 7hr 10min and that was precisely the time we took – a common experience on Swissair.

Convair 990A Coronado

My first attempt to fly in a Coronado ended in failure. I chose to fly Swissair from Athens to Zürich in order to experience the type. The cabin was delightful and beautifully light. We started up, taxied to the 15L runway and got going, but at about V1 there was a swing to port, a straightening up, a lot of reverse thrust and hard braking after which we turned onto a taxiway to cool the brakes. We had lost No. 1 engine and that episode taught me to like runways which were 60m wide.

I had to wait until the Coronados had nearly finished their working lives before I got into the air in one. On that occasion I flew first class and it was rather odd because that small cabin had fewer seats than some of the aeroplanes with which Swissair began in 1931.

It was a pleasant flight; I liked the aeroplane and that time we beat the Zürich – London 1hr 10min flight plan by 4 minutes.

One of Swissair's fleet of Airbus A310-321s.

Douglas DC-10-30

When I made my first DC-10 Atlantic crossing to study the type in operation, Swissair was working at a disadvantage as a result of being first. The airline introduced the DC-10-30 on 15 December 1972 as the first operator of the type and the -30 had not been fully certificated. This does not mean that the -30 was an untried aeroplane but it was working on the somewhat conservative certification of the -20/40, and when I flew on it on 1 February 1973 it was still having to be flight planned on -20/40 figures. This meant that a direct Zürich – New York flight could not be planned;

Swissair scheduled service SR 805
Airbus A310-221 HB-IPD *Solothurn* Capt H Baumann
31 August 1984

Sector	London – Zürich
Stage length	493nm
Flight plan	1hr 10min
Block time	1hr 28min
Airborne time	1hr 9min
Take-off weight (brake release)	109,800kg
Runway	28R
Runway length	3,902m/12,802ft
Runway elevation	80ft
Temperature at take-off	23 °C
Wind at take-off	260/13kt
Take-off flap	15°
V1	134kt
VR	134kt
V2	138kt
Max take-off weight for conditions	143,800kg performance limited / 132,000kg structural limit
Fuel (at brake release)	11,100kg
Endurance	1hr 50min
Basic equipped service weight	79,800kg
Zero fuel weight	98,600kg
Commercial load	19,180kg
Max payload for sector	28,700kg
Cruise level	FL 370
Cruise level temperature	−57 °C
Cruise level wind	280/30kt
Average cruise TAS	460kt
Average cruise Mach	0.80
Cruise technique	Min cost = LRC
Estimated landing weight	103,900kg
Estimated fuel over destination	5,400kg
Actual fuel at end of landing run	5,600kg
Burn-off from ramp	5,700kg
VAT	124kt
Average cruise consumption	2,040kg/eng/hr
Destination weather	280/7kt 10km scattered 5,000ft / 25 °C QNH 1020
Alternate	Basle
Alternate weather	300/10kt 10km scattered 25,000ft 1020
Take-off roll	32sec
Number on board	107 passengers, crew 2+8

Approach was at 130kt

Swissair scheduled service SR 800
Douglas DC-9-81 (MD-81) HB-ING *Glarus* Capt Alfred Kunz
1 September 1984

Sector	Zürich – London Heathrow
Stage length	455nm
Flight plan	1hr 18min
Block time	1hr 42min
Airborne time	1hr 27min
Take-off weight (brake release)	58,400kg
Runway	28
Runway elevation	1,416ft
Runway length	2,500m/8,202ft
Temperature at take-off	14 °C
Wind at take-off	310/4kt
Runway length required	2,100m
V1	138kt
VR	143kt
V2	150kt
Rotation to	17°
Max take-off weight for conditions	59,000kg
Fuel (at brake release)	7,840kg
Endurance	2hr 12min
Commercial load	12,800kg
Cruise level	FL 310
Cruise level temperature	−42 °C
Cruise level wind	−40kt
Average cruise TAS	440kt
Average cruise Mach	0.74
Cruise technique	Min cost
Estimated landing weight	54,400kg
Estimated fuel over destination	3,800kg
Actual fuel at end of landing run	3,050kg
Total burn-off from brake release	4,790kg
Terminal weather	250/8kt 50km 2/25 1017 19 °C/15 °C No sig
Alternates	Stansted 76nm Luton 77nm
Average cruise consumption	1,400kg/eng/hr
VAT	132kt
Number on-board	129 – passengers 122, crew 2+5

6min holding at Biggin Hill Take-off run 37sec
Max t/o weight 63,500kg. Max landing weight 58,900kg
Cat IIIA DH 50' RVR 200m

instead the plan had to be Zürich–Halifax with Boston as alternate and then replanned from Halifax to New York with Bradley as alternate.

There was snow on the ground at Zürich and our take-off was impressive at -1 °C even though we were less than 1,000kg below maximum weight and operating from a runway more than 1,400ft above sea level. Once more the flight plan and airborne times agreed.

I returned on a 747 and the night take-off from New York was so enjoyable that I would like to have gone back and experienced a second one.

Airbus A310-221

Swissair introduced the Airbus A310 on the Zürich–London route on 21 April 1983 and was the first airline to operate the Pratt & Whitney powered version (Lufthansa had started using the General Electric engined A310 on 10 April).

I flew to Zürich on the airline's fourth A310 on the last day of August 1984 and found it a very impressive aeroplane. Admittedly the A310 was light, with only 107 passengers, but the temperature was 23 °C and we had a take-off run of 32 seconds. Climb was impressive and from the front cabin the Airbus seemed to purr uphill; nevertheless, this apparently effortless climb took us from brake release to 37,000ft in 19 minutes including two slight ATC holds. One of the beauties of the A310 is that at European sector weights it will climb straight to 41,000ft.

The EFIS (Electronic Flight Instrument System) and EICAS (Engine Indication and Crew Alerting System) were impressive. The captain simulated a hydraulic failure for me, the warning came up as well as the effect and then the actions required. A steward reported a passenger complaining about an overheated cabin, the crew selected the cabin plan and temperatures and found they

The two-crew flight deck of a Swissair Douglas DC-9, in this instance a -51 model.

had a sensitive passenger – his cabin was 1 °C warmer than the rest!

Having been held at altitude for traffic we made our descent at nearly 6,000ft per minute and it was during this descent that I appreciated the outstanding view from the flight deck. To complete the delightful 1hr 9min flight we made an automatic landing with auto roll out and braking – and we were one minute ahead of flight plan.

McDonnell Douglas MD-81 (DC-9-81)

With the -80 series of DC-9s Swissair was once again the leader, putting the -81 into service on the Zürich–London route on 5 October 1980.

I had long felt that Douglas had gone a bit far in stretching the DC-9 so far (its overall length is 135ft 6in and its wheel-base 72ft 3in) and I had no great desire to fly in it. I should have known better, however; Douglas and Swissair do not often make mistakes.

I flew back from Zürich to London in a -81 after my Airbus visit. Boarding at Zürich through the front door the main-wheels looked a long way back and inside the cabin looked very long, but the crew assured my they had encountered no stability problems – this was just a long DC-9.

This time we overstepped the flight plan by 9 minutes but we encountered strong winds and were held for 6 minutes by ATC, although this gave me several good views of the Thames barrage.

I have yet to sample Swissair's latest, the Fokker 100, but that may be a pleasure to come. □

One of Swissair's Douglas DC-9-32s.

Air Traffic Management in Europe

Peter Cunningham

This article, written in May before the 1989 summer holiday season began, provides an overview of air traffic management. This is a complex and unremitting task requiring long-term planning and execution if today's problems are to be resolved.

Today there can surely be few passengers, travelling either on business or for pleasure, who are unaware of the many problems affecting the punctuality of their flights.

In this article the editor has asked me to outline the causes of delays and what is being done about them. If solutions could not be found, he said, there might be few commercial aviation develop-

Denmark has one of the most modern ATC centres in Europe with a high degree of automation. The controllers are using this to advantage and are currently reorganising their sectors to achieve substantial improvements in handling capacity.

ments to write about in the future! He added that aerospace had always thrived on challenge and indeed had solved more problems than it had created.

Our World Today

Mass holiday travel by air to sunny Mediterranean resorts boomed in the 1970s but, by the end of that decade, newspapers were showing pictures of holidaymakers stranded for hours in airport lounges. The situation improved for a few years but by 1987 it had become headline news again and 1988 was the worst year since the later '70s. 1989

certainly did not start well.

Business travellers using scheduled services throughout the year were relatively unaffected until recently. With only occasional exceptions, commuters on domestic flights, as well as those flying from Britain to the continent, have been able to count on arriving, if not on schedule, then within 20 minutes of the published times. On some routes this is no longer the case. Will the businessman also have to expect hours of delay and uncertainty in the future? Have the airport and airspace planners now reached the point at which it is no longer

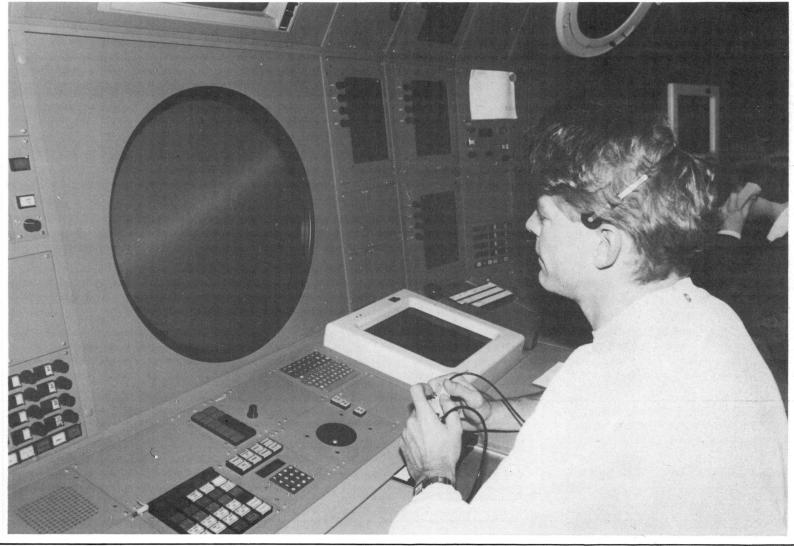

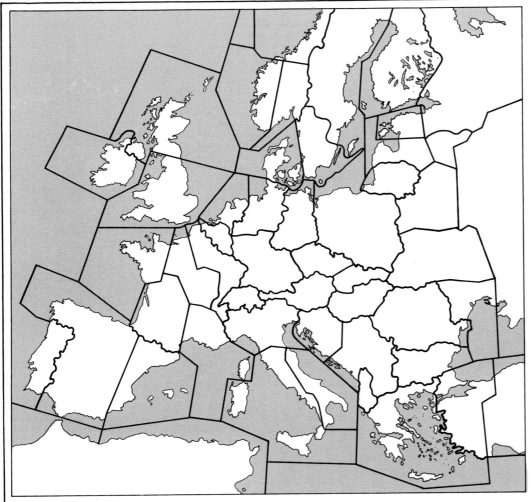

Europe's upper airspace divisions are based largely on international frontiers. Will 1993 lead to a more rational organisation?

The United States bases its airspace divisions on operational factors and is able to control traffic in an area the size of Europe with less than half the number of centres. But the US is equally enclosed by its international frontiers.

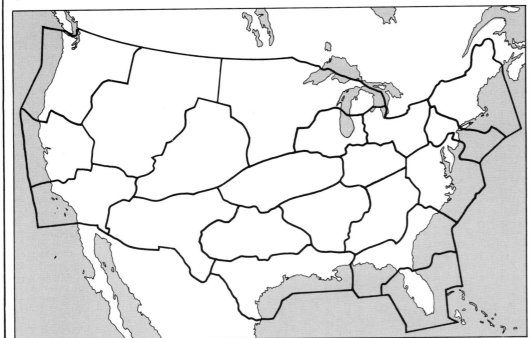

governmental agencies, parliamentary groups, airline organisations, conferences and consultants. Many recommendations have been and will continue to be made. The subject has to be dealt with on a European level so that improvements in one country are not nullified by inaction across a national border. We need all countries to work together to create an integrated air traffic service using the latest technology, planned on a systems, rather than a national, basis and managed in a fully co-ordinated manner.

Is this a Utopian ideal, impossible to achieve in the real world? Perhaps so. Difficulties abound – political, financial, technical, managerial – but a great deal could be achieved if all countries would recognise the need to have an efficient air transport service capable of responding to the public's desire to travel much more than they did in the past and to use aircraft for their holidays, businesses and personal affairs; and, more importantly, provide the necessary infrastructure for this to take place.

The Problems Identified

What, then, are the main problems that have to be overcome? Do they really start at the airports? As far as delays are concerned, the answer to this question is generally no. Airports have easily quantifiable limitations such as the runway and taxiway layout, the number of aircraft stands and the passenger flow through the terminal. Because of this, the airlines co-ordinate their timetables within these constraints.

Airline schedulers perform this difficult and complicated task at all the major airports in the world, dovetailing their programmes into a complex worldwide timetable of air traffic services every six months. This co-ordination process includes the charter operators with their European holiday programmes, as well as scheduled services.

Passengers suffer from many inconveniences at airports, and airlines have their problems, but reasonably efficient airports would work smoothly if delays did not occur for other reasons. It is when such delays snowball that airport lounges start to resemble refugee camps.

The primary causes of delays at airports lie in the organisation and management of the airspace. The various aspects of this are now examined.

The Airspace

There is no European air traffic management system. Instead there are many national systems, each conforming to international standards but at a different level of development and with their own ideas on how to improve their airspace.

possible to meet the demand for air travel in Europe?

The traveller has a right to ask not only why we have arrived at this situation but what is being done to overcome the problems. Cannot the airlines plan their flights more realistically instead of bunching them together? Would fewer, larger aircraft be the answer? Are plans being made to meet the anticipated demand? Are there enough airports and/or are the existing airports big enough? Is Britain's air traffic control system in need of modernisation? Are the other European countries playing their part? What really are the prospects for a better air transport service?

These and other questions have been exercising the minds of many people in recent months: governments, inter-

Formed in Paris in 1973, CORTA was the first ATFM unit in Europe. It is still the busiest.

These airspaces are defined by sovereignty, with national borders projected into the sky, rather than on operational requirements.

If this fragmentation is such a limiting factor, why has it not been dealt with? A brave attempt was, in fact, made some thirty years ago. The need to remove the constraints imposed on air traffic control (ATC) by national frontiers was recognised in the late 1950s. After several years of discussion, Britain, France, the German Federal Republic and the Benelux countries founded the Eurocontrol agency in 1963. Ireland joined the following year.

The original intention was that Eurocontrol would operate ATC centres responsible for handling both civil and military traffic in the upper airspace of its member states. Unfortunately, Britain, France and The Netherlands all opted out, primarily for defence reasons. An excellent centre was constructed at Maastricht in The Netherlands but the airspace it was allowed to control was restricted to a peculiarly shaped area over Belgium, Luxembourg and north Germany. It was not until 1986 that The Netherlands handed over its upper airspace to Eurocontrol. Britain and France have never done so.

The International Civil Aviation Organisation (ICAO) co-ordinates the development of the international civil aviation infrastructure worldwide. This activity has been of the greatest importance in Europe. Nevertheless, ICAO has been powerless in the face of political and military opposition. Many plans have been based, not on a European concept, but on what individual countries have wanted to do; and when some States' delegates have made sensible compromises at international meetings, these have not always been supported at home. Proposals based on solid operational needs have been frustrated.

ICAO encourages, supports and cajoles its member States but has no powers of enforcement. Because of this and a lack of funding in some cases, States have not always fulfilled their obligations to the international community.

Apart from staffing matters, the main problems in European airspace today are the imbalances that exist. Whole streams of traffic are slowed down by differing standards of technical sophistication between one country and another. The availability of radar and the use made of it is a good example. It may surprise many readers to know that there are still areas in Europe through which main trunk routes flow which have no *en route* radar coverage. This means that aircraft separation has to be increased as aircraft cross national boundaries. To accomplish this, ATC has to reorganise the traffic further upstream, thereby reducing capacity even in those countries that can apply minimum separation with sophisticated radar techniques.

Because the route network is based on point-source aids, traffic flows can be throttled by a conjunction of routes over a VOR, or two VORs lying in close proximity to one another. This even occurs in places where there are large volumes of 'empty' airspace nearby. A close examination will reveal that traffic is being funnelled through a narrow channel because of an airspace boundary, a national frontier, a military area, or simply because of the inflexibility brought about by basing a route network entirely on point-source aids sited at convenient locations on the ground. Nevertheless, neither reductions in separation nor additional routes will produce increased capacity if the control system, and the controller himself, is not able to handle the extra traffic.

The ability to handle more traffic depends on many factors such as the ATC equipment, the complexity of the traffic mix, the communications available, the staff and the working environment. Dramatic improvements have been achieved when new automated ATC centres have been implemented. Portugal is an excellent example. A few years ago traffic was severely restricted entering Portuguese airspace. Now there are virtually no restrictions, mainly because of the introduction of the new up-to-date ATC centre in Lisbon.

Capacity can also be enhanced by splitting sectors so that, for example, two controllers will handle the airspace previously handled by one. There is a limit, however, beyond which the reduction in size of the sectors becomes counterproductive because of the workload involved in the rapid transfers required between sectors.

A great deal has been written about the adverse impact of military operations on civil air transport. The civilian complaint is that too much airspace is reserved for military use and its utilisation is comparatively low. It is felt that the number and size of military areas should be reduced and that civil aircraft be permitted to fly through them at peak times when they could provide useful relief routes.

It has to be recognised, however, that the military forces in Europe comprise a large number of units from many countries, equipped with a wide variety of aircraft requiring large envelopes of space for their operational manoeuvres. Defence ministers no doubt see the role of their forces as more important than the transport of sun-seekers to sandy beaches. They may also think that airspace released today might not be retrievable tomorrow. Concessions have been made in some countries, however, and the situation is more flexible than it used to be.

Air Traffic Flow Management

Some ATC sectors started to become saturated at peak times in the late 1960s. When this happened, ATC managers resorted to flow control.

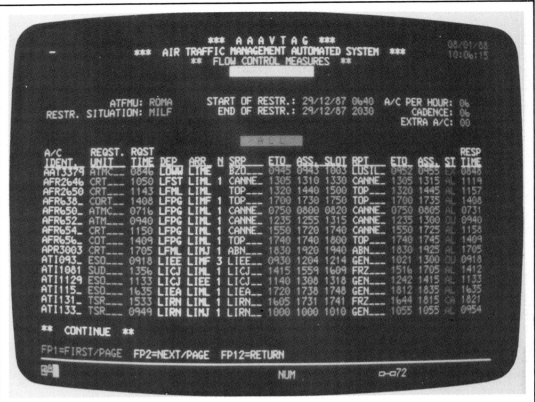

The Rome ATFM unit allocates slots for all departures from Italian airports.

ATC en-route radar display at the new Lisbon Area Control Centre which has dramatically increased the capacity of Portuguese airspace.

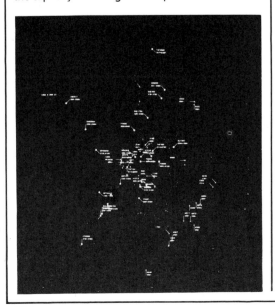

France, which because of its geographical situation was bearing the brunt of the traffic congestion, took the initiative and proposed to ICAO that a meeting of senior air traffic controllers should be held to work out ways of dealing with the problems. Representatives of the airlines were invited to attend as observers. This started a series of meetings that has continued to this day.

The principal remedies that were introduced at the start were a new routeing system, known as traffic orientation, and flight level allocations. These solutions, while helping ATC by simplifying their task, were never popular with the airlines. The early flight level allocation scheme meant that aircraft had to follow one another at 10-minute intervals at the same level instead of being able to operate on the basis of vertical separation. This reduced capacity and caused delays. The traffic orientation made matters worse, giving the operators additional mileage and very little flexibility. Both schemes resulted in fuel penalties.

The flight level allocation scheme was later dropped except where necessary to provide vertical separation at boundary crossings and route intersections or convergencies. Traffic orientation gradually became more flexible as alternative routeings were made available. Nevertheless, as time went on, the benefits of these strategic measures began to be nullified by the disorganised application of tactical flow control on the day.

Acceptance rates, designed for occasional use for short periods to overcome temporary overload situations, were imposed by individual ATC centres without any knowledge of their effect in other areas, and frequently without warning. As this practice spread, it created chaos for the airlines and delays escalated.

Upstream centres took a proportion of each acceptance rate for their own traffic until there was nothing left, after which all other departures became subject to PPO (prior permission only). This permission had to be obtained from units downstream who had not used all the slots they had allocated themselves. As most restrictions in those days originated in southern Europe, the airlines on the northern periphery, ie in Britain, Ireland, Finland and Scandinavia, found the situation intolerable.

Once again, there was no flexibility in the system. When two flows converged over the same point, each would receive a fixed allocation. The actual demand would, however, vary with time so that, when the flow rate on one stream was fully subscribed, the other might be scarcely used. Slots not required on the quiet route could not be reallocated to relieve the busy route. Consequently, even when the potential congestion was at its worst, the capacity in the restricted area itself was not fully utilised. This gave rise to the graphic description by a pilot held on the ground of seeing, as he said, empty slots flying across Europe.

Radical new thinking was called for and, in the 1970s, the concept of a centralised organisation was born. The term flow control was changed to air traffic flow management (ATFM) to emphasise the need to make effective use of the available capacity rather than merely to impose restrictions.

The original idea was to create a single ATFM unit but once again it immediately became clear that some powerful governments did not intend to allow their traffic to be managed by a unit outside their control. The word 'centralised' soon became 'integrated' and it was agreed to establish a small number of sub-regional ATFM units together with national units where required.

While ATC would continue to impose acceptance rates based on their ability to handle traffic, these would be channelled through the new ATFM service where they would be co-ordinated and used to allocate departure slots equitably.

The new arrangements were effective for a few years but gradually broke down as more and more units became involved and the combination of increased traffic and poor communications procedures exerted added pressure. Despite the new term, there was little flow management and unco-ordinated flow control restrictions continued to be imposed. Many State delegates at international ATFM meetings paid more attention to protecting their own areas than to seeking the best ways to expedite the traffic flows across Europe.

Today many delays are once again caused, not by the lack of ATC capacity, but by the inadequacies of the ATFM service itself. In some cases, for example, departure slots requested two hours in advance – the maximum permitted time – have taken so long to process that they have been received after the aircraft should have taken off. The slot request procedure has then had to be started all over again resulting in a two- or three-hour delay.

Recent statistics show that, once again, the northern-based airlines are suffering the most. In 1988, delays on flights to Mediterranean destinations were four times longer from Britain, and three times longer from the Benelux countries, than they were from France.

Now, at last, the need for centralisation has been accepted. After a six-month study in 1988, ICAO agreed that there should be two Central Executive Units with responsibility for Eastern and Western Europe respectively. These units, to be staffed internationally, will be responsible for the planning, co-ordination and execution of all ATFM measures in Europe. They will be in direct communication with the ATC centres in, and adjacent to, the European region; and, of course, with the aircraft operators. Their introduction will be accompanied by the withdrawal of the present twelve units.

Eurocontrol has been commissioned by the transport ministers of the twenty-two member States of the European Civil Aviation Conference (ECAC) to build the central unit in the West. Work is now well under way and is scheduled for completion in 1994. In the interim, the central functions are being jointly handled by five of the existing units: Frankfurt, London, Madrid, Paris and Rome, who will gradually release their responsibilities to Eurocontrol.

A Common Data Base

One of the main problems for ATC controllers and ATFM managers alike has been that they have had a very imprecise knowledge of what traffic to expect. Although Germany established its own system for prior notification, most centres had to rely on past records and intuition.

Until a few years ago, for example, Greek controllers had little idea of the numbers of flights bound for their airspace until the flight plans were received shortly before the aircraft crossed their boundary. Sometimes, flights arrived before the flight plans. This was not too difficult to deal with in the days of comparatively light traffic but it became increasingly problematic over the years.

There were calls for data bases from several countries and it was agreed to establish a central data bank (CDB) in 1980. Eurocontrol took on this task and the CDB, now known as the Data Bank Eurocontrol (DBE), became operational in 1987.

Intended for strategic planning purposes, the DBE can predict end-to-end traffic with considerable accuracy. This is not enough, however, for the complex and peaky traffic patterns in Europe. What ATFM managers need is an accurate picture of the hourly load on each sector and over each critical point for use in the immediate pre-tactical phase.

Consequently, Eurocontrol is now embarking on a further development of their data bank to collect and process flight plans on the day of operation. Incorporated with the western executive unit, this combined facility will provide a Central Flow Management Unit for the ECAC member States.

Rising Demand

For the past fifteen years, Eurocontrol has kept a record of the numbers of IFR movements in the area for which they have been collecting *en route* charges: Austria, Belgium/Luxembourg, France, West Germany, Britain, Ireland, The Netherlands, Portugal, Spain and Switzerland. (IFR traffic, operating in accordance with instrument flight rules, includes the vast majority of all air traffic in Europe. It excludes most private and sports flying, some air taxiing, helicopters and military aircraft flying at low level or within their own operational areas.)

The total numbers of annual and peak day IFR flights and the growth rates for the past fifteen years are shown in the histograms below. Passenger growth is also shown for comparison.

There was a five-year period, 1978–83, when virtually no growth in the number of movements took place; by contrast the past two years, 1987 and 1988, have seen exceptional increases. Therein lies one explanation as to why the planners have been caught out.

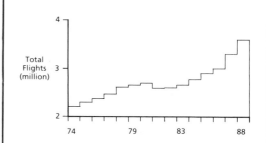

Total IFR flights 1974–88 (Eurocontrol data covering Austria, Belgium/Luxembourg, France, West Germany, Britain, Ireland, The Netherlands, Portugal, Spain and Switzerland).

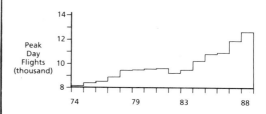

Peak day movements 1974–88 (Eurocontrol data).

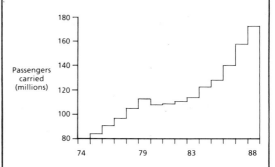

Intra-European passengers, 1975–88 (compiled from data provided by the Association of European Airlines, AEA, and the European Civil Aviation Conference, ECAC).

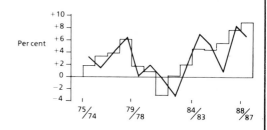

Aircraft movement growth rates, 1974–88 (Eurocontrol data). Blocks show annual total rates, graph line shows peak day annual rates.

Furthermore, the introduction of the integrated ATFM service in the early 1980s, accompanied as it was by a marked decrease in aircraft delays, may unfortunately have acted as a disincentive to ATC planning. This was the period also when, after many European governments introduced a 100 per cent cost recovery programme from the aviation users, the International Air Transport Association (IATA) was exerting pressure on them to curtail expenditure.

As a result of this combination of factors, development plans were slowed down and in some cases, such as controller recruitment, stopped altogether in some countries.

At the other end of the scale, few foresaw the rate of increase in traffic that started in 1987 even though many will now say that the signs were all there.

As five years may be taken as the minimum lead time for any substantial developments, including controller recruitment and training, we are now going through an extremely difficult period.

ATC Improvements

This is not to say that ATC planners have done nothing. By and large they want to provide a good service in their area of responsibility. Unfortunately, they are often hampered by political vacillation and a lack of human resources as not enough national effort is put into the planning function in many countries.

Nevertheless, many ageing ATC centres have been replaced by modern automated centres in the past ten years or so, eg at Brussels, Copenhagen, Geneva, Karlsruhe, Lisbon, Manchester, Prestwick, Rheims, Rome, Vienna and Zürich. Madrid opened its new centre this year. Many others have been, and are being, updated with latest technology.

Radar coverage has been increased, eg in Italy, Spain and Yugoslavia, and, when radar replacement programmes have become due in other countries, the quality has been improved by introducing monopulse radars, automatic radar data processing systems and better displays for the controllers.

The principal ground-based aids to navigation, approach and landing, VOR, DME and ILS, have all been continuously upgraded since they first appeared on the scene. ILS, for example, now has a level of accuracy and performance far exceeding the levels of twenty years ago. As individual facilities are replaced, advantage is taken of new technology.

The route network has gradually been improved over the years through the ICAO planning processes and the military have released some of their areas on weekends, permitting additional route-

Aircraft Size
(Percentage of total IFR flights by category)

Category	Seats	Percentage of total flights 1980	Percentage of total flights 1988
Commuter	Up to 40	10.0	16.2
Commuter	41–70	9.0	9.7
Short-range	71–100	6.8	4.6
Short-range	101–150	9.4	16.2
Medium-/long-range	90–130	4.3	1.8
Medium-/long-range	131–160	15.9	11.4
Medium-/long-range	161–200	18.5	12.4
Medium-/long-range	201–350	5.6	9.3
Medium-/long-range	Over 350	4.0	4.3
Supersonic	–	0.1	0.1
All-cargo	–	1.6	1.0
General aviation	–	9.9	8.4
Helicopter	–	0.3	0.5
Military (on civil routes)	–	4.8	4.1

Source: Eurocontrol

The average size of commercial aircraft operating in Europe has decreased since 1980 despite more than a 50 per cent increase in the number of passengers carried.

ings to be used at these peak times.

ICAO and IATA have made studies of the practical application of area navigation (RNAV). This enables aircraft to fly direct tracks, bypassing turning points on the fixed network. The first and current stage is called Basic RNAV. It involves a track-keeping accuracy of plus or minus 5 nautical miles for 95 per cent of the flight time. This is similar to that attained today using VOR/DME where VORs are less than 100nm apart. The next objective is Precision RNAV with a track-keeping accuracy of within 0.5nm. This is already achievable by aircraft capable of continuous automatic DME/DME position fixing.

IATA's objective is to have RNAV much more widely recognised and used than it is now. This will depend on the co-operation of ATC and on the suitability of airline fleet equipment.

One of the present stumbling blocks to progress is the relatively poor quality of the aeronautical communications system. The fixed network (AFTN) is well behind the state of the art and the implementation of an improved version known as CIDIN has been continually delayed. Unfortunately, the vital importance of the communications element is not always fully appreciated.

Equally important is the need to deal with the small deficiencies. A new telephone link, an additional frequency, an alternative route segment, an updated ATC co-ordination agreement between

adjacent centres can all add small increments to the capacity of the system. For this purpose, Eurocontrol has established a special activity in which an expert team takes a fresh look at ATC operations area by area.

The Airports

Having already dismissed airports as not being primarily responsible for the poor performance of air traffic management and although this article deals mainly with the airspace, it would not be complete without some reference to the problems caused by airports.

The situation as regards airports is fundamentally different from that relating to airspace. External influences such as environmental factors and access by surface transport play a much larger role. There is much less government influence in the planning, management and operation of airports than in the case of airspace. Airports have few international obligations and many have in the past paid scant regard to internationally agreed requirements as set down in the ICAO regional plan.

Airports in one country do not have to be compatible with those of their neighbours and may even compete with them for business. This competitive element also occurs within some countries as in Britain. Many airports are either commercial enterprises or are closely related to commercial activities such as tourism.

Nevertheless, the expansion of capacity to meet demand is very often left far too late. This is partly because of political and environmental constraints, coupled with the long lead times involved in making major developments. The desire of airport owners to maximise their investments before embarking on new projects in a largely monopolistic world also plays a role.

A major problem for the charter operators who carry the bulk of the holidaymakers in Europe is that few passenger terminals were designed to handle this type of traffic. Road access, queueing space at check-in counters and holding areas, etc, are quite often inadequate for this market. Scheduled operators suffer as well. Poor baggage handling systems, for example, limit the effective capacity of a whole airport.

A growing number of international aerodromes have reached the limit of their present capacities and are saturated throughout the day. Thus airport limitations restrict the growth of air travel. The future of commercial air transport in Europe depends as much on the airports as it does on the airspace.

The Operators

IATA has said that delays in 1988 wasted 150,000 flying hours. One European airline alone estimated the cost of its delays that year to be over £30 million.

Much has been written about the inadequacies of airports, ATC, ATFM and governments. Can the airlines be blameless? Could they do more to help themselves?

In a paper written by the British Civil Aviation Authority (CAA) in 1988, it was suggested that the airlines should:

● make better use of available off-peak capacity
● build more recovery time into their schedules
● have more spare aircraft available

BAA, formerly the British Airports Authority, and others have commented on the need to increase the size of aircraft if the present airport congestion problems are to be overcome.

In putting forward their ideas, the CAA pointed out that the costs involved in the last two suggestions would need to be set against the savings that would result from a reduction in delays and the loss of revenue which could result from the deterrent effects on future demand.

Leaving aside the obvious comments on the inherent nature of demand for any form of transport in a competitive market, the fact is that diurnal peaks have already been spread so much that slots for additional services are unobtainable from a number of European airports at any time of the day.

There is room for spreading the load more evenly throughout the week at some holiday resort airports where the number of movements is very high on certain days of the week and very low on others but this is easier said than done.

The package tour business is geared to fixed patterns throughout the season, with accommodation changeovers being made on set days of the week for each tour group.

Building more slack into turnround times does not stand up to close examination. How much slack? ATC delays average 25 minutes but many flights operate on time and some are delayed over two hours. An extra 25 minutes on all turnrounds would jam airport aprons and severely reduce their capacity; and it would do little to help in the case of longer delays. It would also require considerably more aircraft.

Airlines are already buying more aircraft than they would need if the system operated efficiently but this is not so simple a remedy for an airline serving a large number of airports with various aircraft types to match differing markets.

A transport service that constructs its timetable on being late would be a laughing stock. It is a sad reflection on those responsible for providing air traffic services that they should put forward such solutions. Like it or not, the public and the legislators are demanding a better air transport service at the lowest possible fares. This requires efficient airlines and an efficient infrastructure, not an inefficient infrastructure that can support only an inefficient and costly airline operation.

The question of aircraft size is altogether more serious. Many air traffic and runway problems would be solved if there were fewer and larger aircraft. As far as ATC is concerned, each IFR aircraft requires a block of airspace regardless of its size. Runway capacity is reduced when slow aircraft intervene between jet arrivals and departures.

Liberalisation and market forces, however, require a wider range of origin and destination airports than before at an acceptable frequency. Because of this and with the common market approaching, airlines have ordered large numbers of short-range 100/150-seat aircraft. These now account for 16 per cent of the total number of IFR flights compared to 9 per cent in 1980. There is also a growing number of commuter operators and air parcel services (up from 19 to 26 per cent since 1980). Although wide-bodies increased their share from 10 to 14 per cent, there was an overall drop in the numbers of 160/450-seaters from 28 to 26 per cent of the total. By contrast, the number of passengers carried increased by over 50 per cent.

The average size of aircraft operating in Europe has therefore declined since 1980 in contrast to the historical trend when aircraft size and passenger numbers increased roughly in proportion to one another. This turnabout is partly responsible for the very rapid increase in the number of aircraft movements that has taken place over the past few years.

There is no doubt that the average aircraft size will have to increase in the future if forecast demand is to be met. This will ease the strain on the airspace, the runways and the larger airports but it will create problems for the smaller ones. It is much easier for them to absorb a frequent delivery of small batches of passengers than occasional bursts of a few thousand.

Staffing Problems

Of all the difficulties involved in air traffic management – a fragmented airspace structure, bottlenecks, congested airports, poor co-ordination and increased demand – staffing problems constitute perhaps the most difficult. Some observers believe that, if these could be overcome overnight, there would be few delays on the morrow despite all the technical shortcomings.

Today most countries in Western Europe are short of controllers. This is a major source of delays, notably for flights to and from Britain. It adds to the potential overload situation and inhibits capacity improvements.

The diurnal spreading of the traffic, together with the increased use of smaller regional airports, has caused difficulties for ATC in some places because they are unable to provide this more widespread service. When the airlines raised the possibility of reducing the night curfews at an ICAO meeting, one State refused to discuss the subject, not because of environmental issues but because ATC would not be able to man the extra sectors.

The results of the cutbacks in recruitment in the early 1980s have been exacerbated by the increased establishments that have been agreed in some States, eg to relieve stress. The old (1980) staff levels have been reduced by natural wastage and the total work required has increased with the volume of traffic.

This combination of factors has resulted in deteriorating morale, frequent disruptions and a general lowering of the quality of service. It is very regrettable but nevertheless true to say that causing delays to aircraft has become an industrial weapon. It is continually being used.

What Does the Future Hold?

What prospects do we have of seeing major improvements? Will holidaymakers stop dreading those horrendous delays which can ruin a short holiday? Will businessmen be able to count on an efficient and punctual air transport service? Will we be able to plan weekend breaks by air confident that we can get where we want to go without too much hassle and, what is more, get back on schedule? Will the new centralised ATFM system produce results? Will the advent of the common market in 1993 lead to a unification of the airspace?

There are no signs yet that 1993 will affect the organisation of the airspace as such but this may come about gradually in the course of time. In the meantime it will hopefully stimulate other planned international developments such as:

- the centralised ATFM service
- the Eurocontrol common medium-term plan
- the ICAO future air traffic management concept

Centralised ATFM will not increase the capacity of the system, which is what we really want. Nor will it solve the staffing problems. All it can do is to use the available capacity to the best advantage. Nevertheless, this is of very great importance and the centralised system will make an enormous difference if it goes ahead as planned.

The political will exists (at present). The technology exists. There is a working example of a central facility in the United States (albeit in a completely different ATC environment). Eurocontrol is fully capable of building and operating a Central Flow Management Unit in Western Europe with the emphasis on management rather than restrictions. But will the many countries involved co-operate at the working level? Will ATC in the field work in harmony with a central ATFM organisation? Will the existing ATFM units withdraw and hand over their excutive role? Or will this be another grand failure?

A European ATC System

Following an initiative by Eurocontrol a special ICAO working group has developed an air traffic management concept for Europe for the period up to the year 2015. This is based on a number of principles. Those most relevant to this article are summarised in the box below. It will be seen that they deal well with the main faults that now exist.

---

1. The future system shall provide regularity and efficiency commensurate with the demands of all users, with the same or better safety levels than at present.
2. It shall be developed on a systems, rather than on a national, basis but it should be adaptable to the requirements of the various parts of Europe.
3. The design, organisation and operation of the system shall be such as to eliminate the impact of airspace boundaries on the users.
4. The system shall be based on an area control concept rather than on an airway network and shall provide the users with maximum flexibility.
5. The primary emphasis will be on the ability of the ATC system to cope with the traffic demand rather than on the ability of the ATFM system to adjust the demand. Nevertheless, it may not be practicable to provide for the extreme peak levels of demand.

As a complement to this, the Eurocontrol member States have agreed to develop a Common Medium-Term Plan to overcome the problems experienced when each country plans its future in isolation. This plan will go into considerable depth on both technical and operational aspects of the future system.

This is exactly what is needed. With the growing membership of Eurocontrol, it should provide the basis for developing a harmonised Eurocontrol air traffic management system based on the agreed ICAO principles.

Do We Have Time?

The lead time for any significant step forward varies from five to ten years. If we add to that the ten years required to amortise the investment, we should plan now for the period 1995–2010. But forecasting future demand so far ahead is not easy. Forecasts being made today are very different from those made even three years ago.

In 1986 Eurocontrol was forecasting an annual overall growth rate of 1.5–2.5 per cent in total movements. These figures have now been more than doubled. Their present hypotheses for 1995 and 2010 are shown in the table below.

It is numbers like these that have alerted everyone to the magnitude of the problem. Some would suggest that all three hypotheses are now too high. Nevertheless, it is really a question of timing rather than magnitude. If the total of 5 million movements is not reached by 1995 it surely will be a few years later.

As plans invariably take longer than expected to be implemented, it would seem to be wise to plan on the higher rather than the lower side. The worst that could then happen would be that capacity would be ahead of demand: a pleasant change for passengers and airlines alike.

Technical Developments

ICAO has set the objectives, worldwide, for the communications, navigation and surveillance (CNS) elements of the future system. These comprise the technological framework within which air traffic management has to be developed.

The automation of both airborne and ground systems will play a major role and both systems will be integrated with one another. The airborne components will basically operate independently with the ATC elements obtaining precise information from the aircraft through an automatic data link interface. The control function will, however, remain with ATC on the ground.

ATC will be greatly improved with better workstations, high definition colour radar displays, advanced data processing, conflict resolution software, automatic track deviation warnings, etc. Controllers will be able to provide a management service rather than simply ensuring safe separation.

Routine communications between ATC and aircraft will be by data link. This will primarily be achieved through SSR Mode S, the latest type of secondary surveillance radar that will be introduced in Europe at the turn of the century, or via satellite. VHF is not so suitable for data exchange but will still be required for voice communications.

The new CIDIN (Common ICAO Data Interchange Network) should be in place by the end of the century. In the meantime, SITA, the airline network

---

Forecast Growth 1988–2010
(Aircraft movements per annum (million)

Hypothesis	1988 Actual	1995 Number	1995 Increase	2010 Number	2010 Increase
Low	–	4.6	+30%	7.0	+94%
Most probable	3.6	5.3	+50%	10.0	+177%
High	–	5.8	+60%	13.0	+260%

Source: Eurocontrol

Eurocontrol forecast a 50 per cent increase in aircraft movements by 1995; 177 per cent by 2010.

which is already very efficient worldwide, will be implementing its own aeronautical satellite data system.

Navigation will be based on Precision RNAV with aircraft having to satisfy a 'required navigation performance capability'. RNAV computers will be able to interrogate up to five latest technology DMEs to determine an aircraft's position with the greatest accuracy. There will also be a global satellite navigation system. VORs will become largely redundant and be gradually withdrawn.

Flight Management Systems (FMS) will have been further developed and will be standard fits for the majority of business and commercial aircraft. They will be linked to the RNAV computers to provide extremely accurate navigation. Speed control techiques will be introduced and flight profiles will be flown far more precisely than they are today in all four dimensions (4D), ie including the time factor. These improvements will enable reductions to be made in the tolerances required by ATC to ensure safe separation.

Surveillance will be undertaken both by SSR Mode S and satellite, the latter being particularly useful to cover the less densely flown areas.

If an efficient and reliable workforce can be assured, these technological developments will provide a quantum increase in the capacity of the airspace.

This will place great pressure on the airports which may well become the elements that limit the whole system in the future.

National Plans

So much for the next century. What about the 1990s? This is where the Eurocontrol common plan, based on the initiatives taken nationally, comes in.

In Britain, the National Air Traffic Services (NATS) has developed a new concept called the Central Control Function. Restructured sectors will cover flight paths rather than limited boxes of airspace. Controllers in terminal areas will handle flights throughout elongated one-way tubes, reducing the need for controller-to-controller co-ordination when flights are handed over from one box to another in the traditional way.

This will soon be introduced in the London ATC centre. When fully implemented in 1995, it is claimed that it will increase capacity by at least 30 per cent and perhaps considerably more. If this is so, it is an innovation that could well be copied elsewhere. As far as ATC organisation is concerned, many planners believe that improved sectorisation such as this, coupled with the application of the latest technology, offers one of the

best hopes for capacity improvements in the medium term.

NATS is also working on a very advanced *en route* centre which should be operational in the late 1990s. In the meantime, the Manchester sub-centre will be expanded and equipped with a new radar data processing system. The Scottish centre is next on the list for redevelopment. These will all add further to the capacity of the airspace.

Denmark is currently reorganising its sector following the introduction of its new centre in 1988. This is also expected to bring about major improvements.

Swiss ATC planners are working on another idea. This is to use flexible sector boundaries which can be varied at comparatively short notice according to the traffic characteristics. This would reduce the need, for example, to hold arriving aircraft. By expanding a sector horizontally or vertically for a period of time, flight profiles could be varied to make maximum use of the limited airspace in the terminal area minute by minute.

France has been very much in the forefront of technical developments in ATC. They introduced their conflict alert facility 'safety net' nearly ten years ago. Their fourth generation ATC computer has been in operation since 1984 and sub-systems are being introduced all the time. A completely new computer will come on-line in about five years.

Some interesting computer-based descent and approach sequencing systems are under development in France, West Germany and Eurocontrol. These trajectory management techniques take account of the aircraft performance, the upper winds and temperatures, etc, and automatically meter the arriving traffic to optimise use of the landing runways.

Other research and development is taking place in many fields such as airborne collision avoidance systems, data link technology, data processing techniques, flight data displays and on the important man-machine interface.

At the other extreme, however, good operational master plans developed by the ATC services in some parts of Europe gather dust waiting for political decisions to be made or implemention action to be taken.

The Bottom Line

The question now arises as to whether the developments described in this article will result in an efficiently managed air traffic service in Europe capable of handling future demand.

There may come a time in the future when the total system is completely saturated and human ingenuity is unable to enhance capacity any further. But this

is not the case now, nor will it be for many years to come.

The achievement of a sound air traffic management service is possible but it will need concerted and continuing determination by all concerned. A few could negate the efforts of the many. All elements of the system must be developed so as to be in balance with one another. This is no time for sheltering behind other people's problems for, when these have been resolved, the complacent, the lazy and the indecisive will find themselves with a five- or ten-year backlog to make up.

The cost of making the required changes cannot be ignored but the mood of the airlines at the present time is that the delays and waste are costing them far more than the price of the improvements discussed in this article. Funding should not be the problem. Money is available for such projects, the costs of which can be recuperated from the operators through *en route* charges. This does not mean that the airlines are giving *carte blanche*. Cost benefit studies should be a routine part of all planning processes and the operators should be involved.

Governments must fulfil their responsibilities to the public. They have a duty to live up to their policies of liberalisation by providing the necessary resources. □

WE NEED:

- integrated European planning
- continued political support
- firm decisions
- resolution of industrial problems
- more controllers
- full radar coverage
- latest technology ATC centres
- state-of-the-art communications
- area navigation
- centralised flow management
- balanced airport planning
- military co-operation

The Rise of Embraer

David Godfrey

Brazil, with a pioneering tradition going back to Santos Dumont, has since 1969 through Embraer turned its attention to producing aeroplanes of world class for specialised and growing markets at home and abroad. The foundations, however, were laid in the 1940s with training programmes for aeronautical engineers backed by an institute for research and advanced studies.

Concepts that have no direct competition place Embraer* in unique positions in two separate world markets, one military and one civil, typifying the Brazilian company's far-sighted approach. If these were mere projects with little chance of realisation, they would be worth only their weight in brochures, but Embraer has demonstrated the knack of putting its ideas into practice for both domestic and international customers in military and commercial applications, witness the successes of the Bandeirante, Tucano, Xingu and Brasilia.

So both the Embraer/FAMA CBA-123 twin-pusher pressurised feederliner and

*Empresa Brasileira de Aeronáutica SA.

the new MFT/LF fighter trainer/light fighter stand excellent chances of carving new market niches. Before looking at these latest developments, let us consider how Embraer came into being and how it developed to its present rank as a world-class aircraft manufacturer.

In the Beginning

Before Embraer was founded in 1969, there had been a long history of Brazilian aviation activity going back to the

Some buttresses of the business. The Tucano trainer, Xingu pressurised VIP transport, Bandeirante nineteen-passenger commuter aircraft and the thirty-seat pressurised Brasilia for a similar role. All are powered by Pratt & Whitney Canada propeller-turbines.

pioneering work of the celebrated Alberto Santos Dumont. One of Brazil's national heroes, he flew his airships around the Eiffel Tower in 1901, and in 1906, also in Paris, demonstrated controllable heavier-than-air flight in his 14bis tail-first aeroplane. Since all his flights were made before large crowds and were officially observed in a foreign country, it was not surprising that the barely witnessed flights of the Wright brothers in 1903 were given scant credence in Brazil. Also, the Wrights used a take-off ramp and, although they used assisted take-off only in later flights, it was wrongly assumed by many that the first flights also had the benefits of such a launching method. Thus, the claim for Santos Dumont to have made the first *unassisted* take-off and successfully controllable flight persisted and remains an article of faith for Brazilian historians to this day.

In 1907, Santos Dumont demonstrated what today would be termed an ultralight aeroplane, the tiny Demoiselle monoplane, a replica of which starred, over half a century later, in the film *Those Magnificent Men In Their Flying Machines*.

The suburban airport in Rio de Janeiro is named after Santos Dumont. The Brazilian Aerospace Museum at the former air force aeronautics school at Campos dos Alfonsos, near Rio de Janeiro, contains reproductions of the Santos Dumont 14bis and Demoiselle,

Men of Embraer

Ozires Silva, chairman.

Ozílio Carlos Da Silva, chief executive.

Guido Fontegalante Pessotti, technical director.

Irajá Buch Ribas, production director.

Heitor Fernandes Serra, commercial director.

with many items of personal memorabilia. His ingeniously designed summer house (the Enchanted House) is preserved as a very popular museum in Petropolis, near Rio, and there is a Demoiselle restaurant in a São José dos Campos hotel, near São Paulo. Not only was Santos Dumont the father of aviation in Brazil, but he is said to have invented the wristwatch by requesting such a device from Cartier in Paris as a convenience when flying exposed to the elements, and to have invented the aircraft hangar and other devices including the bathroom shower.

Over the years, several attempts were made to develop training and sports aeroplanes in Brazil, happily some of which survive in the aerospace museum which opened in 1976. The first aeroplane to be put into series production in Brazil was the M7 two-seat aerobatic training biplane designed by Col Antonio Guedes Muniz, manufactured between 1935 and 1943. The Ypiranga EA201 tandem two-seat high-wing monoplane flew in 1941 and was built later as the Paulista CAP-4 Paulistinha before being taken over by Neiva.

A very familiar aeroplane in the museum is the Fairchild PT-19B Cornell trainer of which 232 were built by the Fábrica do Galeão to supplement 170 delivered to the Brazilian Air Force in 1942/43 by the US Army Air Force. Several North American T-6 Texans are on display including one of eighty-one built by the Lagoa Santa Aircraft Factory between 1946 and 1951. Another licence-built primary trainer shown is the Fokker S.11, ninety-five of which were made by Galeão as the T-21 after assembling five from kits. Fifty nosewheel-undercarriage Fokker S.12s were built as the T-22.

Neiva (which was to become a subsidiary of Embraer in 1980) constructed the L-6 liaison derivative of the Paulistinha in the mid-1950s. The Aerotec T-23 Uirapura trainer on show was used by the air forces of Bolivia, Paraguay and Brazil. Another trainer at the museum is the Neiva T-25 Universal, designed to replace the Fokker S.11 and S.12, and the North American T-6. The T-25 is particularly interesting as an ancestor of the Embraer EMB-312 Tucano because both were designed by teams led by Ing Joseph Kovacs, the T-27 Tucano being a replacement for the Cessna T-37C. Neiva is also represented by the L-42 Regente liaison version of the C-42 utility aeroplane. The L-42 replaced Neiva L-6s and Cessna O-1s in the Brazilian Air Force. It was historic in being the first all-metal Brazilian aircraft.

There are many aircraft of foreign origin in this most attractive museum at Campos dos Alfonsos, including the sole

surviving example of a Focke-Wulf Fw 58 Weihe twin-engined bombing and navigational trainer. But the most significant aircraft built in Brazil is the Bandeirante. This is a YC-95, the first of three prototypes designed as the IPD-6504 light-twin transport at the Centro Técnico Aeroespacial by a team led by the French designer Max Holste (who was earlier responsible for the Broussard and the Super Broussard that was developed into the Aérospatiale Nord 262, Aérospatiale Frégate and Mohawk 298).

Founding a National Industry

With formation of the Brazilian Air Ministry in 1941, it was obvious that a training centre was needed for aeronautical engineers, backed by an institute for research and advanced studies. Planned in 1945, the heart of the Brazilian aerospace industry is the Centro Técnico Aeroespacial (CTA) at São José dos Campos. The first of five research institutes for aeronautical technology (ITA – Instituto Tecnológico de Aeronáutica) was completed in 1950.

The other four CTA institutes are for research and development (IPD – Instituto de Pesquisas e Desenvolvimento), space activities (IAE – Instituto de Ativadades Espaciais), industrial coordination and development (IFI – Instituto de Fomento e Coordenação Industrial), and advanced studies (IEAv – Instituto de Estudos Avançados). There is also a civilian institute for space activities (INPE – Instituto de Pesquisas Espaciais).

Undergraduate and graduate studies plus basic research are done at ITA while programmes of applied research and development are the responsibility of the IPD, IAE and IEAv for all activities concerning certification of aerospace products. About 7,000 people work in CTA, 80 per cent of them civilians.

The prime aims of CTA are education, research and development, and technology transfer to industry, a mandate undertaken with great success. The main institute, ITA, provides five-year degree courses plus graduate programmes and some doctoral-level courses. After two years on fundamental subjects, undergraduates spend three more years in professional specialities – aeronautics, electronics, aero infrastructure (civil engineering applied to airports, airfields and facilities), mechanics (propulsion and energy conversion) and computers. Normal student total is about 1,000 of whom 400 are graduate students.

Entry is by competitive four-day examinations held nationally in a dozen major cities and administered by ITA. Careful screening of applicants ensures that the very best are selected for award of complete scholarships – all tuition, food, on-campus housing, medical and dental care. Most of the 180 professors also live on campus. There is a similar total of administrative and support staff. ITA uses the 'bootstrap' concept of developing good bachelor-degree students as instructors and encouraging them to obtain higher qualifications to achieve the level of assistant professor, some going abroad for doctorates although an increasing number pursue the highest education in Brazil.

Very close contact is maintained between students and faculty in a tutorial manner, the ratio in later years being about five students to one professor. However, average class size for lectures is about twenty-five and in the first two years of fundamental subjects the classes are of sixty to eighty students who have thirty-two hour weeks for sixteen-week semesters. The senior thesis involves a full-year project as was the Bandeirante. Although ITA belongs to the Air Ministry, only about 10 to 15 per cent of students enter military service as a career. Students have to complete one to two years of officer-training corps service.

Before the advent of ITA, Brazilian education was based on the classic French academic pattern. ITA is deliberately different in that great emphasis is placed on practical laboratory-type classes including machine shop, fitting shop, welding and other workshop courses. Data processing was introduced as a subject in the late 1960s and, more recently, CAD/CAM was designated a 'frontier' area. The graduate-degree programme was instituted in 1961 and includes an air transport degree.

The higher levels of education involve close contact with Embraer and, in turn, ITA provides the aircraft company with most of its technical personnel. This helped the Bandeirante to get going as a national project by ensuring that there was an adequate flow of human resources from ITA to industry. Avibras, the company concentrating on missiles, rockets and other weapons, benefited similarly.

Overall, there is close integration and co-operation between the CTA institutes on research, training and certification for military and civil requirements.

Embraer is Created

A 1965 CTA market survey indicated a gap in the range of transport aircraft then available. The airlines were quick to take up each new type as it was introduced and it was clear that the development of air transport was limited by the speed and size of equipment available.

It appeared that there were several market niches of speed/size combinations and that one such niche, for what is now termed a feederliner, was vacant. Thus was born the IPD-6504 (fourth project for 1965 of the CTA research and development institute, the IPD) project design which was to evolve into the Embraer EMB-110 Bandeirante. It was aimed at developing a transport aeroplane for small regional airlines to feed passengers to larger carriers for long-haul flights. In parallel, there was an obvious domestic application within the vast area of Brazil for a small and fairly fast aeroplane that would be really useful for both civil and military purposes. In the latter role, it would replace the Beech C-45 Expediter, the military development of the Beech Model 18.

Ozires Silva, who was in charge of the IPD-6504 project and later became Embraer chairman, explains that there was some criticism for choosing propeller-turbine engines rather than turbojets, but time and the energy crisis proved the wisdom of the chosen powerplant and aircraft configuration. The design was developed methodically and at one point enlarged and extensively revised. This explains the period between first prototype flight in 1968 and first flight of the definitive production configuration aeroplane in 1972.

The EMB-201 Ipanema agricultural aeroplane had been designed to specifications of the Brazilian Ministry of Agriculture by the IPD, the prototype flying in 1970 while an improved model flew in 1977. In 1981, production was transferred from CTA to the Neiva division of Embraer.

To replace Lockheed T-33 jet trainers for the Brazilian Air Force, Embraer started to build the Aermacchi MB-326 as the EMB-326 Xavante in 1977 and used the opportunity to increase its technical experience of building high-speed aircraft. In all, 182 Xavantes were manufactured. Silva is the first to acknowledge the tremendous amount of assistance provided by the Italian company which also helped to prepare for Bandeirante development in terms of overall technology transfer. This relationship also helped later to further the AMX international programme.

Rather than launch a new line of general aviation aircraft in a market already dominated by a few large companies, Embraer decided in 1974 to build under licence several Piper designs at its Neiva division.

As its first pressurised design, Embraer developed the EMB-121 Xingu corporate and military VIP propeller-turbine transport which flew in 1976. More than 100 were built for civil and military customers in several countries and a great deal was learned that would be invaluable for the Brasilia.

Development of the EMB-312 Tucano propeller-turbine trainer was the next major programme, with first flight in 1980. Also in 1980, an EMB-120 mock-up was exhibited in Rio de Janeiro. A pressurised twenty-four-seat successor to the unpressurised nineteen-passenger Bandeirante was projected in 1974 to have Pratt & Whitney Canada PT6A-45 propeller-turbine engines with five-blade airscrews, but was shelved because of pressure of other work. Under the same designation, a completely new design for a thirty-seat aeroplane was projected in 1979 to take advantage of advances in propulsion, aerodynamics, materials and structures, and to meet the latest certification standards. This became the EMB-120 Brasilia.

The latest Embraer production programme is the AMX strike fighter.

Embraer at Twenty

The visitor to Embraer at São José dos Campos, 50 miles from São Paulo and 220 miles from Rio de Janeiro, sees a bustling, extremely clean and neat plant employing more than 12,000.

When Embraer was founded in 1969, the Brazilian Government owned 82 per cent of the shares and proceeded to offer them instead of 1 per cent of the income tax owed by other Brazilian companies. Over 240,000 companies invested in Embraer and have received dividends since 1974. By 1983, federal ownership of stock was only 5.7 per cent, although a majority of voting shares was retained. Negotiations are proceeding with financial institutions to increase the capital. Debt conversion will be used to transfer ownership to the banks. Future expansion may depend on public share offerings to raise equity.

Perhaps the simplest explanation of Embraer's success is the statement by chairman Ozires Silva that it was very deliberate and based on realistic planning and careful execution, leading to the establishment of a range of production aircraft.

In its twentieth year, indeed, Embraer has a remarkably diverse range of aircraft in production, both civil and military; a military international programme; a civil international programme; a new project that may well become another international venture; and a number of other aerospace activities.

Bandeirante

In production since 1972, the EMB-110 Bandeirante nineteen-passenger feederliner has two 750shp Pratt & Whitney Canada PT6A-34 propeller-turbine engines. Nearly 500 have been delivered to customers in twenty-seven countries, the largest single operator being Westair of

Landmarks for Brazil. In the background, Rio de Janeiro's Santos Dumont city airport with the celebrated Sugar Loaf mountain only 2 miles from the parallel 02 runway thresholds. In the foreground two of the three Embraer Bandeirante prototypes, all of which have been preserved and are on display, reminders of Brazil's rich aviation heritage. With 500 built, production of the Bandeirante ended in 1989.

A firm foundation for Embraer has been the EMB-110 Bandeirante. Of some 500 sold in some forty countries, approximately 25 per cent have gone to the home market, a good basis for export success. Pictured is the latest commercial model, an EMB-110P1A with dihedral tailplane and improved cabin sound-proofing, operating with a mining company in the north of Brazil.

The P-95 (EMB-111) is the maritime patrol version of the EMB-110 Bandeirante used by the Força Aérea Brasileira.

Designated C-95C, this military Bandeirante is used to combat smuggling. It has a partial 'glass cockpit' with digital avionics.

All photographs are by courtesy of Embraer except where stated.

Fresno, California, with thirty-five 'Bandits', as they are known popularly because they 'hold up' well in service. The C-95C military model is the first variant to have a partial 'glass cockpit' digital avionics display. Like the newest commercial Bandeirantes, the C-95C has a dihedral tailplane.

The EMB-111 version of the Bandeirante has been ordered by the Brazilian Air Force for maritime patrol missions. It will have the partial 'glass cockpit' with EFIS-74 electronic flight instrument system, ADI-84 altitude display indicator system and APS-65 automatic pilot – all by Collins. Other electronics equipment includes a Super Searcher radar by MEL of the United Kingdom, a DR 2000 Mk.II/DALIA 1000A Mk.II electronic support measures system by Thomson-CSF of France and an Omega CMA 771 Mk.III navigation system by Canadian Marconi.

This new electronics suite gives the EMB-111 the capability to search, detect, identify and track several different radars and maritime targets; to process the data; and to present data graphically at the tactical operator's station for over-the-horizon targeting and vectoring of friendly vessels, shore-based surface-to-surface missile batteries or aircraft.

Ipanema

The EMB-201 Ipanema crop-duster has a 300bhp Avco Lycoming IO-540-K1J5D piston engine. More than 600 of these aircraft have been built and are used in Bolivia, Uruguay and Brazil. The newest model has several improvements including better cockpit ventilation, new spray gear and a three-blade airscrew.

The propeller-turbine EMB-121 Xingu for corporate use was Embraer's first with a pressurised cabin. A total of 105 were built of which fifty were exported, including this example for use by the French Navy.

Tucano

The first military programme for Embraer, the EMB-312 Tucano went into production in 1982 for the Brazilian Air Force. It was later ordered by Honduras, Egypt and Iraq. In 1985, the Royal Air Force ordered 130 of a more powerful and strengthened version to be built by Shorts in Belfast. Later orders were placed with Embraer by Venezuela, Peru, Argentina, Paraguay and Iran, bringing the total to 608 aircraft as at March 1989.

The Tucano is produced in two forms, the Embraer EMB-312 with 750shp Pratt & Whitney Canada PT6A-25C and the Shorts S.312 with 1,100shp Garrett TPE331-12B engine. The higher-powered RAF Tucano has structural reinforcing for higher manoeuvring loads and longer fatigue life, a different cockpit instrument layout to match that of the BAe Hawk, a ventral air-brake and other modifications. Shorts have also sold the S.312 Tucano to Kuwait and Kenya.

Brasilia

Fastest and also most economical in the 30/40-seat class of commuterliners, the EMB-120 Brasilia cruises at 300 knots, a little above the predicted performance, which is a tribute to the combination of smaller fuselage diameter (three-across seating), curved windshield (like the Lockheed TriStar), minimal wetted area and generally clean aerodynamic design. The 1,590shp PW115 engines of the prototype were replaced by 1,800shp PW118s for production aircraft to cope with inevitable weight growth.

Describing Brasilia design development, Embraer technical director Guido Fontegalante Pessotti recalls that talking with Bandeirante operators showed that a quick-change layout was desirable for the new aircraft, but a proposed large cargo door was not needed and thirty seats were enough. The keynote was to be passenger appeal with a quiet, comfort-

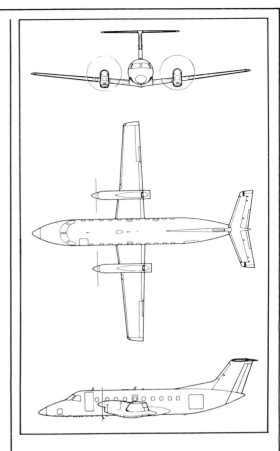

General arrangement of the EMB-120 thirty-seat Brasilia.

able cabin combined with high performance and fuel economy to produce an aeroplane that could be turned around rapidly for maximum productivity.

By selecting a thirty-passenger capacity with seats in rows of three, a small cross-section gave a high cruising speed, yet the 31in seat pitch gave full airline comfort. During the project stage, the engines were moved outboard 10in to increase propeller/fuselage clearance

Fast Feeder One. This EMB-120 Brasilia, the one-hundredth delivered (October 1988), was the twenty-ninth delivery to Atlantic Southeast Airlines, the Brasilia's launch customer. The airline provides feeder services for Delta Air Lines' hub at Atlanta, Georgia, from more than twenty-five cities in six states. A similar hub and spokes operation is centred on Dallas/Fort Worth.

and reduce cabin noise by 5dB to 77dB, increase propeller/ground clearance to 19.7in. A mock-up displayed at various shows was modified as the result of suggestions made by potential customers. The cabin is pressurised to 7lb per sq in differential at 32,000ft when the cabin is at 8,000ft pressure altitude. (The Xingu had a pressurised differential of 6lb per sq in and so was a useful starting point in design of the Brasilia cabin.)

While the Bandeirante is of conventional built-up construction, the Brasilia structure makes wide use of chemically milled panels with stretch-formed skins for double-curvature panels. The wing box has three spar shear webs and is built in one piece from tip to tip with chemically milled and shot-peened panels. The semi-monocoque fuselage has polished Alclad skin and is built in three sections.

The roomy flight deck is very modern to minimise pilot workload because of the emphasis on short lengths in commuter flying. Military experience led to choice of a curved, laminated windshield of glass/polycarbonate construction. This combines low drag with an excellent view and maximum resistance to bird strikes. To give the best view of cathode-ray tube (CRT) electronic instrument displays, yokes are provided instead of control wheels.

Although the Brasilia is very clean aerodynamically, the wing profiles are the well-proven NACA 23018 (modified) at the wing roots and 23012 at the tips. Pessotti considers that advanced-technology aerofoil sections are too expensive to manufacture for this type of aircraft, whereas the NACA-profile leading edges are relatively insensitive to transient shape changes due to ice build-up, pulsating pneumatic de-icing boots, dead insects and mud. However, one refinement is flattening of the sections between wing roots and engine nacelles to reduce drag. Wing spoilers are not needed for low-speed manoeuvring since the Brasilia is not a STOL aircraft.

An unusual feature is a fly-by-wire flap system in which each segment of double-slotted area-increasing flap is driven by a single hydraulic jack that is computer-controlled. A potentiometer located in each jack measures flap position and supplies this information to the computer to ensure that each inner and outer left-and-right pair of flap segments is balanced. An additional safeguard against asymmetric flap operation is provided by the fact that the Frise-type ailerons are sufficiently powerful to counteract the effects of flaps extending fully on one side only. There are two sets of mechanical connections for each control surface.

Fast Feeder Two. Britt Airways provides feeder services to Continental Airlines' hubs in the United States. Fourteen Brasilias, painted as 'Continental Express', form a key part of the Britt operation.

Although the Brasilia is a low-wing aeroplane, it has a T-tail to get the tailplane well away from flap wash and propeller slipstream effects. (The earlier Xingu also had a T-tail.) There are tandem rudders (as proved effective by de Havilland Canada on the Buffalo, Dash 7 and Dash 8 high-wing aircraft). In combination with a long moment arm, this type of tail unit provides excellent control and minimises the size and weight of the vertical surfaces because the tailplane acts as an endplate to increase the effective aspect ratio of the vertical stabiliser and rudders.

The flight deck has five CRT displays and an automatic flight control system may be installed above the glareshield. The electronic flight instrument system and radio equipment are Collins Pro Line II equipment.

Since a Garrett auxiliary power unit (APU) is standard in military Brasilias, its installation is available for those regional airlines and corporate owners wishing to be independent of airport ground supply carts.

Late in 1987, the 'hot and high' version of the Brasilia became available with both reduced structural weight and PW118A engines of 1,800shp flat-rated to 108 °F (42.2 °C). Empty weight is reduced by 858lb to 15,554lb by increased use of composites, thereby increasing range by 320nm for the same take-off weight. The latest Brasilia has passenger seats made of carbon-fibre/Kevlar; the floor and partitions of a carbon-fibre and Nomex-core sandwich; and side panels and cabin ceiling of glass-fibre, Kevlar, Nomex and carbon-fibre sandwich materials. Leading edges and tips of all flying surfaces are of Kevlar/glass-fibre as are wing/fuselage fairings, nose radome and the tailcone for non-APU equipped aircraft. Flaps are of carbon-fibre, while titanium replaces stainless steel on engine cowlings and exhausts. About 10 per cent of the basic equipped empty weight of the Brasilia is composite materials.

AMX – an Italo-Brazilian Programme

A 1975 Brazilian requirement for an Xavante replacement, the AX project, was not officially encouraged at first, says Silva. Various configurations were studied and evaluated including derivatives of the Xavante, but it was clear that the greater potential of a new design was desirable. In parallel, the Italian Air Force considered a development of the Fiat G.91 with a new wing and improved payload/range performance, and the possibility of upgrading the Macchi MB.339 trainer. In 1978, Aeritalia and Aermacchi joined forces in Italy on configuration development and Embraer became involved. The first contract for the definition phase of the AMX was finalised in May 1979. Brazilian participation became official in July 1981. First flight of the first prototype was in May 1984.

The main task of the AMX is close interdiction against ground forces on or behind a battlefield area. So, in addition to a useful payload/range capability, the AMX was designed for good airfield performance, high penetration speeds

A high-subsonic speed strike aircraft, the AMX is an international venture involving Embraer with Aermacchi and Aeritalia in Italy.

with military loads, low-level flight capability, high navigation and attack accuracies, post-failure operability and electronic warfare equipment combined in a highly versatile weapon system.

The flight controls are so designed that for each axis the function is shared between electrical/hydraulic and mechanical/hydraulic operations that drive independent control surfaces. The tailplane, spoilers and rudder are electrically

The Embraer EMB-326 Xavante was the Italian Aermacchi MB.326 built in Brazil under licence.

controlled through two computers, while the ailerons and elevators are hydraulically operated by mechanical circuits and have manual reversion.

The navigation/attack avionic system has two main computers connected to the sensors and displays via data bus bars. Redundancy is applied to increase the probability of mission success. Thus, for example, in the attack mode the inertial platform may be substituted by the attitude-director/standby-attitude heading-reference system, and the radar ranging substituted by the radio or barometric altimeter to generate back-up reversionary modes in case of failure of the main mode. The same is true for the navigation sub-system where autonomous and assisted modes are present.

The AMX fuselage volume is large to ensure operational flexibility by providing accommodation for various palletised payloads of reconnaissance and other equipment. The fuselage also contains electronic countermeasures equipment with infra-red/radar warning sensors, dispensers for flares and chaff, and either one 20mm M61A1 Gatling gun (Italian) or two 30mm DEFA 553 cannon (Brazilian). Externally, in addition to wingtip mountings for air-to-air missiles, there are five hardpoints for up to 8,378lb of external stores including AAMs.

Basic performance parameters are short take-off and landing runs; a penetration speed above Mach 0.7 while maintaining adequate manoeuvring margins; and a ground-level escape speed of Mach 0.85. The AMX has a single Rolls-Royce Spey RB168-807 non-afterburning turbofan of 11,030lb thrust. The first licence-built Spey Mk. 807 was completed by CELMA in Petropolis in February 1989.

Embraer has 30 per cent of the AMX programme, including design, development, testing and manufacturing of all wings, air intakes, external ordnance pylons, external fuel tanks, main undercarriage and reconnaissance pallets. Total Italian and Brazilian procurement is to be 317 aircraft, including 51 two-seaters. Two prototypes have been built in Brazil and three in Italy. Series production at Embraer started in 1987 and is now at the rate of five ship sets a month. Deliveries to the Italian Air Force have begun and were due to start to the Brazilian Air Force in mid-1989.

A special Embraer-developed item is the rocket-sled fuselage section for Martin-Baker to use for testing zero-zero crew ejection seats for both the single-seat strike fighter and the two-seat combat trainer/special missions versions of the AMX.

CBA-123 – Brings Together Brazil and Argentina

In a joint-venture programme with FAMA (Fábrica Argentina de Materiales Aeroespaciales) of Cordoba, Embraer is developing the twin-pusher pressurised nineteen-passenger CBA-123 (Co-operation Brazil Argentina) feederliner. Embraer has two-thirds of the programme and FAMA one-third. First flight is expected in March 1990 followed by certification and entry into service the following year.

Technical director Pessotti describes how the new aeroplane's configuration evolved. It began with a lengthened Xingu cabin with two-abreast seating. Seats for nineteen passengers meant that there had to be a long fuselage ahead of the wings balanced by the engines at the extreme tail, and canard surfaces up front for adequate longitudinal control

Novelty in the marketplace. The CBA-123 is due for certification in 1991.

because of the short tail moment arm. However, use of a foreplane created significant design penalties with such a three-lifting-surface layout, so the next configuration considered was based on a shortened Brasilia cabin with three-abreast seating in five rows and a final row with four seats at the back of the cabin. With this layout, which located the galley, wardrobe, toilet and optional flight attendant's seat at the front of the cabin, all loading conditions could be accommodated with a centre-of-gravity range between 10 and 15 per cent of mean aerodynamic chord, using a variable-incidence tailplane of reasonable area and dispensing with canard surfaces. This also eliminated the need for an elevator trim tab.

As shown in model form at the 1987 Paris Air Show, the CBA-123 had an upswept rear fuselage while the engine nacelles had prominent exhaust pipes. However, design refinement resulted in reshaping of the rear fuselage to optimise cross-sectional area distribution (area ruling) to reduce drag in the region of the nacelles. And the exhaust pipes were eliminated by venting the engine efflux directly rearward around the propeller spinners, reducing base drag and increasing thrust.

The new Garrett TPF351-20 propeller-

turbine is a 1,300shp engine with free power turbine aft of the gas generator and a gearbox capable of absorbing 2,000shp to allow for power growth. Specific fuel consumption at 30,000ft is 0.421 when the engine is delivering 1,050shp for a 351 knot cruising speed, although the aircraft can cruise at up to 40,000ft. New, six-blade Hartzell propellers have forged solid-aluminium blades, rather than the composite construction so popular today, to resist damage from ice shed by the wings. The propellers counter-rotate by means of an extra gear so that the engines are otherwise identical and do not rotate in opposite directions.

Pessotti agrees with the suggestion that the CBA-123 has no direct competition in terms of new-technology designs, now that the de Havilland Canada/ Shorts NRA90 projects have been abandoned. It is, he says, sometimes frightening to be the only one in the marketplace, but Embraer is sure that, once again, it is on the right track.

In addition to the CBA-123 being an international programme in terms of risk-sharing, the reactions of the Brazilian and Argentine civil and military markets have been very good. Also, there should be a large market in Europe because of airline deregulation as the

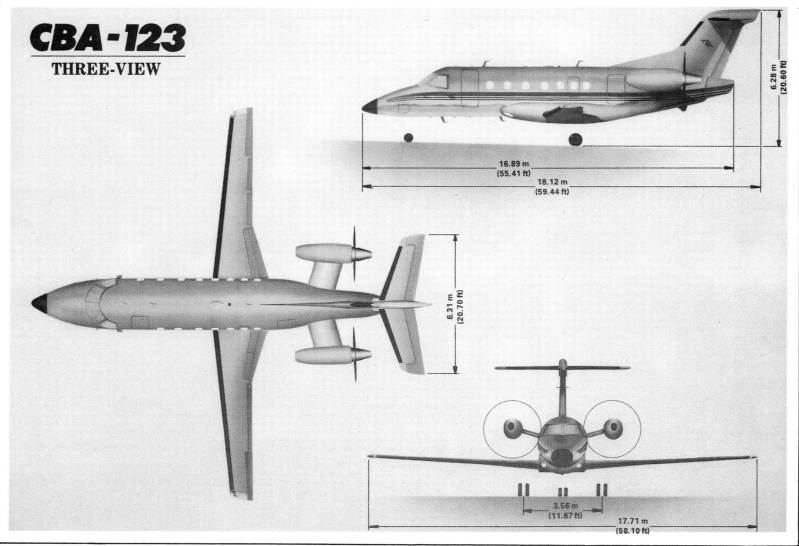

CBA-123
THREE-VIEW

6.28 m (20.60 ft)

16.89 m (55.41 ft)

18.12 m (59.44 ft)

6.31 m (20.70 ft)

3.56 m (11.67 ft)

17.71 m (58.10 ft)

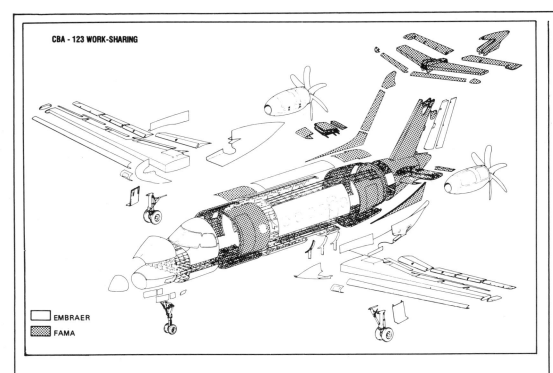

CBA - 123 WORK-SHARING

☐ EMBRAER
▨ FAMA

Programme share CBA-123: Embraer, Brazil, two-thirds; FAMA, Argentina, one-third.

CBA-123 will be able to operate from a runway of only 1,100m under FAA Part 25 rules, and 80 per cent of European runways are shorter than 1,200m. The 753nm range (with all nineteen seats occupied) should make the aircraft very attractive to regional airlines. The high block speed and low cabin noise level with Brasilia-sized cabin cross-section and aisle headroom combine to make a very appealing feederliner. Good operating economics, particularly high cruising speed, make the CBA-123 very productive, based on clean aerodynamic design and suitable engines.

Stressing the attention paid to systems, Pessotti says the objective here is

This artist's impression shows the pressurised nineteen-passenger CBA-123 commuter aircraft, intended as a Bandeirante replacement and due to fly in March 1990.

to ensure maximum reliability based on simplicity. Avionics systems are covered by a contract with Collins that includes four electronic flight instrumentation system (EFIS) displays, dual engine instruments and crew advisory system (EICAS) displays, dual air-data computers, dual attitude-heading reference systems (AHRS), dual autopilots, dual flight directors, dual navigation/communication sets and single DME, transponder, ADF, radio altimeter and weather radar. Engines have full-authority digital electronic control (FADEC).

Avionics inputs are integrated to inform flight and ground crews of conditions for push-button|or|read-out\troubleshooting. Use of advanced systems concepts, high rate of climb and high cruising altitudes make the aeroplane very competitive for corporate use, despite its relatively large seating capacity and cabin diameter. If nineteen seats are not needed, very comfortable accommodation may be installed for ten executive passengers

while retaining the capability to use maximum seating whenever needed for larger corporate teams or groups of customers. With ten passengers, range is 1,800nm while ferry range, using long-range cruise techniques, is 2,200nm. The cabin pressurisation is 8.2lb per sq in.

New Design Techniques

Asked about computational aerodynamics instead of wind tunnels, Pessotti is enthused and says he has 'complete confidence' in the new technique, using wind tunnels only for final confirmation of flow characteristics. Use of computer-aided design with real-time motion added has resulted in development of a computer which has wind tunnel representation on its screen. Digital aerodynamics, he says, is developing so fast that the only limitation is the size of mainframe computer installed in a plant. Only in the regimes of viscous flow and stalling are there problems because of the complications of simulating detached flows, since there is as yet no adequate theory available. However, a suitable digital programme could be devised. Even so, nearly 2,000 hours of wind tunnel testing have been accumulated on the CBA-123, in both Brazil and Argentina.

Data generated in wind tunnel tests were embodied in a software program and used in a Brasilia simulator at Embraer's Florida plant to allow company pilots to 'fly' the CBA-123. During thirty hours of this simulator work, the flight envelope was completely explored to confirm the integrity of the design and permit its refinement before the configuration was frozen. A new wind tunnel model was then built to validate the modified features. With the refined data from the new model, new software generations were produced – one of the CBA-123 in frozen configuration and one of a Brasilia designated EMB-120-3. The latter had a third engine installed at the rear like one of those on the CBA-123 to check the airflow in the region between the two rear-mounted pusher engines and the tail unit. The two configurations were 'flown' on the Brasilia simulator for about twenty hours in August 1988 to prove that both the CBA-123 and the EMB-120-3 had no handling or performance problems.

Additional CBA-123 design evaluations on the Brasilia simulator are taking place as more refined data becomes available from later wind tunnel tests. Eventually, a CBA-123 flight simulator will become available much sooner than would otherwise have been possible, with consequent significant savings in time and cost for flight testing and certification phases of development.

Asymmetric powered Brasilia provides data for CBA-123. A third engine mounted on a stub wing under the tail of the Brasilia prototype represents the slipstream effects on the flying controls of the CBA-123. After take-off with all three engines operating, the port Pratt & Whitney Canada PW118 is stopped. Although not a pusher, the test propeller plane is in the same relative position as in the CBA-123 which has a shortened Brasilia fuselage. The test installation is a Garrett TPE331-12B with four-blade propeller which is the engine on which the TPF351-20, with six-blade pusher propeller, is based.

Tucanos in final assembly. (*David Godfrey*)

Major component jigs – Brasilia fuselage sections. (*David Godfrey*)

An important sub-contract to Embraer is the carbon-fibre outboard wing flaps of the McDonnell Douglas MD-11. The Brasilia also has carbon-fibre wing flaps.

New Materials

Turning to materials, Pessotti says that Kevlar/Nomex sandwich and other advanced composites used on the CBA-123 amount to more than 15 per cent of airframe weight. Composites are used for ailerons, flaps, ground spoilers, rudders and elevators in addition to the more common applications to wing and tailplane, leading edges and tips as on the Brasilia. Composite work with the Brasilia and recent experience with the 29ft-long carbon-epoxy outboard flaps for the McDonnell Douglas MD-11 built by Embraer have preceded increased use of advanced composites in the new feederliner. One example is graphite undercarriage doors for the CBA-123, components which will also be developed for the Brasilia. Pessotti notes that adoption of any new material, whether advanced composites or one of the new aluminium-lithium alloys, 'takes time'.

Missiles and Space

Early in 1987, Órbita Aerospace Systems (Órbita Sistemas Aeroespaciais SA) was formed for the design, development and marketing of guided weapons and missile systems, sounding rockets and launchers for civil space applications. Its founders were Embraer, the armoured vehicle manufacturer Engesa (Engenheiros Especializados SA), Imbel (Indústria de Material Bélico do Brasil), Esca (Engenharia de Sistemas de Controle de Automação SA) and Parcon (Participações e Consultoria SC). In this group, Embraer represents the Brazilian Air Ministry and Imbel the Army Ministry.

Current programmes are the MAA-1 infra-red-guided air-to-air 'fire and forget' missile for the AMX strike fighter; MSA-3.1 shoulder-launched ground-to-air missile; and MSS-1.2 anti-tank weapon. The MAA-1 replaces the venerable Sidewinder. It has a single-shot kill probability (SSKP) of better than 70 per cent and a range of 6 miles. The MSA-3.1 has a similar SSKP, uses automatic command line-of-sight, has a range of 4 miles and reaches a speed of Mach 3.9. The MSS-1.2 is a beam-riding ground-to-ground weapon.

Commercial Developments

Certification of the CBA-123 is due in 1991. Once this has been achieved, the same 'glass cockpit' instrumentation display on the flight deck is to be made available for the Brasilia, says Pessotti. At that time, other Brasilia improvements will include a higher speed of about 340 knots; a cabin stretch to increase the three-abreast seating from thirty to forty-two; cruising altitude raised to 35,000ft; and the ability to operate from 1,200m runways.

Current concentration on the Brasilia is to build to meet substantial orders on hand and to improve the already excellent dispatch reliability still further, since some operators have demonstrated that 99.9 per cent reliability is possible. Engineering support is focused on certain vendor-supplied items, there being no problems with such major items as the PW118 engines, Hamilton-Standard propellers, air-conditioning, undercarriage, and so on. The Collins avionics suite has been 'outstanding', says Pessotti with enthusiasm. At this stage, the emphasis on the Brasilia is 'sustaining engineering', with new requirements for major customers. An example is the Sundstrand APU specified by Texas Air, although a Garrett APU is standard.

Key to the Brasilia's success in world markets, says commercial director Heitor Fernandes Serra, is not merely high performance but the fact that it was better than estimated. This pleased the airlines because the combination of high cruising speed (300 knots) and fuel efficiency meant greater productivity per unit cost. This had been matched by very high despatch reliability well before 100 aircraft had been built. The choice of a thirty-seater capacity with three-abreast seating for optimal drag proved wise for success in launching the programme.

Pointing out that the nineteen-passenger Bandeirante has been in service in Europe since 1977 and a year less in the United States, Serra sees the new CBA-123 as its logical successor. Although the feeder segment of the market is shrinking, between 25 and 28 per cent of commuterlines worldwide will be operating nineteen-seat aircraft at least until 2006, according to Embraer. Also, the corporate and business market potential is appreciable, so the combined probable market justified embarking on a new programme. At the same time, Embraer wanted to ensure that the basic design was a suitably versatile platform for the varied duties it would inevitably perform during its production life – feederline, business/corporate and military transport. High block speed and far greater fuel efficiency would also make the CBA-123 attractive as a replacement for some of the older business jets.

Continued co-operation with FAMA of Argentina depends on how well the relationship with Embraer works out during co-production of the CBA-123, says Serra, noting that FAMA has not previously been exposed to the demands of international markets and so has to make a transition to entrepreneurial activity, but there is no reason why this should not be successful. Pessotti says about eighty FAMA technical staff are at Embraer and that all has gone smoothly with the collaborative technical effort.

Important international contracts such as the McDonnell Douglas MD-11 composite wing flaps are only the start of such work on major programmes in other countries. Not only is this an excellent market opportunity, but it generates much-needed hard currency and permits acquisition of new technology.

Aspects of Manufacturing

About 90 per cent of Embraer's newest machine tools, mainly numerically controlled Cincinnati Milacron five-axis millers, are used for the AMX programme, so machining for the Brasilia and CBA-123 is contracted to US companies to avoid over-expansion of the plant. Explaining this, production director Irajá Buch Ribas says that Embraer is encouraging companies in the São José dos Campos region to establish specialised machining facilities. At least 200 different parts for the Brasilia are now produced on five-axis machine tools by sub-contractors.

There is a chemical milling installation in the Brasilia assembly building because nearly all the skins for this aeroplane – for wings, fuselage and tail – are profiled in this way. The CBA-123 wing has double-curvature panels that are shot-peened after taper machining and chemical milling.

There is some difficulty in obtaining some types of new machine tools, says Ribas, and Embraer has found that used equipment in the United States is usually in good condition at attractive prices.

A second and larger autoclave has been added to increase curing capacity since both the MD-11 flap and CBA-123 programmes involve use of advanced composite materials. Also, other companies are increasing their requests to Embraer for composite components.

Since volume does not justify processing of specialised, high quality aircraft raw materials in Brazil, these come mainly from the United States and Europe. A change in business procedures is the introduction of import taxes, so Embraer has a 'drawback' arrangement whereby aircraft sold abroad are not taxed. Because major US suppliers have worked with the company for a long time, there is no problem with deliveries, although careful long-term planning is required, says Ribas. Obviously, application of 'just in time' supply is limited in view of the ordering lead time needed and the distances involved to the suppliers, not only for materials but also for standard fasteners, seals, bearings and chemicals. A special problem is posed by chemicals because of their short shelf life, so they must be bought in small quantities, refrigerated and used as soon as possible. The logistical solution is to fly in the chemicals.

Manufacturing research is conducted by a dedicated process-development group, concerned mainly with new composites, to simplify fabrication procedures. For metal cutting, research continues into tool design and materials with the objective of reducing machining time. Aluminium-lithium alloys are being evaluated, but have not yet been adopted as standard materials. Titanium machined items are used in the Brasilia and AMX to save weight for highly stressed parts, and titanium is used for some engine nacelle sheet-metal work where heat resistance is required.

An essential part of production planning is training, says Ribas, because there is not a large pool of high-technology personnel on which to draw. Embraer has invested in a new, comprehensive training centre which takes technical school graduates and gives them specialised skills for aircraft production. Not surprisingly, other industries (notably automotive) try to raid Embraer of its staff, but the excellent working conditions, employee benefits and *esprit de corps* prevent this from being a problem.

Computer-integrated manufacturing (CIM) is used to load machine tools for complete management control of production work. Material-resource planning is used to integrate financial planning with utilisation of materials in a new CAD/CAM/CIM system known as ECIM (Embraer CIM). The CBA-123 is entirely a CAD/CAM programme with no paper drawings. Instead, all information is stored in a central, computerised data base for total control of design, tooling, manufacturing, inspection and warehousing. The ECIM system was due to be fully implemented by mid-1989.

Embraer production departments employ about 6,000, says Ribas, and this total is increasing slightly for AMX assembly. The tooling department has about 600 people.

The Brazilian aerospace industry has to be vertically structured because there are not the tiers of sub-contractors, suppliers and vendors as in Europe and North America. Thus, Embraer has to be self-sufficient at almost all levels. It would be much easier for the company – and allow it to be smaller – if other, specialised companies would take on machining, tooling and parts fabrication which require high-technology techniques but are not necessarily restricted to the aerospace industry.

Personnel Training

Recruiting and training personnel are prime concerns. Embraer works with universities and technical schools in Brazil and other countries to prepare prospective new employees for the aerospace industry before they join the company. Special problems have been presented by rapid expansion – 3,000 were hired between July 1986 and September 1987, a 42 per cent increase in total employment. Most of these people were not experienced in aircraft work, so extensive training was necessary. In 1986, some 7,000 people were given courses and in 1987 this rose to 19,000, some employees taking more than one course. The total for 1988 was also about 19,000. The current level of employment, over 12,000, is rising slowly but is not expected ever to exceed 15,000.

Training courses are given mainly in-house by Embraer instructors with some outside specialists. Some courses use video instruction and informatics techniques. However, an instructor is always necessary because of the very varied and dynamic nature of aeronautical engineering.

Equipment Division

In 1984, Embraer established an equipment division in a separate plant. General manager of this division, Roberto Negrini Pastorelli, explains that the main reason for this was the AMX

Originally at the main Embraer plant, production of the EMB-201 Ipanema was transferred to the Neiva general aviation division.

Embraer Current Aircraft Programmes

EMB-110P2A Bandeirante
19-passenger feederliner, unpressurised. Span 15.33m; length 15.10m; wing area 29.10sq m. Empty weight 3,516kg; max take-off weight 5,670kg. Max speed 248kt; max cruising speed 223kt. Service ceiling 2,440m.

EMB-201A Ipanema
Single-seat agricultural aircraft. Span 11.20m; length 7.43m; wing area 19.94sq m. Empty weight 1,011kg; max take-off weight 1,550kg. Max speed 124kt; max cruising speed 115kt.

EMB-312 Tucano
Two-seat basic trainer. Span 11.14m; length 9.86m; wing area 19.40sq m. Empty weight 1,810kg; max take-off weight 2,550kg. Max speed 242kt; max cruising speed 222kt. Service ceiling 3,050m.

EMB-120 Brasilia
30-passenger commuterliner, pressurised. Span 19.78m; length 20.00m; wing area 39.43sq m. Empty weight 7,070kg; max take-off weight 11,500kg. Max speed 328kt; max cruising speed 300kt. Service ceiling 9,085m.

AMX
Single-seat strike fighter. Span 8.87m; length 13.58m; wing area 21.00sq m. Empty weight 6,700kg; max take-off weight 12,500kg. Max and max cruising speeds Mach 0.86. Service ceiling 13,000m.

CBA-123
19-passenger feederliner, pressurised. Span 17.72m; length 17.77m; wing area 27.20sq m. Empty weight 5,640kg; max take-off weight 8,500kg. Max cruising speed 350kt. Service ceiling 12,190m.

programme. It was necessary to create the equipment division because there were no companies in Brazil with the capability to produce aircraft-quality items under sub-contract and the only alternative would have been to procure such items from the United States or Europe. So, it was necessary to create a new plant with vertical organisation to cope with all facets of engineering and manufacturing needed by Embraer aircraft programmes, particularly the international AMX. Also, the high-precision nature of the work made it desirable to have physical separation from the main plant with self-sufficiency as to laboratories, quality control, salt-spray and sand/dust environmental test chambers, surface heat-treatment facilities, and a range of numerically controlled machine tools for milling, boring and other precision operations.

For the AMX programme, the equipment division manufactures the main undercarriage under licence from Messier-Hispano-Bugatti, wheels and brakes with anti-skid system, hydraulic and fuel system valves, and ordnance-release ejection units for the underwing pylons. Other products include the standard and the heavier Shorts versions of the Tucano main undercarriage, underwing pylons for the Tucano, and airstair door dampers for the Bandeirante.

The Embraer equipment division anticipates receiving sub-contract work from major international aerospace corporations. Not only will this be useful in earning money, but it will provide technology transfer for future programmes. The first application of this will be the CBA-123 main undercarriage which is being developed by Embraer in collaboration with Messier-Hispano-Bugatti. The workload is increasing steadily as AMX production gets into top gear and employment in the division reached 600 in 1989.

MFT/LF

A proposal for a modern fighter trainer/light fighter was submitted to the Brazilian Air Ministry in mid-1987. Precise details are naturally restricted, and in any case the configuration is subject to change, but the implications are significant, given Embraer's success with other projects that have matured into production programmes. This applies both to strictly Brazilian designs and to international programmes such as the AMX and CBA-123.

The MFT/LF is an Embraer proposal for an uncompromised basic airframe/engine/systems aerial platform to be developed in parallel as both a modern fighter trainer and a light fighter, the two versions differing by their role-dedicated equipment. One engine of about 6,000lb thrust dry and 9,000lb with afterburning would give a maximum thrust to weight ratio of around unity to provide a high energy level for manoeuvring agility and a top speed of about Mach 1.8. The airframe would be mainly of advanced composite material for low weight.

State-of-the-art avionics systems and cockpit displays would be used, but not a sidestick controller. So, the MFT version would be complementary to advanced combat aircraft such as the European Fighter Aircraft, Dassault Rafale and Advanced Tactical Fighter. A flight simulator would be used for operational missions, including intercepts, to maximise cost-effectiveness and safety. Tandem seating would be used, as in the Tucano, two-seat AMX and the majority of advanced trainers. The LF version would be a single-seater, although production of a two-seat variant would be no problem because of parallel development of the trainer airframe. The light fighter would have similar performance, depending on ordnance and fuel loads carried for different roles as an interceptor, strike or close-support aeroplane.

As proposed, the MFT/LF generally resembles the AMX except for different air intakes to suit supersonic speeds. There is a high-set wing with slight sweepback, curved-back tips and no dihedral. The wing has leading-edge flaps, large inset trailing-edge flaps, and spoilers for lateral control. At the wing roots are leading-edge extensions forward to near the lips of the fuselage-side air intakes. The all-moving horizontal tail surfaces have anhedral. There is a single, large fin and rudder. Missiles are located at the wingtips while two hardpoints under each wing are available for a variety of ordnance stores and long-range fuel tanks.

In Brazil, the MFT would be an ideal replacement for the Xavante, while the LF would be a long-overdue successor to the Dassault Mirage and Northrop F-5. Similarly, many air forces would be interested in a mixed-force acquisition of both versions. Some armed forces would doubtless be attracted to the MFT to match their own high-technology fighters, while others would find the LF both a cost-effective fighter replacement and an upgrading of equipment from old-technology aircraft of many types.

Technical director Pessotti says the MFT/LF must be both affordable and reasonably expendable. Once the programme has a go-ahead, the aeroplane must be developed quickly – within six years – which is a big challenge both technically and financially. Therefore, the programme should be an international joint venture with another country which has similar requirements, as with Italy for the AMX.

General Aviation Programmes

At Indústria Aeronáutica Neiva SA, the Embraer general aviation subsidiary, more than 2,000 aircraft have been built at Botucatu where about 530 are employed. Of the ten aircraft manufactured each month, half are EMB-201 Ipanema crop-spraying aeroplanes and the remainder are various models of single- and twin-engined aircraft derived from Piper designs, plus sub-assemblies for the Tucano. The Piper-based aircraft are EMB-711S Corisco II (Turbo Arrow IV), EMB-712 Tupi (Archer II), EMB-720D Minuano (Saratoga), EMB-810D Seneca III and NE-821 Carajá (PT6-powered Schafer Comanchero 500).

View from the Bridge

Chief executive officer Ozilio Carlos da Silva is, like all Embraer executives, steeped in aviation experience. As an aeronautical engineer trained at the Instituto Tecnológico de Aeronáutica, he started his career at the aircraft certification commission of the Centro Técnico Aeroespacial and headed it until 1964 when he went to France to take specialised courses. On returning to CTA, he worked on the IPD-6504 project design that became the Bandeirante.

As a founding member of Embraer in 1969, da Silva was production director until 1979 when he became commercial director. In this capacity he was responsible for the policy of establishing a subsidiary in the United States (Embraer Aircraft Corporation at Fort Lauderdale, Florida) and a Paris office (Embraer Aviation International at Le Bourget). He also conducted some of the most significant negotiations for Embraer in the export market. In July 1986, he became chief executive officer when the chairman, Ozires Silva, took on an additional important government posting at the request of the President of Brazil, José Sarney.

The most likely candidate for a new military programme at Embraer is the MFT/LF, says da Silva. An international partner is essential, but first must come a requirement from the Força Aérea Brasileira (Brazilian Air Force) for a supersonic trainer since it is essential to have a domestic order as the basis of the programme. As to the conjecture that the AMX partners Aermacchi and Aeritalia might also become the MFT/LF partners, this clearly depends on Italian requirements, although it would be a very logical progression.

For the AMX, a comprehensive international export-market survey has been made with potential sales and areas

The Embraer NE-821 Caraja was derived from the Piper Navajo via the Schafer Comanchero 500.

allocated to the three partners. There is a good market, says da Silva, for a naval attack version of the AMX to replace such older aircraft as the Dassault Super Etendard and McDonnell Douglas A-4 Skyhawk for which there are no modern equivalents, whereas the AMX weapons, avionics and systems are completely up to date and extremely effective.

The Tucano continues to be a very successful international programme and several additional countries are seriously interested in the propeller-turbine trainer as a replacement for jet types. One example is Canada where agreements with Spar Aerospace would be the basis of an agreement such that the cost of procuring the Tucano would be offset by the value of work on the Brazilian telecommunications satellite programme. For the Canadian Armed Forces, the Tucano would have a 1,110shp Pratt & Whitney Canada PT6A-67 engine instead of the 750shp PT6A-25C of the standard aeroplane, so that its performance would be similar to that of the version built by Shorts for the RAF.

A minimum of fifteen years' work is planned for the CBA-123 programme with FAMA of Argentina, including derivatives. Business/corporate and military versions of the new feederliner would be the natural replacement for both the pressurised Xingu and unpressurised Bandeirante, says da Silva.

The success of the Brasilia was not a surprise, says da Silva, recalling that de Havilland Canada had been regarded as the most serious competition, rather than Saab, because although the Saab

340 was closer in size than the Dash 8, the Canadian company had extensive experience of developing short-haul aircraft and had built a strong customer base. In the event, the thirty-six/forty-seat Dash 8 with four-abreast seating was obviously destined to be stretched and so would not compete directly with the thirty-seat Brasilia in most cases. In deciding to concentrate on a thirty-passenger aeroplane, Embraer knew it had to provide the highest speed in the class and had achieved this. The proposed cabin stretch to forty-two seats is now being discussed with those Brasilia customers who need a larger-capacity

These EMB-312 Tucano propeller-turbine trainers are from the air forces of (left to right) Peru, Brazil, Argentina and Venezuela.

aircraft but want to maintain fleet commonality. More than 200 firm orders have been placed for the Brasilia with options on another 150.

Since general aviation activity is still not very active, says da Silva, Embraer decided to sell off the Neiva division, but no buyer came forward so a five-year programme is being implemented to invest more capital and rationalise production on a reduced range of models to make Neiva more attractive to buy.

A new and significant part of Embraer corporate strategy is the creation of Órbita for the development of armaments, missiles and launching rockets for scientific purposes, says da Silva. Space work is also an objective and its precise role is the subject of negotiations. Embraer owns 40 per cent of Órbita. □

Success Through Perseverance: the Rolls-Royce RB.211 Engine

J R Cownie

Our contributor records his belief that two-thirds of the life and times of the RB.211 may yet be to come. Brought out from the abyss in 1971 to scale new heights of technical excellence and commercial appeal in the late 1980s, the RB.211 represents an object lesson on the need to persevere with technology capable of bringing long-term rich rewards.

The story begins in the early 1960s when the concepts which led to the RB.211 were developed in the Rolls-Royce project office at Derby. Since then the RB.211 programme has been fascinating at a number of levels. Firstly, it has been a major export success, having been designed from the beginning for aircraft produced outside the United Kingdom. Nearly 2,900 RB.211s were on order or had been delivered by 1 May this year – a great deal of business when a single RB.211 for the Boeing 747-400 costs well over £4 million.

Throughout its history the RB.211 has depended on human determination and preseverance, often in the face of difficult odds. The engine has always faced fierce competition. Rapid technical development of the RB.211 has been driven by market forces, leading to major advances in performance – more power, better fuel consumption, reliability improvements and lower overall cost of ownership.

Throughout the engine's life there has been a struggle to assure its commercial success and to gain a greater share of the market for large turbofan engines, which Rolls-Royce shares with General Electric (GE) and Pratt & Whitney (P&W). There have been constant efforts to increase the number of airframes for which the RB.211 is specified, endeavours which have been much more successful recently than in the early years of the engine programme.

Following the engine's launch to power the Lockheed L-1011 TriStar, determined efforts were made to ensure that no other manufacturer's engines were installed in that aircraft. This was achieved: the TriStar is the only Western wide-bodied airliner to have been powered by a single type of engine. But this success was won at some cost in fighting off bids by other engine companies and by airlines.

Rapid and costly upgrading of the RB.211-524 was needed, for example, to persuade Pan American World Airways to buy the TriStar with RB.211s rather than with P&W JT9D engines. The airline had been the launch customer for the Boeing 747 and its JT9D engine; it had powerful reasons for wanting its TriStars to have the same engine.

The bid to widen the application of the engine led to the first order for RB.211-powered Boeing 747s in 1975 by British Airways. This followed the launch of the 48,000lb thrust RB.211-524 in 1974 for extended-range L-1011-200 TriStars ordered by Saudi Arabian Airlines.

Since 1975 the Boeing 747 has been a key market for successive versions of the engine. The latest Boeing 747, the long-range -400 version, has been ordered by a number of operators with the 58,000lb thrust RB.211-524G/H engine following

The RB.211-524G/H in cutaway. This engine entered service in 1989 and higher-thrust versions are also being developed.

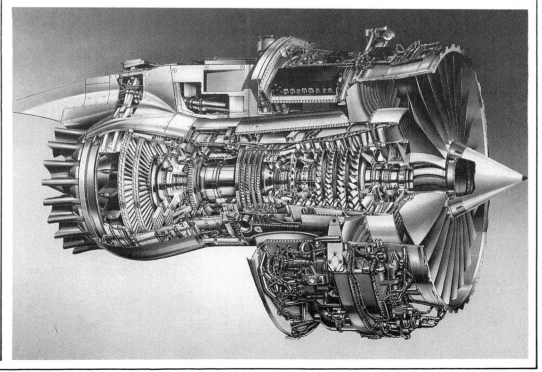

their satisfaction with earlier RB.211-powered 747s.

Boeing 757 Requirements

The next advance was the launch of a smaller version of the engine for a Boeing airliner to succeed the 727, which was at that time the best-selling jet airliner ever produced. A considerable success was achieved in 1978 when the derivative RB.211-535C was selected by Eastern Air Lines and British Airways, the launch customers for the Boeing 757. It is rated at 37,400lb take-off thrust.

This was the first time that a Rolls-Royce engine had been the launch engine for a new Boeing airliner. Success seemed to be assured, particularly when GE withdrew from the competition after trying to obtain part of the 757 market.

But it was not to be; P&W came into the picture with a new and more fuel-efficient engine, the PW2037, which it offered for the Boeing 757. This was launched with an order from Delta Air

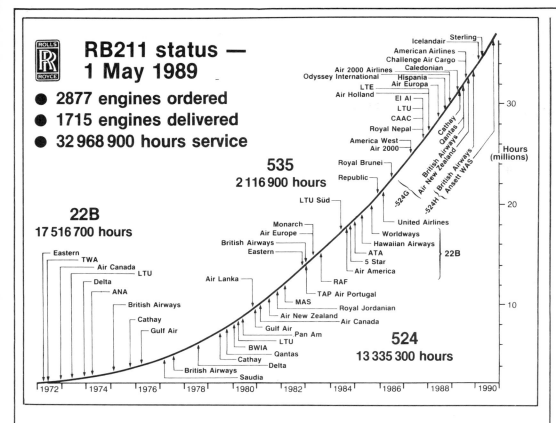

**RB211 status —
1 May 1989**

● **2877 engines ordered**
● **1715 engines delivered**
● **32 968 900 hours service**

22B
17 516 700 hours

535
2 116 900 hours

524
13 335 300 hours

An Eastern Air Lines RB.211-powered Lockheed TriStar carried its first fare-paying passengers on 26 April 1972. In the seventeen years that followed (to 1 May 1989) Rolls-Royce delivered 1,715 RB.211s (of 2,877 ordered) and 32.97 million engine hours of service were accumulated in Lockheed TriStars, Boeing 747s and Boeing 757s.

The diagram also shows the engine hours and operators of the -22B, the -524 and the -535 variants. Dates of initial airline orders are also shown.

Lines and Rolls-Royce had to respond by rapidly developing a much more efficient RB.211-535, the 40,100lb thrust 535E4.

In order to compete with the PW2037, Rolls-Royce pulled out all the stops and put a great deal of its advanced technology into the RB.211-535E4. The smooth and rapid development of this engine was one of the great successes of the RB.211 programme. With its excellent performance and outstanding reliability, the 535E4 has proved a worthy competitor to the PW2037, which suffered from

the lower reliability inevitable when introducing a new engine.

Initial sales of the Boeing 757 were slow because the aircraft was somewhat larger then the deregulated airline market originally needed. Each order was fiercely contested by Rolls-Royce and P&W but today 75 per cent of customers have selected the Rolls-Royce engine and it has been specified for well over half the 757 aircraft ordered. Demand for the Boeing 757 has increased in recent years, with orders and options now placed for over four hundred 535-powered aircraft.

The RB.211 Family

Up to 1987 there had been three main variants of the RB.211: the 42,000lb thrust RB.211-22B for medium-range TriStars; the RB.211-524 family for

RB.211 engines entering service from 1972 to 1993.

longer-range TriStars, Boeing 747s and more recently Boeing 767s, with thrusts from 48,000 to 60,600lb; and the RB.211-535 family for Boeing 757s.

Technical improvements had permitted take-off thrust to be increased and specific fuel consumption (sfc) to be reduced without going to a larger engine pod. This was made possible by use of a wide-chord fan, introduced on the 535E4 and later installed on the most powerful 524 engines.

But clearly a larger engine was needed to provide thrusts reaching from 65,000lb towards 80,000lb. These are required for the latest generation of long-range wide-bodied aircraft – the McDonnell Douglas MD-11, Airbus A330 and possible further developments of the Boeing 767 – and maybe later the Boeing 747 as well. As a result Rolls-Royce is developing a new larger-diameter engine, the RB.211-524L, based on experience with earlier RB.211s and incorporating the same high-pressure core as the RB.211-524G/H.

The launch order for 524L engines to power MD-11 airliners was announced by Air Europe early this year. Soon afterwards Cathay Pacific Airways announced the first order for Airbus A330s with this new development of the RB.211. All these new airliners will enter service early in the 1990s.

In June this year the company announced that this engine was to be known as the Trent – the third in the Rolls-Royce line of engines to bear this river name. The first was a propeller-turbine version of the Derwent turbojet; two of these Trents powered a Meteor testbed aircraft in 1945. The second Trent was Rolls-Royce's first three-shaft turbofan, an engine in the 10,000lb thrust class which was intended for a development of the Fokker F.28 airliner to be produced by Fairchild. This engine was test-run in the late 1960s, but did not go into service.

With its work on the new Trent, Rolls-Royce is now manufacturing or developing versions of the RB.211 for five types of large commercial aircraft: the Boeing 747, 757 and 767; the MD-11; and the A330.

Background

In the early 1960s the United States developed the propulsion and airframe technology needed for large wide-bodied aircraft. Both P&W and GE received government funding to design and build demonstrator versions of large high-bypass engines. Boeing and Lockheed were given contracts for the study and development of large airframes.

These contracts were applied to develop the technology needed for the giant

RB.211 Engine Data

	RB.211-22B	RB.211-535E4	RB.211-524G/H	RB.211-524L (Trent)
In-service date	1972	1984	1989	1993
Take-off thrust	42,000lb	42,100lb	60,600lb	67,500lb*
Certification	Feb 1971	Oct 1984	Mar 1988	1992
Bypass ratio	4.8	4.3	4.1	5.1
Pressure ratio	24.5	25.8	34.5	34.0
Length	119.4in	117.9in	125in	147.6in
Fan diameter	84.8in	74.1in	86.3in	97.5in
Basic weight	9,195lb	7,264lb	9,874lb	N/Av

* For Airbus A330 the engine has growth potential to 80,000lb.

Aircraft Performance

	Lockheed L-1011-1	Boeing 747-100	Boeing 747-400
In-service date	1972	1970	1989
Wing span	155ft 4in	195ft 8in	211ft 0in
Length	178ft 8in	231ft 10in	231ft 10in
Height	55ft 4in	63ft 5in	63ft 5in
Max take-off weight	426,000lb	735,000lb	870,000lb
Max cruise speed	Mach 0.85	Mach 0.84	Mach 0.84
Design range	2,878 st miles	4,600 st miles	8,233 st miles
Passenger accommodation	Max 400	385 (two class)	386 (three class)
Powerplant	Three RB.211-22B	Four P & W JT9D-7A	Four British Airways RB.211-524G
Engine take-off thrust	42,000lb	46,950lb	58,000lb
Total take-off thrust	126,000lb	187,800lb	232,000lb

Some wide-bodied jets. Variations in installed power, take-off weight and design range in the period 1972 to 1989.

C-5A military transport. The airframe contractor chosen in 1965 for this aircraft was Lockheed and its large TF39 turbofans were ordered from GE.

This technology was later applied in the civil field by the losing contractors in the C-5A competition, Boeing and P&W. They worked together to launch the Boeing 747 with its large JT9D turbofan engines. The launch order for the Boeing 747 was placed by Pan American in 1966.

In the early 1960s Rolls-Royce did not receive as much government research and development funding as the US companies, nor did it gain their early design and development experience with large turbofan engines. It had been studying the best type of engine to succeed its Conway turbofan, which was rated at up to 21,800lb thrust.

In July 1961 Rolls-Royce made proposals for the 25,000lb thrust RB.178, a two-shaft engine for long-range subsonic airliners. In April 1965 the main board was asked to authorise manufacture of an RB.178 demonstrator engine rated at 28,500lb thrust.

By September 1965, market studies showed Rolls-Royce that there would be a large enough demand in the next decade to support one, and possibly two, large engines. At this time the company began to favour its revolutionary three-shaft concept and was soon convinced that it could offer a three-shaft engine with a much higher thrust than the RB.178 demonstrator.

This demonstrator first ran in July 1966 and, after initial problems, achieved 93 per cent of its design thrust. Mechanical defects were found and test running was limited because the engine cost more than its budget. The demonstrator programme was then dropped owing to shortage of finance, although continued running could have helped development of the RB.211, particularly its high-pressure module.

Some experience was gained later with the smaller three-shaft Trent engine, cancelled in July 1968. Many believed that some of the problems later experienced in RB.211 development could have been solved earlier if running had continued with the RB.178 demonstrator.

While the US companies were gaining experience with large high-bypass engines, Rolls-Royce obtained only limited test information from its lower-bypass RB.178. Its choice of a relatively low-bypass ratio was influenced by the company view at that time that large-diameter engines with high-bypass ratios (5 to 1 and above) would be so heavy and would have such a large installed drag that this would nullify the benefits of the low testbed specific fuel consumption provided by using high-bypass ratios. Higher propulsive efficiency is produced with high-bypass ratios as each pound of thrust is produced by accelerating a larger mass of air to a lower exhaust speed.

This changed when Rolls-Royce discussed its new engine proposals with Boeing, which took the view that it knew how to install high-bypass turbofans to minimise their drag. Rolls-Royce bypass ratios would have to go up to around 5 to 1 if its engines were to be competitive for the 747. Rolls-Royce then submitted an engine proposal for the Boeing 747, but the P&W JT9D was selected.

Meanwhile, Rolls-Royce had been competing with Bristol Siddeley Engines on a range of projects and Bristol Siddeley had ambitions to enter the big-engine field. The European Airbus was at its early design stage and in June 1966 Bristol Siddeley and SNECMA of France took an option from P&W to manufacture and sell the JT9D in Europe. Rolls-Royce did not welcome this move, which could have facilitated the launch of a big US engine in Europe.

This move by Bristol Siddeley accelerated discussions which had begun in March of that year and led to Rolls-Royce taking over the company. It did this in September 1966 by acquiring The Bristol Aeroplane Company Limited, which owned half of Bristol Siddeley, and by buying the other half interest in the company from the Hawker Siddeley Group Limited.

The Department of Trade and Industry report of 1973 states that 'it is difficult to escape the conclusion that the acquisition was primarily defensively motivated' and that 'Rolls-Royce paid about £22 million for goodwill but profits earned fell far short of forecast'. Thus the

Maiden flight of the RB.211-powered Lockheed TriStar prototype on 17 November 1970.

The first RB.211-22 production engine, the ring of snubbers clearly visible on the early narrow fan blades.

JT9D indirectly led to some problems for Rolls-Royce even before the launch orders were placed for the RB.211.

After it failed to sell the RB.211 for the initial version of the Boeing 747, Rolls-Royce became progressively keener to win a major order for the engine. It was becoming clear that large engines would account for the biggest share of the future orders for commercial engines, as they do today.

Rolls-Royce forecasts showed that only the Spey would contribute significantly to its civil turnover after 1970. In February 1967 a board committee was given figures showing that sales of ex-

The Rolls-Royce solution, as shown here on the RB.211-535E4B, as fitted to the Boeing 757s of American Airlines, is the wide-chord fan blade, without snubbers, giving more thrust and better fuel efficiency for any given fan diameter. A wider chord for each blade gives a reduced number of blades per fan – a reduction from 33 to 22 on the -535 for instance.

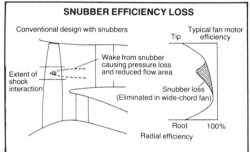

With the engine running the snubbers, fitted to stabilise the fan blades, prevent flutter and fatique failure but reduce the engine's efficiency.

isting engines would fall from £58.9 million in 1969 to £3.5 million in 1975, with spares revenue falling from £36.5 million to £31.9 million in the same period. The company would lose its position in the major league of aero engine suppliers if it could not launch a large new transport engine.

Launching a Big Engine

At this time Rolls-Royce was proposing two large new engines, both of three-shaft design: the RB.207 and the RB.211. The RB.207 design was the larger of the two and was eventually rated at thrusts of well over 50,000lb before it was abandoned. It was specified for the European Airbus in the late 1960s and proposed for the BAC Two-Eleven project and the twin-engined US airbuses.

In February 1967 a Rolls-Royce board committee decided that offers of the RB.207 and RB.211 engines should be made for the European Airbus and Lockheed L-1011 aircraft. By May 1967 US studies indicated that an engine of 33,000lb thrust would be needed if a

Fresh market for a developing product. The RB.211-524L Trent will be the first Rolls-Royce engine to power an Airbus Industrie product. The artist's impression shows the Airbus A330-300 to be operated by the launch customer Cathay Pacific.

three-engined US airliner went ahead instead of the twin-engined types initially proposed. In June American Airlines decided in favour of a trijet rather than the big twin-engined design.

Lockheed was a potential supplier and in response to the request for a proposal Rolls-Royce offered the RB.211-06 engine in June 1967. Douglas was also considering a three-engined aircraft which became the DC-10.

Rated at 33,260lb thrust on a 90 °F day, the RB.211-06 was an ambitious design offering high thrust, a reduction of 25 per cent in specific fuel consumption over earlier engines and very low noise levels. It incorporated many technical advances including its three-shaft design, use of new materials such as the Hyfil carbon-fibre composite and an annular combustion system. At that time Rolls-Royce had only limited running experience with the RB.178 demonstrator and test running had not yet started with the three-shaft Trent.

The thrust of the RB.211-06 was over 50 per cent greater than for any previous Rolls-Royce engine. It was planned to design and develop the engine for service entry in 1971, about four years ahead. When the offer was made the company had begun design and development work on the engine and the first run was planned for August 1968.

At a main board meeting in July 1967 Rolls-Royce authorised final bids for the RB.207 to power the European Airbus and the RB.211 for the US trijets and for possible application in the BAC Two-Eleven. In the same month engineering data and commercial information on the RB.211-06 were given to Douglas. In September of the same year the aircraft companies were told that Rolls-Royce was pursuing a parallel programme to provide solid titanium fan blades for the RB.211 in case these were needed in place of those planned with Hyfil carbon-fibre composite, abandoned later.

After a period of hectic and competitive negotiation the launch order for the RB.211 was announced on 29 March 1968. Lockheed ordered 150 'ship sets' of the RB.211 (450 engines) and at the same time TWA and Eastern Air Lines announced orders for a total of 94 TriStars, while Air Holdings of the United Kingdom ordered fifty aircraft.

These fifty helped to meet the political need that the US balance of payments should not suffer through a large offshore purchase of British engines. In the event these fifty aircraft were progressively taken by the non-US airlines which later ordered TriStars.

When Lockheed and Douglas made bids to sell a wide-bodied airliner with transcontinental range to large US airlines, it was assumed that only one type would be ordered. In the event two of the launch customers, American Airlines and United Air Lines, selected the Douglas DC-10-10 with GE CF6 engines and the other two chose RB.211-powered Lockheed L-1011-1s.

This had adverse consequences for all the manufacturers involved because the launch and follow-on orders for these aircraft and their engines had to be shared between the manufacturers. It was unfortunate for Lockheed, because the TriStar represented its re-entry into the civil market following its experience with the propeller-turbine Electra.

Douglas was able to sell a long-range version of the DC-10 soon after it first sold the medium-range version, but Lockheed did not launch the smaller long-range TriStar L-1101-500 until much later. The result was that Lock-heed became the first, and so far only, manufacturer to abandon production of its family of wide-bodied airliners when it concluded the TriStar programme after 250 aircraft had been sold.

Development Task

Because of the tough negotiations and increases in aircraft performance which preceded the launch of the TriStar and DC-10, the take-off thrust required for the initial production RB.211-22 engine rose to 40,600lb by March 1968.

This compared with the 33,260lb of the RB.211-06 offered by Rolls-Royce less than nine months earlier; the manufacture of development engines to this standard was under way. This increased thrust had to be provided by an engine of the same size as the original RB.211-06.

The engineering staff at Derby faced an immense task. The three-shaft RB.211 which was scheduled to enter service in late 1971 now had to provide nearly twice the take-off thrust of the largest engine the company had produced previously. Much of the technology it depended on was as yet unproven. Compared with the RB.211-06 the specification had changed in many ways which included its cost, weight, noise guarantees and use of advanced noise-attentuation materials.

The company was heavily loaded with the RB.211 programme and wanted to avoid diversion of its effort on too many programmes which did not generate useful production orders. This point was made by Sir Denning Pearson of Rolls-Royce in January 1970 in a letter to the Ministry of Technology saying, 'We are not prepared to provide an engine for launching another BAC aeroplane (the BAC Three-Eleven) without adequate stretch built into the original engine. The sale of seventy-nine Tridents and 171 BAC One-Elevens has required from Rolls-Royce six quite different ratings of engine, each requiring substantial development and type testing. When you compare this with the fact that the Boeing 727 has sold well over 500 aircraft with exactly the same P&W engine you can understand why we find it difficult to compete and make profits, even with our lower labour rates'.

A good production run was achieved with the RB.211-22B, over 600 of which have been produced and 560 of which continue in service today on 143 aircraft. But for competitive reasons it has not been so easy for later versions, although the RB.211-535E4 is doing well.

The task of developing the RB.211-22

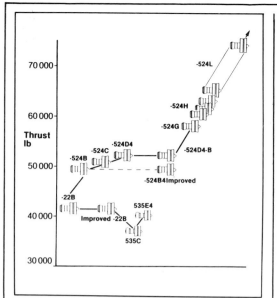

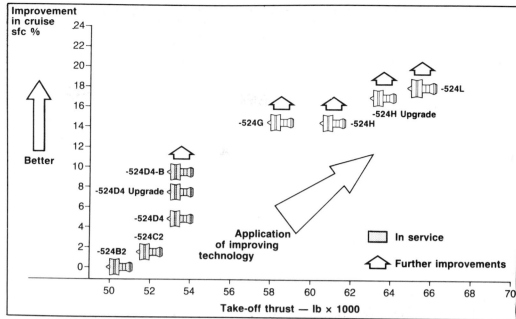

The RB.211 family from -22B to -524L Trent and beyond.

RB.211 development to increase thrust and efficiency.

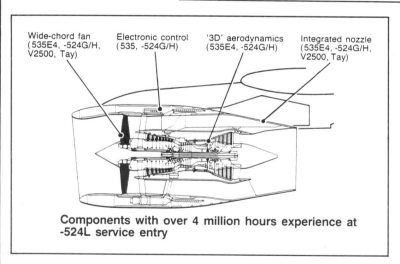

-524G/H (1989)
58 000-60 600 lb

-524G/H (1991)
58 000-65 000 lb

-524L (1993)
65 000-75 000 lb

− 2% sfc

− 2% sfc

To pace market requirements for powerplants for medium- to long-haul jet airliners – almost half of the total civil jet engine market – is the objective for RB.211-524 performance development. More power, with better specific fuel consumption while at the same time improving reliability and cost of ownership (fewer inflight engine shutdowns, fewer engine removals and longer component lives) are the benefits to be derived from a policy of continuous development of the modular -524, itself a stepped power improvement from the basic -22B.

There is now to be commonality of core development for the three standards of RB.211 engine, the -534E4, the -524L and future -524G/H.

Wide-chord fan (535E4, -524G/H, V2500, Tay)   Electronic control (535, -524G/H)   '3D' aerodynamics (535E4, -524G/H)   Integrated nozzle (535E4, -524G/H, V2500, Tay)

Components with over 4 million hours experience at -524L service entry

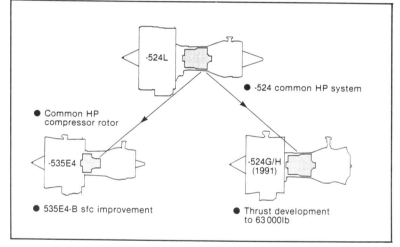

-524L

● -524 common HP system

● Common HP compressor rotor

-535E4

-524G/H (1991)

● 535E4-B sfc improvement

● Thrust development to 63 000lb

The RB.211-524L (Trent) will derive component in-service experience benefit from earlier engines in the series as well as from the smaller Tay and the international V2500 engine, all fitted with Rolls-Royce wide-chord fans.

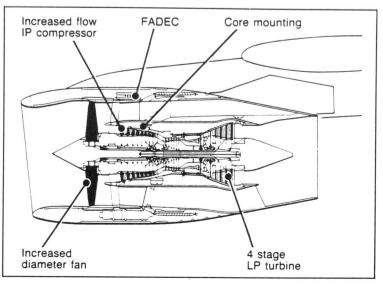

Increased flow IP compressor   FADEC   Core mounting

Increased diameter fan

4 stage LP turbine

The changes from the RB.211-524G/H of the 1991 standard to the -524L (Trent).

When big also means beautiful. In production at Derby in 1989, the -524G of 58,000lb thrust for the British Airways Boeing 747-400. Also in production is the -524H of 60,600lb for the British Airways Boeing 767. By the end of 1989 these two RB.211 variants will be mechanically identical, enabling the airline to use the same engine pool for both aircraft types.

to the committed timescale was formidable. It was made no easier in May 1970 when it was agreed with Lockheed that by March 1972 the take-off thrust of the engine would be increased to 42,000lb on an 84 °F day. The reason was that the weight of both the TriStar and the engine had increased.

Testing of the RB.211-06 engine began in August 1968. It soon showed that changes would be needed to obtain the necessary thrust at acceptable operating temperatures. The engine's specific fuel consumption was also too high.

Engine development continued against a background of great urgency and increasing development costs. The British Government had agreed to provide launch aid of up to 70 per cent of the engine's launch costs with maximum government payments totalling £47.13 million. Costs rose above the initial forecasts and the company's financial position deteriorated. Top management changes were made.

The first flight of the TriStar was preceded by engine testing in a VC10 with two of its four Conways removed and replaced by a single RB.211. The TriStar first flew on 16 November 1970 with RB.211-22 engines giving 33,500lb take-off thrust. Efforts continued to meet the engine's specified performance and the production programme which would allow the TriStar to enter service in December 1971 as planned.

In the event Rolls-Royce went into receivership on 4 February 1971 and its gas turbine activities were subsequently taken over by the State-owned company Rolls-Royce (1971) Limited. Later it regained its independence as Rolls-Royce plc.

Ironically, on the morning of 4 February test results were sent to London showing that modifications to the engine were proving effective and thrust levels were approaching those required. After a period of uncertainty the RB.211 contract was renegotiated and the TriStar went into service in April 1972, five months later than originally planned. The engineering achievement was very great in view of the difficult targets to be met. The official report on events leading up to the receivership states, 'The performance of the company on the project was remarkable. The extent to which countless individuals contributed exceptional personal efforts to the task should not go unrecorded'. That dedication has continued since then through the ups and downs of the RB.211 programme.

Expanding the Range

In common with the other large turbofans, the RB.211 did not at first achieve the levels of reliability targeted for these engines. This is not surprising as any totally new engine requires a period of years before it becomes mature and its operating costs and inflight shutdown and unscheduled removal levels are reduced to the outstanding levels now achieved.

This involved considerable effort with the RB.211, both to improve the operating lives of key parts, such as its high-pressure turbine blades, and to develop the engine in service to overcome the initial problems which arose. These included two early fan-disc fail-ures which led to intensive technical investigation and rapid replacement of the fan-disc material used.

As with all engine families, the later derivative versions have proved much more reliable when first introduced. The second, third and fourth generations of an engine benefit from all the operating experience with earlier versions and also incorporate improvements in technology as advances are made. For example, the RB.211-535E4 engine for the Boeing 757 established record levels of reliability and low ownership costs as soon as it entered service.

As the 42,000lb thrust RB.211-22B was being established in service efforts were made to develop and sell more advanced versions, both for new versions of the TriStar and for other types of aircraft.

Technology Benefits

In the two decades since the early days of the RB.211 great strides have been made in the technology of the engine business. This does not apply only to the technology which gives the engines higher performance, better fuel consumption and improved reliability; it applies across all aspects of the business because of the introduction of much cheaper and more effective computing power.

Engine components are now designed using computerised three-dimensional gas-flow techniques which have led to new component shapes and much better performance. Computerised design and manufacturing systems have greatly increased efficiency and led to the introduction of automated factories. The computer has made a great impact across every aspect of the business.

The management of technical advance has also made great strides, with advanced engineering programmes which demonstrate new technology and prove its effectiveness before it is incorporated in new or established designs of engine.

The steady improvement of the RB.211 engine has benefited from its modular construction. It has proved possible to incorporate technical improvements in individual engine modules and thus to upgrade their performance so that operators can embody these advances in their existing engines. This technique has been applied very successfully to the RB.211-524 engine family.

As a result of these improvements operators of RB.211-powered Boeing 747s have been able to upgrade the performance of their engines and to improve the payload/range performance of their aircraft. This has been particularly valuable for long-range operators such as Cathay Pacific Airways. Improvements in cruise specific fuel consumption have permitted the profitability of their aircraft to be greatly improved over long-distance non-stop routes such as the direct flight from Hong Kong to London.

Advances have been made in RB.211 performance through improvements in high-temperature technology, materials and manufacturing techniques. The life of RB.211-22B high-pressure turbine blades has been transformed with the use of directionally solidified turbine blades. As a result RB.211-22B engines can now remain on the wing of aircraft they power for up to 20,000 hours. Later engines make use of single-crystal cast turbine blades.

One of the most significant advances pioneered by Rolls-Royce is the wide-chord fan blade, introduced on the RB.211-535E4 engine nearly five years ago. When used in conjunction with a common exhaust nozzle this provides a reduction in engine specific fuel consumption of between 4 and 5 per cent.

The fan on a high-bypass engine acts like an efficient propeller to provide nearly 75 per cent of the engine's thrust. It is driven by a shaft providing 50,000hp or more and driven by the low-pressure turbine. Any advance which improves the efficiency of this fan is very valuable.

After a lapse of some twenty-five years a Rolls-Royce turbine engine, the RB.211-524L, has been chosen to power a McDonnell Douglas commercial aircraft, the MD-11 trijets on order for Air Europe (artist's impression).

Conventional solid fan blades have snubbers which act in effect as a reinforcing ring when the engine is viewed from the front and stabilise the blades. Unfortunately, the snubbers also reduce the airflow through the fan disc and lead to an efficiency loss where shockwaves springing from them restrict the flow. Rolls-Royce has eliminated this difficulty by developing wider and stiffer fan blades of fabricated construction which do not need snubbers. These blades are much more aerodynamically efficient and also permit a larger airflow to pass through a given fan area. Their installation on later engines of the RB.211-524 family has permitted take-off thrust to be increased without using a larger-diameter fan.

Other benefits of the wide-chord fan are its better resistance to foreign-object damage and the noise reduction which it provides. This has helped operators of Boeing 757s with RB.211-535E4 engines because they can operate at night into some US airports where the same aircraft with the competing engine cannot meet the tight noise limits imposed.

Following its success on the 535E4 engine the Rolls-Royce wide-chord fan is used in the Tay turbofan and the International Aero Engines V2500 as well as later large RB.211s.

It is not always easy to appreciate the high power loadings of components in advanced engines such as the RB.211. A single wide-chord fan blade on an RB.211-535E4 engine, for example, requires some 2,000hp to drive it – more than double the power of a Merlin engine on an early Spitfire – and provides most of the power needed to fly five passengers at speeds well over 500mph.

Similarly, the technology of high-pressure turbine blades is impressive. One RB.211 turbine blade is 4in long. It operates in a gas stream which is hotter than the melting point of the metal from which it is made and extracts 500hp from this gas stream. It is prevented from melting by an internal air cooling system plus a system of surface cooling by air supplied from small holes running along its surface. These provide a continuous thin film of cooling air which helps to insulate the blade surface so that it is not directly exposed to the hot combustion gas which drives the blade. At the same time the heated blade revolves at about 10,000rpm and is subjected to a centrifugal load equivalent to a London bus suspended from it. In these operating conditions the blade must run effectively and reliably for many thousands of hours during which the aircraft it helps to power will fly at least 10 million miles.

The development of such technology has taken many millions of pounds and thousands of man-hours of effort. It indicates why products such as the RB.211 are ideal for Britain, with their high added value based on technical knowledge and experience gathered over many years.

Continuing Improvement

Development of the RB.211 family is continuing to increase thrust, reduce fuel consumption and enhance the commonality between engine variants.

The RB.211-524G/H already provides a 14 per cent better cruise specific fuel consumption than the initial RB.211-524 engine. A further 2 per cent improvement in cruise sfc will be achieved by late 1991, plus increased take-off thrust capability

to 63,000lb. This will come through the introduction of a new high-pressure compressor.

The RB.211-524L (Trent) will provide a further 2 per cent sfc in 1993 through its increased bypass ratio and advanced intermediate-pressure compressor and low-pressure turbine. It has a thrust capability in the 65,000 to 75,000lb range, with development potential to 80,000lb thrust.

The engine benefits from advanced features proven in other RB.211 engines, the Tay and the V2500. These will have over 4 million hours of operating experience behind them when the 524L engine enters service. The 524L follows the derivative design approach applied to current RB.211s and which has resulted in outstanding levels of reliability in service, including the best engine reliability in the 747 for the 524D4, according to Rolls-Royce.

Core commonality of the 524L with 524G/H engine will be introduced from 1991. All of these engines will have the same core, consisting of the HP compressor, combustor and HP and IP turbines. In addition the HP compressor rotor drum will be common with the 535E4 from 1992.

A new version of the 535E4 entered service this year. Known as the 535E4B, it has an increased take-off thrust of 43,000lb and will power the Boeing 757s ordered by American Airlines. Aircraft deliveries began in summer 1989.

The Future

As indicated, the RB.211 family of engines has probably two-thirds of its lifespan still to come. It enters the 1990s with more aircraft applications than ever before and the prospect of future development to at least 80,000lb take-off thrust.

Throughout its life Rolls-Royce has had no international partners in the RB.211 programme, apart from a re-latively short period when Rolls-Royce and GE took stakes in each other's programmes – GE in the RB.211-535E4 and Rolls-Royce in the CF6-80C2.

Today, Rolls-Royce is going ahead to develop the large new Trent without any launch aid from the British Government but with risk- and revenue-sharing shares in the progamme taken by two Japanese companies, Kawasaki Heavy Industries with 4 per cent and Ishikawajima-Harima Heavy Industries with approximately 5 per cent.

The future of the RB.211 engine family is likely to be at least as interesting as the past, with continuing competition from the engines produced by Pratt & Whitney and General Electric. As ever, human resolution and dedication will be as important as technology in ensuring continued success. These engines bring valuable export earnings to Britain and are standard bearers for British engineering excellence throughout the world. □

A miscellany of RB.211-powered commercial airliners

1972. TWA operated their first TriStar services in June.

1972. In April Eastern Air Lines became the first airline to operate the Lockheed L-1011-1 TriStar powered by RB.211-22B engines.

1975. A Lockheed L-1011-200 TriStar of Saudia – Saudi Arabian Airlines demonstrates its ability to ferry a spare RB.211, slung inboard of the aircraft's normal starboard engine on a special mounting.

1977. A Boeing 747-200 (four RB.211-524) of British Airways.

1979. The long-range Lockheed L-1011-500 TriStar (-524 powerplant) of British Airways.

1987. A Boeing 757 (two RB.211-535E4) of the United Kingdom carrier Air 2000.

1989. A Boeing 747-400 (four RB.211-524G) of Cathay Pacific.

1989. The first Boeing 767 (two RB.211-524H each of 60,600lb thrust) for British Airways.

Power House. The design, development and test facilities of Rolls-Royce at Derby, as seen from a TriStar.

The Great Flight to Australia, 1919

Stanley Brogden

While the names of Ross Smith and Keith Smith are imperishable, not so well known are the others who bid to be first to fly to Australia from Europe. The author recounts the qualities of these Australians and one Frenchman, in the spirit of the age in which they flew. Immediately following the end of the First World War, a struggle in which Australia, vast in extent and half a world away, with a population of only some five million, had sent her sons to fight, some of these men, brave and well qualified, were determined to return home by air.

Australia is perhaps the one country that was held back by what historian Geoffrey Blainey has called the Tyranny of Distance. There were in fact two tyrannies – the distance from Britain and the great empty stretches at home. Five million Australians were huddled in the southeast corner of an empty continent of 3 million square miles. Northern and Central Australia were weeks away from the cities; a stockman in the Kimberleys of Northwestern Australia who broke a leg might be 2,000 miles from help, as an extreme example. Overseas trips took months just to arrive and return.

In 1919 William Morris Hughes, Prime Minister of the Commonwealth of Australia, knew all about the tyrannies and he was one of the first statesmen to use an aeroplane to cross the English Channel for the Paris Peace Conferences. He had spent the previous Christmas Day at a military hospital at Conham Hall, Kent, with some Australian Flying Corps (AFC) pilots who told him they would love to fly home. He at once seized the idea, took notice of the £10,000 prize offered by the *Daily Mail* for an Atlantic flight, and cabled his Cabinet in Melbourne to approve a £10,000 prize for the first flight to Australia.

His Chief of the General Staff, Gen Legge, was an aviation enthusiast, first military officer to fly in Australia, before the war, and founder of the Australian aero club movement. Legge advised Hughes to restrict the entrants entirely to Australians. This was agreed.

Setting the Scene

The most decorated pilot in the AFC, Capt Ross Smith MC, DFC, AFC, read about this in Calcutta. He was the second son of the manager of Mutooroo sheep station near Broken Hill – a true Austra-

lian! He enlisted in the 3rd Light Horse Regiment in August 1914, had three stripes within weeks and left for Egypt in the first convoy as regimental sergeant-major. For a twenty-one-year-old public school boy son of a well-off Adelaide family, this was a certificate of leadership, for the Light Horse were a tough breed. The only other public school boy I know of with a similar distinction, who was even younger, was, curiously enough, Hudson Fysh, co-founder of Qantas, later Sir Hudson Fysh.

Smith, Fysh, George Jones (later to become Chief of the Air Staff for a record ten years, 1942-52) and Kingsford Smith all fought as dismounted Light Horsemen on Gallipoli against the Turks. Ross Smith was commissioned in The Penin-

Seen at Brooklands with their new Vickers Vimy. Lieut Keith Smith, Capt Ross Smith, Sgt J M Bennett and Sgt W H Shiers. (*Vickers*)

sula, a distinction in itself, then was invalided to England. He returned to fight in Palestine in such cavalry battles as Romani before going over to the AFC's first unit, No. 1 Squadron, when it came to Palestine in 1916.

Smith was first an observer, like Fysh, then a pilot, wounded four times, awarded two Military Crosses and three Distinguished Flying Crosses. His commanding officer, later Air Marshal Sir Richard Williams, who founded the Royal Australian Air Force, told me years later that only an error by an RAF clerk had robbed Smith of a DSO which Williams recommended in 1918.

Smith stood out in No. 1 Squadron, a very impressive achievement in a squadron that produced six knights and a brace of air marshals, as well as the only Australian air VC of the war (Lieut F H McNamara, later Air Vice-Marshal). Nor was it easy to be a leader in the First Australian Imperial Force (AIF), whose other ranks expected officers to be leaders without apology.

Williams was the senior Australian air officer in London when the £10,000 England–Australia flight was being organised. He did not regard it as a race because the entrants left days apart and the most obvious winner for a time was 'a French fellow, Poulet', as he put it. Yet Ross Smith had the experience of having

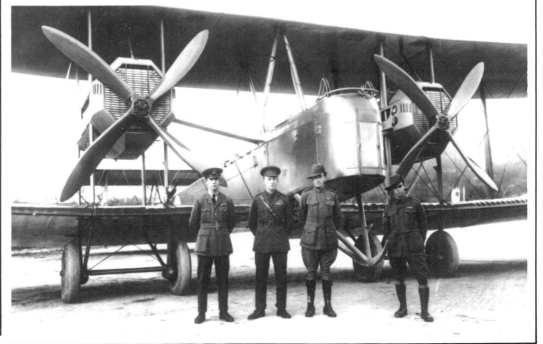

As it was in the beginning. London to Australia, 1919 fashion.

flown over a considerable part of the route to Australia which, in the event, no other pilot had done. Furthermore, he had the basis of a crew in Sgts J M Bennett and W H Shiers, both highly experienced aircraft engineers with pre-war experience. Bennett had won the Meritorious Service Medal with No. 1 Squadron, in which Shiers also served.

Smith is praised at some length in *The Seven Pillars of Wisdom* as T E Lawrence's pilot in the Handley Page O/400 and other aircraft. A Handley Page O/400 was flown out from England by Brig-Gen A E (Biffy) Borton who, at the war's end, sold the idea of flying it to Baghdad, then RAF HQ Middle East, and on to Calcutta for the Viceroy's Cup, and possibly to Australia for the Melbourne Cup in October. But first things first. Smith, the two engineers, Borton and Maj-Gen Geoffrey Salmond, who commanded the RAF in Mesopotamia (now Iraq), left Cairo on 29 November for Calcutta via Damascus and Baghdad.

They flew to India via Basra, Bushire and Bandar Abbas to Karachi and then on to Delhi. Borton, who commanded the aircraft on the flight from Cairo to Calcutta, was the first pilot ever to fly from Britain to India. The Cairo–Calcutta flight itself was the longest ever made up to the end of 1919. Borton seems to have been a character over-looked to some extent by historians. According to the Australian NCOs he was also one of the few senior British officers who understood the balance between Australian officers and NCOs.

Borton became serious about going on to Australia. He secured Whitehall approval to charter a vessel to survey the route east of India in February 1919. The survey was held up when the ship was burned out at Chittagong; the Handley Page's crew were lucky to survive that, but it merely whetted Borton's appetite.

The Indian Government provided another ship for three months' work surveying facilities, or the lack of them, down to Borneo. When the team returned to India, their O/400 had been

In 1922 Sir Ross Smith's book *14,000 Miles Through the Air* was published by Macmillan and Company Limited. The table reproduced here repays study. After reaching Darwin, the crew flew on by short stages to Melbourne; after a month they continued to Adelaide, in all a further 52hr 25min flying time and 3,290 miles past Darwin. Ross Smith's quiet but triumphant statement to complete the book will be understood and applauded by all who have attained flying goals much less ambitious: 'At the request of the authorities I flew the machine on to Adelaide, my native city, and thus realised to the full my ambition of flying from London to my own home'.

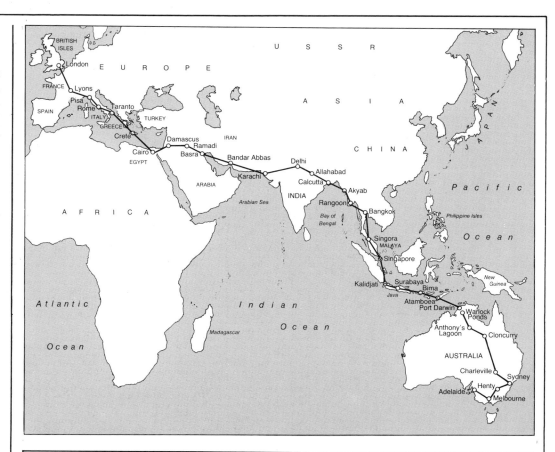

TIME TABLE WITH DISTANCES

Date	Hour	Route	Time in Air Hrs	Time in Air Mins	Distance in Miles
12/11/19	0905	London to Lyons	6	20	500
13/11/19	1006	Lyons to Pisa	4	45	380
15/11/19	1000	Pisa to Rome	3	20	180
16/11/19	0904	Rome to Taranto	2	35	260
17/11/19	0815	Taranto to Suda Bay (Crete) . . .	5	40	520
18/11/19	0812	Suda Bay to Cairo	7	20	650
19/11/19	1024	Cairo to Damascus	4	10	450
20/11/19	1015	Damascus to Ramadie	6		420
21/11/19	1315	Ramadie to Basra	3	30	350
23/11/19	0635	Basra to Bundar Abbas	7	40	630
24/11/19	0700	Bundar Abbas to Karachi	8	30	730
25/11/19	0740	Karachi to Delhi	9		720
27/11/19	1020	Delhi to Allahabad	4	25	380
28/11/19	0830	Allahabad to Calcutta	5		470
29/11/19	0830	Calcutta to Akyab (Burma)	4	45	420
30/11/19	0730	Akyab to Rangoon	4	15	330
1/12/19	0655	Rangoon to Bangkok	6		400
2/12/19	0745	Bangkok to Singora	6		470
4/12/19	1015	Singora to Singapore	6	20	480
6/12/19	0700	Singapore to Kalidjati (Java) . . .	9		640
7/12/19	0735	Kalidjati to Surabaya	4	20	350
8/12/19	1200	Surabaya to Bima (Sumbawa) . .	5		420
9/12/19	0945	Bima to Atamboea (Timor)	5	30	440
10/12/19	0835	Atamboea to Port Darwin (Australia)	6	30	470
13/12/19	1023	Port Darwin to Warlock Ponds .	4	20	220
14/12/19	0900	Warlock Ponds to Cobb's Creek .	5	30	300
17/12/19	1800	Cobb's Creek to Anthony's Lagoon		15	20–
18/12/19	1005	Anthony's Lagoon to Brunette Downs	1		50
19/12/19	1105	Brunette Downs to Avon Downs	2	45	180
20/12/19	0745	Avon Downs to Cloncurry	3		230
22/12/19	0650	Cloncurry to Longreach	4	40	300
23/12/19	0705	Longreach to Charleville	3	40	330
12/ 2/20	1100	Charleville to Bourke	4		260
13/ 2/20	0900	Bourke to Narromine	3		230
14/ 2/20	0700	Narromine to Sydney	4	15	200
23/ 2/20	1005	Sydney to Cootamundra	4	15	240
24/ 2/20	1000	Cootamundra to Henty	1	10	80
25/ 2/20	0600	Henty to Melbourne	3	5	220
23/ 3/20	0700	Melbourne to Adelaide	7	30	430
		Total flying time and distance, London to Adelaide	188	20	14,350

In addition to the above, several hours were spent in the air making test flights at various places.

Some of the other contestants

The Sopwith Wallaby G-EAKS. (*Roger Jackson*)

The Martinsyde Type A Mk.I G-EAMR leaving Hounslow. (*Roger Jackson*)

The Blackburn Kangaroo G-EAOW after its forced landing in Crete. (*Roger Jackson*)

committed to the Afghan War and destroyed on the northwest frontier, so Borton and the Australians went off to London by ship.

Borton rang Vickers to get Smith an aircraft only to find that Salmond had already discussed the idea with Vickers executives hopeful of some commercial interest in a Vimy development.

Ross's elder brother (by two years), Keith, was not a strong fellow and, rejected by the AIF, had paid his fare to London and joined the RFC after having the operation they recommended, also out of his own pocket. With 500 hours' flying experience and considerable expertise in aerial navigation, he joined the team. While he was not in Ross's class as a pilot, he played a very great part in the subsequent success of the venture as co-pilot and navigator.

As international regulations had only then been thought of, the task of securing over-flight permission from the countries *en route* to Australia was a considerable chore for Williams. The Royal Aero Club drew up the rules in liaison with the Australian Aero Club in Sydney.

The Australian Government ruled that the winner must reach Australia before 1 January, while Williams and Hughes agreed on 8 September as the first take-off date, however nobody was ready to start by then. Vickers did not enter the Vimy until 18 October.

The Royal Aero Club ruled that aircraft must take-off from Hounslow, west of London, or, in the case of a flying-boat entry, Calshot Seaplane Station. There was to be only one control station on the whole route to Australia, at Singapore. Aircraft would be allowed one engine change. They could be towed up to 100 miles in emergency. There must be no replacement of aircraft for any reason and a minimum range of 500 miles was set. If the pilot was not a qualified navigator then one must be included in the crew. Fanny Bay at Darwin was the selected Australian destination, the Fanny Bay landing strip being near the gaol.

Back in Sydney, Reginald Lloyd announced he would fly a Handley Page

Snow in November at London's Heathrow Airport is virtually unknown, but on 12 November 1919 at Hounslow Heath aerodrome, within a couple of miles of the threshold to Runway 27L today, the snow had settled as G-EAOU was being prepared for take-off. (*Vickers*)

aeroplane and start a regular air service by October. What he actually did was to become the first man to ride a motorcycle across the continent. More important, the Australian Aero Club in Sydney organised the first survey of facilities across Australia, which meant finding level places on which to land. Lieuts Hudson Fysh and P J McGinness, ex-No. 1 Squadron, set off in a motor car for a really tough pioneering venture in itself. They were soon to found Qantas.

In Australia, public opinion was mixed before the race. The Melbourne *Age*, in the then-capital city of Australia, described it as a circus but changed its tune after the event. Some MPs appealed to the government to withdraw such a scandalous offer of taxpayers' money.

Among all the publicity given to the England–Australia race in the last seventy years, the forgotten achievement was a great feat of endurance and flying by Capt H N Wrigley and Sgt A W Murphy, who left Point Cook, near Melbourne, on 16 November and made the first transcontinental flight in Australia in a B.E.2e. Murphy had flown T E Lawrence in Arabia and later became an RAAF air commodore. Wrigley rose to become an air vice-marshal and ran RAAF Overseas HQ in London for most of the Second World War. Their flight was in many ways an epic achievement, for it was done to provide information for the race competitors who might continue on from Darwin and need data on flying conditions over areas where there had never been any flying before the pioneering B.E.2e effort.

The race, therefore, had very great effects in Australia even before the Smiths arrived. Nobody had thought of flying across the continent as early as 1919 as a serious proposition. That was now to be done twice, first by the B.E.2e and immediately after by the Vimy, a

On 8 December at Soerabaya within 1,200 miles of their goal, Darwin, it appeared that the prize might still elude them as the Vimy sank repeatedly into the boggy ground. A more stable surface was constructed from pieces of bamboo matting, pegged down. (*Vickers*)

combination which was almost as great a revelation to Australians as the three-year-old East-West Transcontinental Railway to Perth.

Back in England an AFC lieutenant, Frank L Roberts of Melbourne, would have been first off the line if A V Roe, for whom he was a test pilot, had agreed to install an Eagle VIII in an Avro 504K, as he disliked rotary engines. Frank Roberts was an airline pioneer back home.

First Away

The first man off was the French crack pilot, Etienne Poulet, in a Caudron G.4. He had been flying since January 1912 and was by far the most experienced and well-known pilot of them all. He actually made the flight to raise money to help the widow of a close friend, Jules Védrines, another veteran French pilot who had been organising a round-the-world flight through Melbourne and up the west coast of Asia to the United States and home to Paris. Védrines, however, flying since 1910, was killed in a crash in April 1919. He must have been quite a man, for Poulet gambled his entire life savings on the venture; neither Caudron, for whom he flew, nor France in general would help.

Poulet and Jean Benoist took off in the Caudron G.4 from Paris soon after dawn on 12 October, flying down the Lyons railway line, struck fog, returned to Villacoublay, and started off again on the fourteenth, a week ahead of the first Australian. Poulet, of course, was not a race entrant.

Race Entrants

The two pilots who were to make the most noise in the history of aviation, H J L Bert Hinkler and Charles Kingsford

Smith, were not in the race. The widely held view that Billy Hughes refused to allow Smithy to compete because in conversation he found he and his crew had no navigational experience or skill has been denied by Williams in his autobiography. Whatever the truth, that crew did not fly the Blackburn Kangaroo they proposed.

Sopwith entered a Dove to be flown by Hinkler, a fine RNAS pilot, subject to money being found elsewhere, which it wasn't – another 'if' of aviation history. He was by far the most experienced pilot, though how a civil version of the Pup fighter with an 80hp Le Rhône rotary engine would have got on with a single-man crew, even with ten hours' tankage, is hard to judge. Moreover, he had organised all this before the £10,000 prize was announced.

Sopwiths entered their official Sopwith Wallaby six weeks later, to be flown by Capt G C Matthews, another former Light Horseman, who had a master mariner's certificate, with Sgt T D Kay, a motor mechanic who had secured his pilot's licence after the war while working for Rolls-Royce. The famous Harry Hawker test-flew the Wallaby the day Poulet left Paris and told Matthews he ought to 'eat the Caudron'.

Matthews took off on 21 October, nine days after Poulet, receiving a telegram from Hughes to say he should ignore the thirty days' limitation on the Hounslow–Darwin trip because whatever time he took would be a wonder of the world anyway, and the first pilot home would win the prize regardless. Billy never worried about toeing the official line. All competitors received the same telegram.

The Wallaby was damaged landing at Cologne and Kay was hurt, but later they got away in bad weather to be forced

10 December. Taxiing up to the fence at Darwin after landing. (*Qantas*)

down eventually 100 miles from the new nation of Yugoslavia. There they had adventures resembling a TV serial for boys; arrested and imprisoned as Bolsheviks, they struggled on only to crash in Bali (where better?) on 17 April 1920. Aircraft and mechanic were badly hurt and both, with the pilot, completed the journey by sea. Matthews later started his own small airline back home, in the 1920s, after flying for Qantas.

Third off was the Vimy on 12 November, but it is a lesson in achievement to describe the opposition first. Williams considered that the aircraft and its crew made the Vimy unbeatable well in advance of the race. He was right.

Fourth off, on 13 November, a day after the Vimy, were Capt R M Douglas, AIF infantryman on Gallipoli who won the Military Cross and a Distinguished Conduct Medal at Polygon Wood, and Pozières who became a pilot in 1918 and instructed in the AFC training squadron. Both he and Lieut J S L Ross qualified as navigators before the race. Ross had been badly wounded flying an S.E.5 with No. 2 Squadron AFC in France.

They flew an Alliance P.2 Seabird named *Endeavour*, a now forgotten biplane with a single Napier Lion 450hp engine, designed specially for long-range operation with a cabin, no less, and upholstered seats for the crew. Compared with all the other entrants, this belonged to a later age. They even had a radio which, during tests, picked up transmissions from Panama.

This team would have got away days before the Smiths but for an accident which the media never heard of, plus bad weather. They fully intended to break a world record by flying non-stop to Brindisi at an average 100mph, which would have made the Smiths look silly. But spectators saw the aircraft spin down from about 1,000ft following engine failure and crash in Surbiton. The sequel was horrendous, for Douglas' fiancée stood up at the coroner's inquiry and demanded justice against the manufacturers, which began a very sad sequence of events. Many pilots blamed the enclosed cabin for affecting the pilot's view.

On 21 November the Blackburn Kangaroo left Hounslow commanded by Capt Hubert Wilkins, later knighted for his Arctic and Antarctic accomplishments, as navigator, and Lieuts Valdemar Rendle and D R Williams as pilots, with Lieut Garnsey H M St Clair Potts as mechanic. Rendle had started a flying school in Queensland at the age of eighteen on the strength of gliding experience, with a Caudron. After the race he planned, but never flew, full-scale aerial services between London and Sydney and a pre-Kingsford Smith transpacific flight. He ended up selling kitchen equipment. Both Williams and Potts were experienced pilots so, as Wilkins was also qualified, this was a four-pilot crew. The Kangaroo was a bomber of more than 8,000lb gross weight powered by a brace of 275hp Rolls-Royce Falcons. As mentioned, it had been the aircraft designated for Kingsford Smith, with Rendle and two others.

This aircraft struggled through to Suda Bay in Crete after many very bad experiences, only to be forced to abandon the attempt when one engine became irreparable away from England. Nothing had gone right.

Second only to Ross Smith for gallantry awards was Capt C E Howell, an AIF infantry sniper who transferred to the RAF, as did Smithy, and won the DSO, MC and DFC in September-October 1918 flying Sopwith Camels and shooting down thirty-two German aircraft. He should have been the star. His mechanic,

10 December, Darwin. Hudson Fysh (back to camera) greets Ross Smith. (*Qantas*)

who had studied navigation for the flight, Sgt G H Fraser, had another distinction in being the oldest man in the race at forty, for most of the crews were in their early twenties.

These two flew a Martinsyde Type A Mk.I with a single Rolls-Royce Falcon, taking off on 4 December. They struggled through Dijon, Pisa, Naples and Taranto in weather that became worse as they

10 December, Darwin – the perfect 'I was there too' picture. In white, resplendent (centre), the governor of the Northern Territory and the chief justice. Flanking them the Vimy's crew – Shiers, Bennett, Ross Smith, Keith Smith and the mayor. Even the customs officer, E P Geraghty (extreme right), seems satisfied, while near the left of the picture Hudson Fysh, co-founder of Qantas, appears in bush hat. (*Qantas*)

flew, and crashed into the sea off Corfu, for no known reason, on the way to Athens on 9 December. Howell's father rejected the verdict of drowning and in Melbourne called a number of public meetings because he believed his son had been murdered on the beach for the money he carried. Howell's fiancée was on a ship bound for Australia and near Corfu at the time of the crash.

Last to take off and the only other aircraft to reach Australia was the de Havilland D.H.9 flown by Lieuts Ray Parer and J C McIntosh, who left on 8 January 1920 and reached Darwin on 2 August, then flying on to Melbourne. This trip has been almost folklore cult in Austalia over the years and the subject of at least one full book – and why not, when the age of flight was being ushered in with miracles? Parer became one of the main aviation pioneers on the goldfields of New Guinea while his Scottish-born mate was to die in one of Australia's first commercial airline crashes with the first Australian airline, West Australian Airways, with whom Smithy was a pilot.

The Smith Brothers

It was a miracle that the Vimy ever got to Australia considering the state of the art and the primitive conditions over most of the route. However, the winning team included three men who had flown together over much of the route. They were the unique factor. And it could be argued that they flew the best combination of airframe and engine available at the time, given the endurance required.

Each competitor had to organise his own ground support, with fuel and oil at each stop. The Smiths had the best in Shell fuel and C C Wakefield oil, as both firms had long aviation experience.

The standard Vimy that the Smiths took over at Vickers was to be fitted with four tanks to allow 516 gallons of fuel, forty of oil and ten of water. About 800lb of spares, tools and even fish-hooks were carried to cope with emergencies in the remote areas of the flight. They would take off 500lb lighter than the 13,500lb Vimy that flew the Atlantic seven months before, as they needed less fuel and oil. Their maximum range would be 1,040 miles, which they hoped would never be needed. In the Dutch East Indies an enthusiastic governor-general had actually suggested and was laying down two new landing fields to reduce what might have been one or two long stages.

The aircraft was stripped down at Vickers, who supplied staff to help Bennett and Shiers. Some parts of the fuselage were re-covered with tougher material. Designer Rex Pierson himself suggested replacing all fuselage tacking with screws, which was done with aplomb by factory girls. There was even a shock-absorber on the control column to ease the strain on Ross Smith's shoulders and back – he would even fly hands-off.

There was no wireless, however, for few radio transmitters east of Central Europe could provide navigational assistance. Keith navigated over land by map and over water by Admiralty compass, groundspeed indicator, drift indicator, experience and gut feeling. It worked. As Norman Macmillan put it, airmen in those days were on their own. Met forecasts were, if not a joke, either primitive or non-existent.

The weather in that particular November was dreadful. Sir John Alcock made the first test flight and Ross had to wait days for a second. The new-fangled system of registrations supplied them with G-EAOU, which was derisively interpreted as 'God 'elp all of us'. They thought of Poulet flying into the Middle East with all that hot sun. Ross considered him the main enemy by now.

Few people got out of bed to see the Vimy leave Hounslow at 0905hrs on the morning of 12 November, the brothers having agreed that the bad weather should be defied if they were to catch up with that sentimental Frenchman.

They left the snow on the Hounslow grass to face bitter conditions that would have forced many of the aircraft of the day to give up. Ross had to climb to within 1,000ft of his maximum limit to get above the snow as he crossed the Channel into France, then wing and fuselage icing forced him down to 1,000ft to see where they were. Even without the icing they were well above the design maximum take-off weight of the Vimy as a type. They had to remove their iced-up goggles and suffer severe pain from the snow. All three cockpits were full of snow, the four men like snowmen.

Overnight at Lyons, the French were not too quick to help, and a delayed take-off at 1000hrs meant landing at Pisa instead of Rome. They blessed the French as they risked taking off from the water-logged Pisa field, which then closed down for a solid week. Dreadful weather affected one engine that morning and they force-landed at Venturina to fix its gauge, but they reached the Italian Air Force base at Centocelle, Rome, on the second day of the flight. Ross had counted on twenty-four landings altogether, but it now looked as if it would be nearer thirty, and so it proved.

They uttered prayers of thanksgiving to find the RAF at Taranto next day and more *en route* to Crete, for they had to fly so low that an island looming up out of the murk almost caught Ross by surprise. That was close to disaster. The 650-mile, 7½-hour flight from Suda Bay to Cairo was no better, Keith navigating along the coast via Sollum.

By now the Australians had hoped to have sun on their faces, but the leg to Damascus was wicked with storms and pelting rain that soaked them all to the skin. A night's rest and more rain made their take-off a very risky business indeed as the wheels fought the deep mud. They had suffered headwinds a good deal of the way so far and on this leg they were forced to land at Ramadi, which Ross knew so well, to escape a night landing at Baghdad, 63 miles further on. And then the storm that night was of such severity that the crew and thirty airmen had to rush to the field to keep the Vimy on the ground. All the following morning was needed for repairs.

When they reached Shaibah, Basra, that afternoon they were exhausted, but the two engineers set to, spending a whole day on the engines. That set the pattern for the weeks ahead, several hours' hard work after landing, and back-breaking toil refuelling from tins by hand. At least the four had the comparative luxury of the officers' and sergeants' messes at Shaibah.

When they reached Delhi on Day 13, they were told that Poulet had left that very morning; he was now something of a worry, as the only threat. They intended to meet him at Allahabad but, too exhausted, they actually reached Akyab in Burma on 27 November when he was at Calcutta. Exhausted indeed – they worked up to six hours a day on the ground in the Indian heat, after flying in open cockpits, almost deafened.

They met at Akyab because it was race day at Rangoon. When the Vimy landed first at Rangoon, though Poulet had left an hour before, they were staggered by the 40,000 people who greeted them. The Vimy was the first aircraft ever to land at Rangoon; two in one day was a miracle.

The kite hawks were not so friendly. One smashed into one of the propellers on take-off next day and his mate into a wing, without grounding the aircraft, and, while dodging more of the birds, Ross Smith cleared the surrounding trees by what he later was told was a bare foot margin.

The mountain ranges between Calcutta and Rangoon rose to 7,000ft and the storms were dreadful. The Vimy was to fly slowly to accompany Poulet's slower Caudron as a safety measure, but one of the Caudron's engines was in real trouble. Ross had to climb to his maximum 11,000ft against headwinds, then made an error in rudder control that resulted in further proof of the Vimy's stability.

Flying blind, they came down to 4,000ft, praying, and were lucky to have passed the main range. In clear air, they flew across lush green river flats to the Siamese Air Force base at Don Muang. They had survived the first days of the monsoon. Poulet was not so lucky, being forced back to Rangoon, and when he tried to take off the next day he suffered two punctures. Luck had turned its face away. The third attempt the following day brought engine trouble in flight, forcing him to return again. Not until 8 December did he manage the 100 miles

Just over a day out of Darwin and still with 3,000 miles to go, the Vimy was forced to make a precautionary landing with a cracked port propeller. In Ross Smith's words they were '150 miles from a telegraph and 450 miles from the nearest railway'. An ingenious repair by Sgt Bennett made a take-off possible three and a half days later. Meanwhile, as this picture of his fellow crewmen taken by Bennett shows, life had to go on. Left to right: Keith Smith, Wally Shiers, Ross Smith. (*Qantas*)

to Moulmein, on his fifty-sixth day. The next day Poulet, without the power the Vimy had, could not get over the mountains. A cracked engine piston, broken propeller and heavy fog forced him back to Moulmein and on 10 December he heard of the Vimy's arrival at Darwin and gave up. He now passes out of this story, although he had done much more flying.

Ross Smith had planned on making the journey to Darwin in eighteen days'

On 23 December Keith Smith in pith helmet does his share of refuelling at Charleville. (*Qantas*)

flying time at the rate of 600 miles a day. He could not better 450 miles and hence twenty-four flying days, for reasons quite beyond his control. The weather was one – unpredictable, undodgeable. The effect of the weather on their eyes was a major factor on some stretches of the journey.

He had counted on a non-stop flight to

Singapore, some 1,000 miles before he heard *en route* about Poulet's arrangements at Singora on the way. When the Vimy got there, the crew found tree stumps more numerous than gallons of petrol, of which there should have been 500 gallons. There were in fact 150 litres, for Poulet. Where there weren't stumps there was water and in dodging the stumps Ross broke the tailskid mounting. A local junk shop provided a steel shaft, which was worked on a man-powered lathe in a ricemill at midnight. The rest of the night they spent protecting their aircraft against storm damage. Ten inches of rain fell. They secured fuel from Penang while 200 convicts were organised to improve the airfield.

Poulet had said the big Vimy would never be able to land on the Singapore racecourse and the Australians succeeded only when Bennett climbed down the outside of the fuselage to bring down the tail while landing. They now had 2,500 miles to cover in eight days before the 30 December midnight curfew. If they landed at Darwin at 0005hrs on New Year's Day they would have lost the £10,000. They then might not be as broke as Poulet, but it would hurt just the same.

They flew 650 miles to Batavia in nine hours in fine weather and the next day luck ran out as the Vimy broke through what appeared to be a firm surface at Sourabaya. The immense crowd wrecked the rest of the airfield surface, which was just mud under a firm top. A conference with local Dutch engineers brought out an acre or so of bamboo mats which made a reasonable surface within eighteen hours. While the two sergeants worked on the engines that night, the Smiths refuelled the tanks. Not only had they to handle the cans, they had to filter the contents through chamois leather, a painstaking and exhausting job that took six hours for more than 300 gallons. That was the way of it at most of their stops, a duty now forgotten in this age of unhandled fuel. In tropical heat, even by night, it was a chore.

That morning the problem of the wash from the Vimy's propellers blowing away the matting was cured by conscripting more mats from homes and pegging them all down in two narrow lines for 500 yards, the two lines accommodating a pair of mainwheels each. By contrast, and thanks to the Dutch governor-general's personal interest, the Sumbawa airfield was 'perfect' and the Timor field at Atamboea scarcely less appealing.

Dutch Air Force co-operation was in the same class, fuel and oil and every need being met without asking. But the field was marginal, the Vimy's undercarriage scraping the palm trees on the way out.

The last stage was over the Timor Sea, the leg which all subsequent record-chasers in the 1920s and into the 1930s feared. Five hours of lonely waters were in this case relieved by Billy Hughes' order that HMAS *Sydney*, the Australian cruiser which sank the *Emden*, should be cruising halfway across. The cheering sailors brought the first whiff of Australia to the Vimy crew. And the news flashed by wireless to Darwin had the pioneer port's small and multi-coloured population running out past Fanny Bay gaol to the airfield.

The Vimy landed at 1500hrs on 10 December, just twenty-seven days and twenty hours out of Hounslow, the first aerial vessel to cross the world. There were fifty-two hours to spare for the £10,000. The crew was whisked off to the gaol for the rather unbelievable entertainment of afternoon tea, Hudson Fysh told me later. He had been first to shake Ross Smith's hand. Wrigley and Murphy, the Australian transcontinental fliers, arrived in time for a long conference with the Smiths before the take-off for the south on the thirteenth. Those two intervening days and nights were devoted largely to work, the two mechanics slaving to counter the effect of 135 hours on engines never designed to cope with the experience of such a flight. They

Removing the port engine for overhaul at Charleville. A new propeller was made by the Queensland government in their railway workshops at Ipswich. (*Qantas*)

needed a week's work, but the 'Wet' was due and if the aircraft was to be seen in Melbourne then it must get off as soon as humanly possible.

After a very difficult series of flights across the country, the Vimy arrived at the Smiths' home town, Adelaide, on 23 March, and that was the end of the great adventure. The Smiths were knighted and Bennett and Shiers awarded bars to their AFMS and later commissioned in the Australian Army Reserve.

Sir Ross Smith and his brother soon decided on a round-the-world flight and by early 1922 were in England testing a Vickers Viking amphibian. Bennett, who had invested his share of the £10,000 (they all got shares in the prize, Australia-fashion) in a car repair business in Melbourne, sailed off to London as soon as the Smiths called him. He and Ross were killed on a test flight, Keith arriving just in time to see them killed, on 13 April 1922 at Brooklands.

Sir Keith Smith told me years later that his brother was a genius, and who would deny it? He was certainly a leader, for he created and led a four-man team that is unique in aviation history. 'We never had

At Adelaide today. Alone, apart, revered, G-EAOU and the sculpted figures of the four pioneers who operated her.

a cross word all the way from Hounslow', Shiers told me. 'Ross was a wonderful fellow'. Fysh told me Ross might have changed aviation history in the 1920s.

Sir Keith was Vickers representative in Australia until he died at the age of sixty-five in 1955. He had been on the boards of Qantas, Tasman Empire Airways and British Commonwealth Pacific Airlines by that time, always a salesman for British airliners, though with little success. Shiers went barnstorming in a Ryan monoplane for some years and was then chief engineer for Airlines of Australia before joining George Mills' Light Aircraft Company in Sydney. Mills was a dentist who flew with the AFC and afterwards started a firm to make parachutes under licence. Shiers died in 1968, aged seventy-seven.

Hudson Fysh was right: Ross Smith and his men did change Australian history. Although the first fare-paying passenger ever to land in Australia (at Darwin by Imperial Airways) did not arrive until 1934, the Vimy flight changed Australian public opinion of aviation. If the little B.E.2e and the larger Vimy could fly across Australia, there must be some truth in what the mad aviators claimed. And the idea of a full air service to Old England just might not be mad, after all.

West Australian Airways began operating up the West Coast in December 1921 and Qantas operated in the lonely stretches of Queensland. The entire style of life in Australia began to be transformed, the result of a dream that youngsters of the AFC had as the First World War ended. □

Regional Airport: Birmingham International

David Smith

Sixty years in the life of a would-be key United Kingdom regional airport are reviewed by our contributor. Today, Birmingham International Airport, with an enviable network of scheduled services to business and holiday destinations, has all the right ingredients for further expansion.

Birmingham International Airport
5227N 0145W

ICAO location indicator: EGBB
IATA location indicator: BHX
Visited on: 24 April 1989
Population resident within 50km: 3½ million
Runways: 15/33 2,405 x 46m, 06/24 1,315 x 46m
Operational: 24hrs
Aids: ICAO Category II Approach and Runway lighting
ILS Category I on Runway 15, Category II on Runway 33
NDB
Surveillance radar approaches down to ½ mile on Runways 06, 24 and 33 and 2 miles on Runway 15
Owned and operated by: Birmingham International Airport plc
Airport director: R R Taylor MBE
Movement totals for 1988: air transport 52,820, other 33,297
Passengers: 2,876,985 (1988)
Cargo: 15,256 tons (1988)
Principal operators at date of visit: Aer Lingus, Air Canada, Air Europe, Air France, Air Malta, Birmingham European Airways, British Airways, Britannia Airways, British Midland, Brymon Airways, Cyprus Airways, Iberia, JAT, Jersey European, KLM, Lufthansa, Manx Airlines, Monarch Airlines, NetherLines, NLM, Ryanair, SAS, Swissair, Tunis Air, Wardair
Scheduled service destinations: Aberdeen, Alicante, Amsterdam, Barcelona, Belfast, Brussels, Cologne/Bonn, Copenhagen, Cork, Dublin, Dubrovnik, Düsseldorf, Edinburgh, Faro, Frankfurt, Glasgow, Guernsey, Hamburg, Hanover, Isle of Man, Jersey, Larnaca, London Gatwick, London Heathrow, Madrid, Malaga, Manchester, Milan, Munich, Newcastle, Oslo, Paris Charles de Gaulle, Pula, Southampton, Stuttgart, Toronto, Valencia, Zürich.

More than fifty years ago the strong entrepreneurial instincts of Birmingham City Council were apparent when, in February 1928, its General Purposes Committee was instructed to 'consider the possibility of establishing a municipal airport .... so that much of the continental traffic might be diverted from London to Birmingham direct, thereby stimulating the present and future trade of the city'. It was decades before this objective was even approached, but it is now well within the airport's grasp and management is looking further afield towards North America.

The terminal area at Birmingham International Airport, a view looking approximately northeast. Curving from the terminal is the MAGLEV automatic shuttle link to Birmingham International railway station and the National Exhibition Centre. The driverless cars operate by frictionless magnetic levitation along a 630m elevated track and provide a 90-second link between airport terminal and railway station.

During the United Kingdom's industrial run-down in the 1970s the West Midlands area, previously a hive of engineering activity, was badly hit. Now, however, the area is booming again with manufacturing industries geared to the times, financial services and distribution complexes, the International Convention Centre and the National Exhibition Centre dating from 1976 and now world-known. The Exhibition Centre with its own associated railway station located on the airport perimeter is Britain's premier location for trade exhibitions. The new airport terminal and the railway station, some 600m away, are connected by the MAGLEV elevated rail shuttle link, operating by magnetic levitation.

With business activity again coming right, with the airport's excellent road and rail links and a very large resident population within 50km, the time had come, by the early 1980s, for Birmingham's International Airport to make a

new beginning. At the very heart of this opportunity were the local communities on the airport's doorstep.

As the airport director Bob Taylor explains, 'When the West Midlands County Council was abolished at the same time as the Greater London Council, the airport was handed over to the seven district councils which formerly made up the old West Midlands county, ie Birmingham, Solihull, Coventry, Walsall, Dudley, Sandwell and Wolverhampton. Under the Airports Act, we were required to form a company (Birmingham International Airport plc) with the seven shareholders being those seven district councils. One hundred per cent of the shares are owned by them in proportion to their ratepaying strength. So Birmingham, for example, holds roughly 40 per cent of the shares and has four board members. Each of the others has one board member and their shares are held in relation to ratepaying numbers and dividends are paid out pro rata on the same basis'.

The Turning Point

Where many airports evolve gradually with traffic expansion pacing their building programme, at Birmingham on the passenger side, little had been done since the creation of the then new terminal in 1939. The decision to create an entirely new terminal in the early 1980s attracted new traffic, a fact which the airport director confirmed. 'Whereas we had quite a strong marketing approach to overseas airlines such as Lufthansa, Air France, Swissair and KLM, to name but a few, it was only when we actually started work on the terminal that they began to sit up and take notice. We were able to use the terminal as a marketing tool and to say, "Now look, we have smart, elegant facilities through which you can travel, this is more your style".'

The new terminal was designed to cope with 3 million passengers a year by 1990. It is a three-storey steel frame structure clad in moulded panels and tinted glass wall units. Two piers serve twenty aircraft stands, eleven of them international, including five for wide-bodied jet transports. The remaining nine are for domestic services, three of them wide-bodies.

Departing passengers check in and are then guided to domestic and international lounge complexes, both of which have easy access to restaurants, shops and other services. Facilities for arriving passengers are designed for efficient passage through the terminal, then onto surface transport from the airport.

The administrative areas are located above the second floor, while a glass-fronted public viewing terrace with separate access gives visitors a good view over the airport. There is also a large open grassed area alongside the southern taxiway which, with the blessing of the airport authority, provides an excellent vantage point for enthusiasts.

The car parks are immediately adjacent to the terminal, a fact which was duly appreciated during our visit as it was pouring with rain. The impressive building is backed by efficient and versatile methods of aircraft handling, most stands being fitted with adjustable airbridges and advanced docking systems. A new taxiway network has eliminated one of the main disadvantages of the original airport layout, namely the need for aircraft to backtrack along the runway. This slowed down the movement rate and caused ATC many problems.

The airport's expansion has also incorporated a new 6,000sq m air cargo centre, provided with the latest ground handling equipment and freight processing systems. Most of the original apron

The relationship of the International airport (centre, top) to the National Exhibition Centre, the Birmingham Metropole Hotel (centre, foreground) and the main rail and motorway system can be seen here.

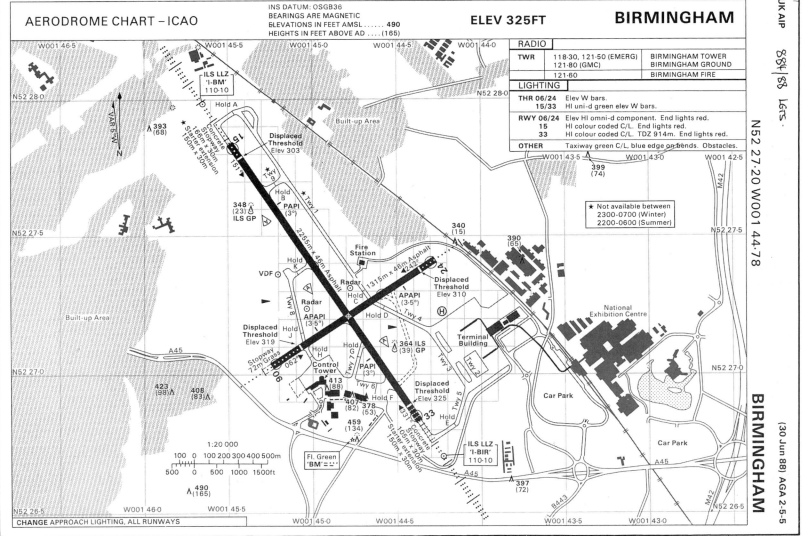

on the southwest side of the airport is now for the exclusive use of cargo aircraft, with its eastern portion reserved for general aviation.

Bob Taylor sees the airport's catchment area realistically as a 50-mile radius but comments that it also goes out via the motorway network to anything up to 100 miles. Obviously, there is some overlap with East Midlands, Bristol, Luton and other airports but, as he says, 'We succeed here by our very extensive scheduled network'.

Asked how successful the airport is in luring traffic away from the London airports and their inevitable congestion, Mr Taylor observed that so far only Air Canada has admitted to wanting to put a schedule out of Birmingham because it could not get the right slots at Heathrow. 'In marketing the airport, I can't tell whether Iberia HQ or KLM, Lufthansa, Swissair, Air France, etc, aren't looking to the future as Heathrow gets congested. They are running their services out of regional airports more and more anyway, but whether that is related to Heathrow as well, I am not too certain. I like to think that it is and that we are attracting those overseas carriers to run scheduled services because people now want to travel via their local airport and be home in 20 minutes.'

A schedule to Canada is not new, however, Wardair having flown between Birmingham and Toronto for several years. During the summer of 1989, the service reached a peak of three trips a week. Air Canada has joined them during 1989 using Boeing 767s or Lockheed TriStars on a twice-weekly basis. For the US-bound traveller, entry can be very much quicker than flying direct to a US airport. With much less queueing, US customs and immigration can be cleared in Toronto. The ease of access to Birmingham Airport and time saved at destination can mean several hours sliced off the real journey time to North America.

On the subject of more transatlantic operations, Mr Taylor commented, 'We are talking to two American airlines whom we would like to see operating through here before too long and I think there is at least one British carrier which is eyeing us with interest. We are certainly feeding it with statistics from Birmingham with a view to early transatlantic operations on a scheduled basis.

'One of the problems is the Bermuda Tour Agreement and the Gateway situations that exist. We've said to the Government, don't just give Gateway status to Birmingham. Why don't you offer every UK airport as a gateway to the Americans and then let us negotiate it through? It is only certain airports that would attract scheduled services, but if

Welcome to Birmingham International.

you offer them all, surely that is a good bargaining point. But the government hasn't seen it our way yet!'

Until 1984, Birmingham suffered from a shortcoming common to a number of Britain's regional airports: marginal runway length for economic long-haul operations. During the planning stages of the new taxiway system the airport director was instrumental in suggesting starter strips at each end of the main runway. This simple and relatively inexpensive measure made only a small difference to the physical take-off run but its effect on the runway length requirements for departure calculations is considerable. He reports, 'British Airways tell me that a TriStar on the old runway

could take off from here for Chicago with a full load of passengers and baggage but restricted cargo. With the 150m starter strip, the same TriStar can depart at maximum all-up weight with fuel reserves'.

The entire northeast sector of the United States and Canada now comes within the reach of a 450-seat Boeing 747 with baggage and up to 10 tons of cargo. With such important destinations as New York, Boston, Washington, Chicago, Detroit and Toronto, the need to extend the runway may not be yet.

If the demand is such that long-haul flights are required from Birmingham, then outline planning approval would be sought and discussion started about land purchase, for which there are still compulsory powers. There would be a public enquiry but a 10,000ft or even 12,000ft runway should not present too many problems, though the A45 trunk road would require a costly tunnel.

Whereas plans for further transatlantic operations are still under review, the expansion of European routes is very much a reality. The most heavily used scheduled services are those in Amsterdam, Paris and Frankfurt, with Brussels and Düsseldorf close behind. These are the parts of Europe with which Midlands people seem to be doing the most business at present; now that Milan and Copenhagen are on jet schedules, this traffic is likely to grow as well.

The Copenhagen route, opened in April 1989, saw SAS operating through Birmingham for the first time. Another national flag carrier new to Birmingham

The new terminal.

is Iberia, which began Madrid and Barcelona services during the past summer. Both airlines use the Douglas DC-9. Air France, Lufthansa, NLM-Dutch Airlines, Sabena and NetherLines have all increased the frequencies of their Birmingham services over the past few years. Some routes, notably Amsterdam, Paris and Düsseldorf, are served by several operators, such is the demand for seats.

British Airways has always been a familiar part of the Birmingham scene and the relationship has now grown to the point where a domestic hub terminal is being planned as a joint venture by the airport and the airline. It will be situated on the extended apron to the east of the present domestic pier, which will eventually have to be demolished.

All BA's scheduled services will be concentrated on the new hub, though other scheduled operators and established charter flights will continue to use the existing main terminal. A possible second stage expansion is planned for the mid-1990s as traffic increases.

With the existing terminal already approaching its theoretical annual capacity, the airport is building a baggage hall extension with additional passenger lounges above it. This will virtually double the size of the present international lounges and create an entirely new first floor unified travel lounge. Once these lounges are operational, the opportunity will exist to replace totally the landside first floor to include expanded buffet and seating areas, shops and other new passenger facilities.

This expansion may not come a moment too soon if the plans of Birmingham European Airways (BEA) come to fruition. Since its formation in 1983 as Birmingham Executive Airways, the company has built a reputation as an exclusive business services airline and during 1988 carried over 50,000 passengers. Newly grouped with Brymon Airways and backed by British Airways and the Danish carrier Maersk, it seems set to emerge as a significant force on the European regional scene.

The reorganised BEA will continue the existing network of routes linking Birmingham with Amsterdam, Copenhagen, Düsseldorf, Frankfurt, Milan, Oslo and Stuttgart, but plans to start a number of new routes within Britain as well as to continental Europe. They include Birmingham–Paris Charles de Gaulle, Birmingham–Newcastle and Birmingham–Southampton. Initially using two BAC One-Eleven 500s leased from BA, BEA will replace smaller types such as the BAe Jetstream and Grumman Gulfstream on the Copenhagen, Milan and Amsterdam routes. Jorn Eriksen, BEA's managing director, declared that

A MAGLEV car nears the airport terminal. From the terminal these cars serve the Birmingham International railway station and from there the National Exhibition Centre.

his company's goal for 1989 would be to double the previous year's passenger total.

Birmingham's share of the holiday charter market continues to increase, with over 150 tour operators featuring flights from the airport in their brochures. All the usual destinations in the Mediterranean, North Africa and the Canaries are covered but, at present, the only long-distance routes are to Canada and Florida. There is obviously scope for further long-distance charter business and the airport authority will not be slow to encourage it.

Does all this traffic create problems with arrival and departure slots, as at Heathrow, Gatwick and Manchester? Mr Taylor says, 'None at all. We get congested at times, of course, particularly around seven in the morning when many schedules are departing and again in the early evening when they return but it isn't causing delays as yet. There are no restrictions on night operations but obviously the bigger and quieter aeroplanes are becoming more and more welcome. Most of the cargo aircraft out of here at night are propeller-turbine types which create relatively little noise'.

The airport's growth rate in 1988 was 8 per cent higher than in 1987, with a total throughput of 2,876,985 passengers. All this had taken place in a year when there was an overall reduction in the charter market. The increase has been in scheduled services and their use by the business traveller. Compared with 1987,

these services have increased by no less than 20 per cent.

In reply to a question about the possible effects on traffic when the European Free Market is introduced in 1993, Mr Taylor said, 'It can only be speculation. An airport like this one, a true regional airport serving the West Midlands region, could well see some benefit. We may see an easing off of the rules governing route licence applications and approvals, especially between some of the smaller cities in Europe. I don't think it will be a magic wand for Birmingham but I feel it will be an opportunity for some of the smaller carriers to bring in business through having access to routes they would not have had before'.

He emphasises that the airport is growing on the healthy basis of 50 per cent scheduled and 50 per cent charter. 'It isn't only new destinations coming in but increased frequency on the scheduled routes. We still see our main business as short to medium haul into Europe and the Mediterranean, along with some long-haul traffic which is slowly growing as the aeroplanes get bigger and operate from smaller runways as well. Having said that, I don't consider our runway to be too short and, as I have shown, it has yet to prove a drawback in promoting business.'

History

The year 1989 marks Birmingham Airport's Jubilee and, with a prosperous future ahead of it, there has been time to reflect on its not always profitable past. Birmingham's original airport was Castle Bromwich, a site which it shared with the Royal Air Force. Although a purely civic airport nearer to the city centre had been suggested as early as 1928, the existence of Castle Bromwich meant that there was no pressure to decide.

Plans progressed at a leisurely pace, land at Elmdon on the southeastern outskirts of the city being selected in 1930 for development. The plan was shelved, however, but revived in 1934 after the Air Ministry hinted that Castle Bromwich could not be used indefinitely as a civil airport.

The Airport Committee formed in 1934 finally approved the new airport on 4 February 1936. It was to cost £500,000 and was planned to have an electric railway and monorail link to the city centre. The latter scheme was abandoned, however, owing to lack of funds. The existence of a main railway line in the immediate vicinity was probably one of the reasons why the Elmdon site was chosen. Indeed, it was considered that the airport be linked with a nearby station at Marston Green, but it was more than forty years before the present Birmingham International Station was built.

The land for the airport proved to be less than ideal, with nearly three years being spent on preparation which included the filling of numerous ponds and the laying of 40 miles of drainage pipes. A layer of solid clay had also to be removed. The first aircraft landed on 20 March 1939, a de Havilland D.H. 90 Dragonfly of Western Airways, one of the aircraft which were to inaugurate a thrice-daily Weston-super-Mare–Bristol–Birmingham–Manchester service on 17 June.

The airport was opened officially on 8 July 1939 by the Duchess of Kent. Elmdon was the last municipal airport to be opened before the outbreak of war and the customary air display was dominated

The passenger terminal today.

by military types, including eight Spitfires and examples of the Whitley, Hampden, Blenheim, Hurricane and Skua.

That summer, Elmdon joined the network of internal air routes with schedules by Railway Air Services from Croydon and on to Liverpool, Belfast and Glasgow. Great Western & Southern Air Lines linked the city with Liverpool, Manchester, Bristol, Southampton, Ryde and Shoreham. While still in infancy, the airport was engulfed by war; civil aviation was suspended, and Elmdon became the home of No. 14 Elementary Flying Training School (EFTS), equipped with Tiger Moths.

Austin Motors at Longbridge switched from the production of Fairey Battles and Hawker Hurricanes to the Short Stirling at the end of 1940. The airfield at Longbridge was far too small for heavy

Ready to be put to work. The original terminal opened in July 1939.

Twenty years ago. Aer Lingus and BEA Viscounts on scheduled services.

bombers so final assembly was undertaken at a factory at Marston Green, connected with Elmdon by a concrete track. The Stirlings and the Lancasters which succeeded them from Austins were towed to the airport for test flying and delivery.

From this operation, Elmdon gained two hard runways, both just over 4,000ft long, and these formed the basis of today's extended layout. Having trained several thousand pilots, No.14 EFTS disbanded on 1 February 1946. Control of the airport was handed over to the Ministry of Civil Aviation on 19 June 1946, but for several years it was little more than a flying club field with occasional charter flights, a situation shared by most British airports at that time.

In 1949, after much persuasion by the city council, British European Airways began a thrice-weekly Birmingham–Paris schedule in association with Air France. In the same year, Aer Lingus began a weekday service to Dublin. During the 1950s, the range of services slowly expanded to include Belfast, Jersey, Düsseldorf and Zürich.

The most significant development for the airport in the early 1960s was its return to city control on 1 April 1960. In the following year, an extension to the terminal was opened for international flights, and other improvements included bigger car parks and better restaurant facilities. In 1966 the main runway was extended to 7,400ft and the apron enlarged.

In the summer of 1970. The control tower is framed by the tail of a BOAC Super VC10.

With its huge catchment area, Elmdon was well placed when inclusive tour operations to Mediterranean resorts became commonplace early in the 1960s.

Regular jet movements were at last possible when the runway extension was completed. A million passengers used the airport in 1973, the first time this had been achieved, and international services increased steadily.

At the official opening in 1939, Prime Minister Neville Chamberlain made a speech which included the observation that those who had been responsible for Birmingham's new airport had planned wisely for the future. Their successors deserve a similar accolade and the airport director summarises today's achievements: 'I believe that the pattern now established is only the start and that we shall continue to see real steps forward in the development of scheduled services at Birmingham'.

I should like to thank Mr R R Taylor MBE, airport director, and his press and public relations manager, Janine Lee, for their help and co-operation. *Aviation in Birmingham* by Geoffrey Negus and Tommy Staddon was an invaluable source of historical information. □

BOOK REVIEWS

Hugo Junkers and his Aircraft
Gunter Schmitt

Transpress, Berlin.
Hardbound, 240 x 280mm, 224 pp,
illustrated. No price quoted.
Distributed by Midland Counties
Publications, 24 The Hollow,
Earlshilton, Leicester LE9 7NA.

Professor Hugo Junkers has been de-scribed as the father of metal aircraft construction and this book certainly confirms it. Readable and carefully re-searched, although suffering somewhat from its translation into English, it describes the Junkers Company and its products during the period (1890s to 1935) when the professor was in control. In the late 1930s, advancing age and the Nazis combined to force his retirement and change the company's character.

In its early years, before turning to aircraft, Junkers was in the business of producing a wide range of gas-fired appliances, including notably a very successful bathroom geyser. In 1909 he patented a new form of wing – not, as previously reported, the design of a complete aircraft. The patent led Junkers into the design of the J 1 'Tin Donkey', its metal construction inspired by the 'corrugated iron duck' of Prof Hans Reissner. Research in his own wind tunnel convinced Junkers that mono-planes with thick cantilever wings of high-aspect ratio were likely to be most efficient. J 1 performed satisfactorily af-ter its first flight on 12 December 1915 but its smooth-skinned, steel structure was obviously too heavy. This led Junk-ers' design office to produce a series of projects and prototypes, built in duralu-min instead of steel and with corrugated skins, which confirmed the soundness of the chosen low-wing metal monoplane configuration and culminated in the J 10 two-seater.

A civil version of the J 10 was followed by the F 13 light transport, a remarkably advanced design for its time, which remained in production through most of the interwar years. It also led to a wide range of civil and military aircraft which sold all over the world and were pro-duced under licence in several countries. Junkers' corrugated-skin duralumin monoplanes competed effectively, and almost alone of their type, against the conventional wood, wire and fabric bi-planes of the rest of the world's aircraft

industries until the modern type of stressed-skin metal monoplane started to take over in the 1930s. From such information as is available, it seems that the Junkers types sold, despite higher costs of manufacture, because of their greater robustness and durability. There was little difference in performance or range and payload. Comparison of the manufacturing costs and prices of the F 13 and a 'conventional' wood, wire and fabric biplane of the same period, such as the Airco D.H.16 (civil conversion of the D.H. 9A), gives the following picture which may well have continued between the wars: according to Gunter Schmitt, the F 13 required 9,000 design man-hours to first flight; the D.H.16 probably required about the same. But production man-hours on the first F 13 seem to have been between two and three times those on the D.H.16. The latter's price (£4,000) compares with £4,500 for the F 13. **PWB**

British Military Airfields 1939–45
David J Smith

Patrick Stephens, Thorsons Publishing
Group, Wellingborough NN8 2RQ.
Hardbound, 235 x 155mm, 294 pp,
illustrated, £14.95.

This superb book is a sequel to the very valuable ten-volume Action Station series (to which the author contributed two volumes) and covers all the military airfields of the United Kingdom and Northern Ireland. This series almost certainly gives the most detailed account of any country's military aerodromes.

To state that David Smith has under-taken massive research and produced an essential addition to British aeronautical history would be an understatement.

I do not pretend that I have completely read this work because it is printed in rather small type and I have yet to find time to undertake this rewarding task, but I have read much of it and studied the large and varied collection of photo-graphs and drawings included between its attractive covers; here I must enthuse over the back cover which has the most beautiful photographs of Hurricanes in a winter scene that looks Antarctic but on reading the caption proves to be Pen-rhos, North Wales, and it was taken in the formidable winter of 1940/41.

Not many people are likely to buy a book simply to read the author's intro-duction but in this case it could be money

well spent, and I certainly learned much that I never knew about aerodromes by reading this introduction very carefully.

David Smith's new book is not specifi-cally about individual aerodromes and flying-boat stations, but about the provi-sion and equipment of these facilities and the runways and buildings thereon.

The work opens with a chapter on 'Airfield development 1912-39', a fine historical survey, and continues with 'Paved runways, taxiways and disper-sals', followed by 'Airfield requirements for the Royal Air Force 1940-42', 'Air-field construction', a chapter headed 'Miscellany: strip aerodromes, emergen-cy landing grounds and other specialised sites', 'The airfields that never were', 'Watch towers, hangars, ancillary build-ings and installation' and... but I am not going to list the other chapters; you can buy the book and find out.

The photographs are extremely in-teresting, ranging from a sunken pillbox and a balloon shed through every type of installation to oblique and vertical views of complete airfields. Drawings include aerodrome plans, location charts and lighting diagrams. There is no question that this is an outstanding book. **JS**

World Aviation in Spain (The Civil War) 1936–1939
American and Soviet Airplanes
J Miranda and P Mercado

Silex Ediciones, Madrid.
Hardbound, 300 x 235mm, 284 pp,
illustrated. No price quoted.

This work was originally published in Spanish and is intended to be part of a much larger work covering all the aircraft involved on both sides in the Spanish Civil War – some 3,200 aeroplanes of 262 types, although the authors can only find photographic evidence for 188 types.

The authors devote more than 150 pages to the US aircraft which found their way to Spain and 100 pages to the products of the USSR. As far as possible the same treatment has been given to all types, with development history, tech-nical data and production lists and details of their Spanish employment.

There are ninety-seven photographs but regrettably many of the types could not be shown with Spanish markings. However, the book contains more than 150 $1/72$nd scale drawings which, apart from showing the aeroplanes, include their-markings and colour schemes both in their countries of origin and in Spanish service. In fact the authors have gone to a lot of trouble to provide information for modelmakers.

Stating, as the publishers do, that there are 153 drawings is misleading. This figure should refer to pages because many pages of drawings contain a number of individual illustrations; for example on one page dealing with the Polikarpov I-16 there are ten complete side views, two partial side views, two partial plan views and four fin badges complete with measurements.

There is even a page of drawings of vehicles which include six versions of the Hucks starter! Some battle diagrams are also included.

This book contains many of the well-known combatants but has surprises for many. In the US section can be found such oddities as the American Eagle A-129, some Bellanca low-wing monoplanes, the Boeing 281, Consolidated Fleetster, Fairchild A-942, Northrop Gamma and even the twin-boom Sikorsky S-38 amphibian. However, there were no surprises in the Soviet section.

An enormous amount of effort obviously went into researching, writing and illustrating this book and it is to be hoped that the rewards will encourage the authors to complete the series. **JS**

Knights of the Air
Peter King

Constable, London.
Hardbound, 240 x 180mm, 530 pp,
illustrated, £20.00.

Sub-titled 'The life and times of the extraordinary pioneers who first built British aeroplanes', the book is 'an attempt to put this handful of men and their achievements, technical and commercial, into perspective.... It is offered as a preliminary study of their lives and motivations in developing the industry in the way they did'.

Acknowledging the help of many, the author is commendably frank in recording that, of these, Sir Peter Masefield and Bill Gunston do not always agree with his interpretation of motivations and events. If he is guilty of a disregard for the pervading atmosphere of the times about which he is writing, then he has also 'been an enthusiastic aviation buff' and 'has tried to balance that enthusiasm against the wider assessment of such facts as are now available'. These caveats provide some armour against the barbs of criticism which are bound to follow from many with personal experience of events chronicled. These criticisms will arise not because they, like 'most of the pioneers, hated objectivity' but because the author has not displayed 20/20 vision in his hindsight and in the interpretation of his extensive bibliography.

Although the author states that his story ends in the mid-1960s when the pioneers had either died or retired, he does continue briefly to the present day with some interesting and stimulating comment. It is a considerable endeavour covering most phases in the development of the industry, both technical and financial, and the interactions of national and personal politics. One knight of the air I would have liked to see included is Sir James Martin. While not a pioneer in the author's chronological definition he was truly a pioneer. His experience with officialdom in his fighter endeavours and his success with ejection seats make an interesting comparison with that of the earlier pioneers.

This book is the author's first in the aeronautical field and although he worked for a comparatively short period in the industry I consider his view that of an outsider. It is thus probably all the more valuable. That he draws the conclusions he does behoves all concerned with the reputation of the industry and of its future to read him with care and heed his plea for further research and the production of a definitive history.

My review copy is an uncorrected proof, but I hope that the reference to the Fairey Battle fighter and the Battle biplane will not survive the proof-reading. **LB**

A Dream of Wings: Americans and the Aeroplane, 1875–1905
Tom D Crouch

Smithsonian Institution Press,
Pennsylvania, USA.
Paperback, 150 x 230mm, 350 pp,
illustrated, $14.95.

This book first appeared as a hardback, published by W W Norton, in 1981, but has been out of print for a few years. As it is one of the classic studies of the aeroplane's nascence, it is good to see it available once more, with a few typographical errors eliminated and some footnotes clarified, but with the basic text unaltered.

Crouch's account begins with the start of Octave Chanute's interest in the possibility of heavier-than-air flight, in 1875, and traces the development of aeronautics in the USA through pioneers such as Langley, Herring, Montgomery, Huffaker, James Means and the Boston Aeronautical Society, and, of course, the Wright brothers.

Keyed into the narrative are the influences of international players in the story: Lilienthal, Mouillard, Hargrave, etc. The account ends with the failure of Langley's Aerodrome and the Wrights' first successful powered, sustained and controlled flights in December 1903. An epilogue provides an enlightening follow-up on the leading characters' careers after 1903.

Notes, bibliography and index round off this excellent study. **PJ**

How we Invented the Aeroplane – an illustrated history
Orville Wright, edited by Fred C Kelly and additional text by Alan Weissman

Constable, London.
Paperback, 210 x 280mm, 88 pp,
illustrated, £5.20.

This book represents a chance to acquire some useful and little-published Wright brothers material. The core of this book is an extract from a paper written by Orville Wright in connection with a 1920 legal action. This was last published in 1953. Added to it is another Orville article, 'After the First Flights', probably of similar vintage; 'After Kitty Hawk: A Brief Resumé' by Fred C Kelly; and the first popular published account, 'The Wright Brothers' Aeroplane', from *The Century* magazine of 1908. The whole is bolstered by a good collection of seventy-six photographs, well reproduced, and with useful captions (in picture 27 I suspect that the 1902 glider is not 'turning to the right', as often stated; the wings are being warped to correct a tilt to starboard). **PJ**

Shuttleworth: The Historic Aeroplanes
David Ogilvy

Airlife Publishing, Shrewsbury.
Paperback, 220 x 220mm, unpaginated,
illustrated, £8.95 (casebound £13.95).

This nicely designed, full-colour book contains a potted history of the Shuttleworth collection, type-by-type descriptions of the aircraft and associated equipment, and the work to keep the aircraft flying (written by chief engineer Chris Morris), and an account of a typical display day.

The colour illustrations, many by leading aviation photographers, are beautifully reproduced, making this an altogether attractive publication. If you enjoy visiting Old Warden as much as I do, this will help allay withdrawal symptoms during the cold months between display seasons. Some of your time could be spent counting the pages, for the enlightenment of this reviewer. **PJ**